W9-CID-373

OT 38
Operator Theory: Advances and Applications
Vol. 38

Birkhäuser Verlag
Basel · Boston · Berlin

Leiba Rodman

An Introduction to Operator Polynomials

1989

Birkhäuser Verlag
Basel · Boston · Berlin

Author's address:

Prof. Leiba Rodman
Department of Mathematics
College of William and Mary
Williamsburg, VA 23185
USA

CIP-Titelaufnahme der Deutschen Bibliothek

Rodman, Leiba:
An introduction to operator polynomials / Leiba Rodman. –
Basel ; Boston ; Berlin : Birkhäuser, 1989
(Operator theory ; Vol. 38)
ISBN 3-7643-2324-8 (Basel . . .) Pb.
ISBN 0-8176-2324-8 (Boston) Pb.
NE: GT

Printed in Germany on acid-free paper
ISBN 3-7643-2324-8
ISBN 0-8176-2324-8

Dedicated to Daniel, Ruth, and Benjamin

PREFACE

This book provides an introduction to the modern theory of polynomials whose coefficients are linear bounded operators in a Banach space - operator polynomials. This theory has its roots and applications in partial differential equations, mechanics and linear systems, as well as in modern operator theory and linear algebra.

Over the last decade, new advances have been made in the theory of operator polynomials based on the spectral approach. The author, along with other mathematicians, participated in this development, and many of the recent results are reflected in this monograph.

It is a pleasure to acknowledge help given to me by many mathematicians. First I would like to thank my teacher and colleague, I. Gohberg, whose guidance has been invaluable. Throughout many years, I have worked wtih several mathematicians on the subject of operator polynomials, and, consequently, their ideas have influenced my view of the subject; these are I. Gohberg, M. A. Kaashoek, L. Lerer, C. V. M. van der Mee, P. Lancaster, K. Clancey, M. Tismenetsky, D. A. Herrero, and A. C. M. Ran. The following mathematicians gave me advice concerning various aspects of the book: I. Gohberg, M. A. Kaashoek, A. C. M. Ran, K. Clancey, J. Rovnyak, H. Langer, P. Lancaster, A. Dijksma, G. Rublein, and C. R. Johnson. Special thanks are due to M. A. Kaashoek who read parts of the manuscript. Finally, I would like to thank J. Riddleberger for speedy and accurate production of the typescript. During the preparation of this monograph, I was partially supported by research grants from the National Science Foundation.

Leiba Rodman
November 1988
Williamsburg, Virginia

TABLE OF CONTENTS

INTRODUCTION

This book is devoted to the study of polynomials of a complex variable λ whose coefficients are linear bounded operators acting in a Banach space $\mathcal{X}$:

$$L(\lambda) = \sum_{j=0}^{m} \lambda^j A_j, \tag{1}$$

where $A_j: \mathcal{X} \to \mathcal{X}$ $(j = 0,\dots,m)$. Such polynomials will be called in short *operator polynomials*.

Three broad topics emerge as the main foci of interest in the theory of operator polynomials. The first one has to do with the idea of linearization, i.e., reduction (in a certain sense) of the generally high degree polynomial $L(\lambda)$ to a linear polynomial of the form $\lambda I-A$, or $\lambda B-A$, for suitable operators A and B. To illustrate the point, the quadratic scalar polynomial $(\lambda-1)(\lambda-2)$ has a linearization $\lambda I-\begin{bmatrix}1 & 0\\ 0 & 2\end{bmatrix}$. In this book we adopt the following definition: A (linear bounded) operator T acting on a Banach space $\mathcal{Y}$ is called a linearization of the operator polynomial $L(\lambda)$ over an open set σ if $\sigma(T) \subset \sigma$, and

$$L(\lambda) \oplus I_{\mathcal{Z}_1} = E(\lambda)[(\lambda I-T) \oplus I_{\mathcal{Z}_2}]F(\lambda), \quad \lambda \in \sigma$$

for some Banach spaces $\mathcal{Z}_1$, $\mathcal{Z}_2$ and some analytic and invertible operator functions $E(\lambda)$ and $F(\lambda)$ defined on σ. The linearization enables us to translate spectral properties of an operator polynomial to the more familiar setting of linear polynomials.

The second topic concerns various types of factorizations and related notions, such as common divisors and common multiples. Thus, one studies factorizations

$$L(\lambda) = M(\lambda)N(\lambda),$$

where $M(\lambda)$ and $N(\lambda)$ are operator polynomials, as well as Wiener-Hopf factorizations (see Chapter 10 for the definition of this notion).

The third topic concerns the problems of multiple completeness of eigenvectors and generalized eigenvectors. Historically, it first emerged in the work of Keldysh [1] (see also Keldysh [2]).

All three topics mentioned above are intimately interrelated. For example, the multiple completeness of eigenvectors and generalized eigenvectors of an operator polynomial is basically equivalent to the completeness of eigenvectors and generalized eigenvectors of its linearization. Also, the problems of factorizations can be studied in terms of certain invariant subspaces of the linearization.

From the very beginning, the theory of operator polynomials drew on many important applications of mechanics and physics (see, e.g., Askeron-Krein-Lapter [1], Zilbergleit-Kopilevich [1], Shinbrot [1], Kreĭn-Langer [1]), and in differential equations (Tamarkin [1], Kostyuchenko-Shkalikov [1], Radzievskii [1], Friedman-Shinbrot [1]) as major sources of inspiration and development.

The subject matter of this book is focussed on linearizations, factorizations, and related notions. The problems of multiple completeness are left out, mainly because they are fully exposed and referenced in Markus [1].

The main tool used in this book is a spectral theory of operator polynomials developed mainly in the last decade. This theory is based, among other things, on the notions of spectral pairs and triples of operators associated with an operator polynomial (1). In the case when $L(\lambda)$ is *monic*, i.e., with leading coefficient $A_m = I$, we associate with (1) the pair of operators

$$[I_{\mathcal{X}} \quad 0\cdots 0]: \mathcal{X}^m \to \mathcal{X} \tag{2}$$

$$(3) \qquad C_L = \begin{bmatrix} 0 & I_{\mathfrak{X}} & 0 & \cdots & 0 \\ 0 & 0 & I_{\mathfrak{X}} & \cdots & 0 \\ \vdots & \vdots & \vdots & & \vdots \\ 0 & 0 & 0 & \cdots & I_{\mathfrak{X}} \\ -A_0 & -A_1 & -A_2 & \cdots & -A_{m-1} \end{bmatrix} : \mathfrak{X}^m \to \mathfrak{X}^m .$$

The operator C_L is the well-known *companion operator* for $L(\lambda)$; it serves as a linearization for $L(\lambda)$ and encompasses (together with X) the spectral structure of the operator polynomial $L(\lambda)$. The pair (X, C_L) will be called the *right spectral pair* of $L(\lambda)$; the epithet "right" is used to stress that it is used to describe the right factor $L_2(\lambda)$ in factorizations $L(\lambda) = L_1(\lambda)L_2(\lambda)$ of $L(\lambda)$ as a product of two monic operator polynomials. A dual construction leads to a *left spectral pair* which can be identified with

$$\begin{bmatrix} 0 & 0 & \cdots & 0 & -A_0 \\ I & 0 & \cdots & 0 & -A_1 \\ \vdots & \vdots & & \vdots & \vdots \\ 0 & 0 & \cdots & I & -A_{m-1} \end{bmatrix}, \quad \begin{bmatrix} I \\ 0 \\ \vdots \\ 0 \end{bmatrix}.$$

The pairs are combined to produce a triple of operators (X,T,Y) whose main feature is the "realization" of the inverse of the operator polynomial:

$$L(\lambda)^{-1} = X(\lambda I - T)^{-1}Y.$$

The notions of spectral pairs and triples relative to a given part of the spectrum of $L(\lambda)$ are developed as well.

It turns out that there is one-to-one correspondence between certain invariant subspaces of a linearization of $L(\lambda)$ and factorizations of $L(\lambda)$ into a product of two monic operator polynomials of lower degrees. Namely, let X and C_L be as in (2) and (3). Then each factorization

$$(4) \qquad L(\lambda) = L_2(\lambda)L_1(\lambda)$$

with monic operator polynomials $L_1(\lambda)$ and $L_2(\lambda)$ corresponds to a unique C_L-invariant subspace $\mathcal{M}$ such that the column operator

$$Q = \left.\begin{bmatrix} X \\ XC_L \\ \vdots \\ XC_L^{k-1} \end{bmatrix}\right|_{\mathcal{M}} : \mathcal{M} \to \mathcal{X}^k$$

is invertible (here k is the degree of $L_1(\lambda)$). Conversely, if $\mathcal{M}$ is a C_L-invariant subspace for which the operator Q is invertible, then $L(\lambda)$ admits a factorization (4) with

$$L_1(\lambda) = \lambda^k I - X(C_L|_{\mathcal{M}})^k (V_1 + V_2\lambda + \cdots + V_k\lambda^{k-1}),$$

where $[V_1 V_2 \cdots V_k] = Q^{-1}$.

This type of correspondence between invariant subspaces and factorizations works as well in the more general framework of operator polynomials $L(\lambda)$ whose spectrum

$$\Sigma(L) = \{\lambda \in \mathbb{C} \mid L(\lambda) \text{ is not invertible}\}$$

is a compact set. This characterization of factorization in the geometric language of invariant subspaces gives new insights into the theory of factorization and admits a treatment of questions concerning existence, perturbations, and explicit representations, for example.

Several applications concern the existence of divisors which have spectrum localized in the complex plane in some way (in a half-plane, or the unit circle, for example). For such problems the geometrical approach admits the transformation of the problem to the construction of invariant subspaces with associated properties. In addition, extensions of the theory to the study of several operator polynomials provide a useful geometrical approach to common multiples and common divisors.

The idea of correspondence between invariant subspaces and factorizations was first developed in the theory of characteristic operator functions (Brodskii-Livsic [1], Brodskii [1], Sz.-Nagy-Foias [1]). In the finite-dimensional case, expositions of the spectral theories of matrix polynomials and rational matrix functions where a correspondence of this kind plays a major role are found in Gohberg-Lancaster-Rodman [2,3,6], Lancaster-Tismenetsky [1], Bart-Gohberg-Kaashoek [2].

Much effort had been put into keeping the exposition in this book reasonably self-contained and accessible to graduate students after (or even in parallel with) the first basic course in operator theory or functional analysis. Of course, compromises are inevitable here (partly due to the rather tight upper limit on the size of the manuscript prescribed by the publisher), and on several occasions the reader is referred to the original sources for the proofs, and several results are presented not in the most general available form. It is hoped, however, that these drawbacks are offset by accessibility of the exposition. Background in the following topics in operator theory is sufficient for understanding the material in this book: Fredholm operators, their perturbations and index; compact operators in Banach spaces; spectral theorem for bounded self-adjoint operators; the closed graph theorem and its basic applications; one-sided and generalized inverses, basics on Banach algebras (e.g., as in Chapter 18 in Rudin [1]). The textbooks Gohberg-Goldberg [1], Schechter [1], Kreyszig [1], Taylor-Lay [1], Weidmann [1], and others can be consulted for the necessary background information in operator theory.

Several topics from the book can be used in a graduate course; to facilitate classroom use, most chapters are equipped with exercises. Many open problems and conjectures are stated in the book; few of them are well known but most are new. The exposition also contains supporting material in operator theory usually not exposed in textbooks, such as one-sided resolvents, inertia theory, basics on indefinite scalar product spaces, hulls of operators, spectrum assignment theorems, etc.

References to the literature are usually done in the notes at the end of each chapter. The list of references is not claimed to be complete, and as a rule only the references known to the author that bear directly on the material presented in the book are mentioned.

We conclude the introduction with a brief description of the contents chapter by chapter. The basic results concerning linearizations (such as existence, uniqueness, special classes of polynomials) are put forth in the first chapter. Chapters 2 to 5 are devoted to the spectral theory of monic operator polynomials and its applications to various problems. In Chapter 2, we introduce the notions and concepts (right and left spectral pairs and spectral triples) that characterize the spectral data of a monic operator polynomial; these notions and concepts form the backbone of the book. The exposition in Chapter 2 is often parallel to the finite-dimensional exposition presented in Gohberg-Lancaster-Rodman [2]. In the same Chapter 2 we also give applications of this material to the problem of density of factorable polynomials and to differential and difference equations with operator coefficients. Further applications are given in Chapters 3, 4, and 5. In Chapter 3 we study common multiples of operator polynomials and some related questions. Here the notion of Vandermonde operators and their properties play a crucial role. We refer the reader to Chapter 3 for the precise definition of the Vandermonde operators; here we indicate only that in case the monic operator polynomials $L_1(\lambda),\ldots,L_m(\lambda)$ are of first degree, so that $L_j(\lambda) = \lambda I-X_j$ $(j = 1,\ldots,m)$, the Vandermonde operators are defined as operator matrices

$$\begin{bmatrix} I & I & \cdots & I \\ X_1 & X_2 & \cdots & X_r \\ \vdots & \vdots & & \vdots \\ X_1^{m-1} & X_2^{m-1} & \cdots & X_r^{m-1} \end{bmatrix}$$

Chapter 4 is devoted to the study of perturbations of divisors of monic operator polynomials. In particular, we introduce the notion of stable factorizations of a monic operator polynomial $L(\lambda)$, i.e., such that any operator polynomial sufficiently close to $L(\lambda)$ has a factorization which is as close as we wish to the original factorization of $L(\lambda)$. It turns out that stable factorizations can be studied in terms of stable invariant subspaces, and this is the approach used in Chapter 4. In Chapter 5 we consider an important class of operator polynomials with self-adjoint coefficients. Here the Kreĭn space structure appears naturally, and the spectral theory and factorizations of operator polynomials are based on the properties of bounded self-adjoint operators in the Krein space.

Starting with Chapter 6, we generally drop the assumption that our operator polynomials are monic. In Chapter 6, we extend the spectral theory based on right and left spectral pairs and triples to the framework of non-monic operator polynomials with compact spectrum. Again, there is a correspondence between factorizations of the polynomial and invariant subspaces of the linearization. In Chapter 7 we solve the inverse spectral problem: construct an operator polynomial with the given right (or left) spectral pair. The solution is based on some notions and results from linear systems theory, namely, completely controllable systems and spectrum assignment theorems. This background is presented in Chapter 7 as well. Application of the spectral theory developed in Chapters 6 and 7 are given in Chapters 8, 9, and 10. In Chapter 8 we study common multiples and common divisors of a finite family of operator polynomials with compact spectra. It turns out that greatest common divisors exist always, in contrast with common multiples (let alone least common multiples) which may not exist. The main thrust of Chapter 9 is towards results on separation of spectrum of operator polynomials (with respect to the unit circle). These results are based on the properties of resultant and Bezoutian operators associated with operator polynomials, and on inertia theorems. Finally, in the last chapter we give necessary and sufficient conditions for existence of a Wiener-Hopf factorization for operator polynomials.

CHAPTER 1: LINEARIZATIONS

A basic approach in the study of operator polynomials of an arbitrary degree is reduction (using analytic equivalence and extensions by the identity operator) to operator polynomials of first degree (i.e., linear), called the linearizations. In this chapter we develop these ideas and study the questions of existence, uniqueness, and special cases of the linearizations.

1.1 Definitions and Examples

Let $\mathfrak{X}$ be a (complex) Banach space, and let $L(\mathfrak{X})$ be the algebra of all bounded operators acting on $\mathfrak{X}$. An *operator polynomial* is a polynomial of the form

$$L(\lambda) = \sum_{j=0}^{\ell} \lambda^j A_j, \tag{1.1.1}$$

where λ is a complex parameter and $A_j \in L(\mathfrak{X})$, $j = 0,\ldots,\ell$. The *spectrum* of an operator polynomial (1.1.1) is the set

$$\Sigma(L) = \{\lambda \in \mathbb{C} \mid L(\lambda) \text{ is not (two-sided) invertible}\}.$$

It is easy to see that $\Sigma(L)$ is closed. However, in contrast with the spectrum of a (linear bounded) operator, the spectrum of an operator polynomial need not be compact or non-empty. For example, let $A \in L(\mathfrak{X})$ be the *quasinilpotent* operator, i.e., $\sigma(A) = \{0\}$ (here and elsewhere in the book

$$\sigma(A) = \{\lambda \in \mathbb{C} \mid \lambda I - A \text{ is not invertible}\}$$

is the spectrum of an operator A). Then

$$\Sigma(I+\lambda A) = \phi.$$

If A is compact with infinitely many non-zero eigenvalues, then $\Sigma(I+\lambda A)$ is not compact. It can happen that $\Sigma(L) = \mathbb{C}$, as the following example shows:

EXAMPLE 1.1.1. Let $S: \ell_2 \to \ell_2$ be the left shift:

$$S(x_1, x_2, \ldots) = (x_2, x_3, \ldots).$$

The operator polynomial

$$L(\lambda) = \begin{bmatrix} \lambda S - I & 0 \\ 0 & S - \lambda I \end{bmatrix}: \ \ell_2 \oplus \ell_2 \to \ell_2 \oplus \ell_2$$

has the property that $\Sigma(L) = \mathbb{C}$ (indeed, $L(\lambda)$ is invertible if and only if both operators $\lambda S - I$ and $S - \lambda I$ are invertible, which never happens). In this example, $L(\lambda)$ is Fredholm for $|\lambda| = 1$ (recall that an operator $A \in L(\mathcal{X})$ is called *Fredholm* if dim Ker $A < \infty$ and

$$\text{Im } A = \{Ax \mid x \in \mathcal{X}\}$$

is a closed subspace of finite codimension). Moreover, the index of $L(\lambda)$

$$\text{ind } L(\lambda) = \dim \text{Ker } L(\lambda) - \text{codim Im } L(\lambda) = 1$$

for $|\lambda| \neq 1$. ■

A very familiar object is an operator polynomial of first degree with leading coefficient I: $\lambda I - A$. This polynomial can be identified with the operator A, thus the study of such operator polynomials amounts to the study of the individual linear bounded operators on $\mathcal{X}$.

One of the basic ideas in the theory of operator polynomials is the relationship between general operator polynomials and operator polynomials of the first degree, which is achieved by means of linearization which will be defined now. It is convenient to introduce the following notation. For Banach spaces $\mathcal{X}$ and $\mathcal{Y}$, we denote by $L(\mathcal{X},\mathcal{Y})$ the Banach space of all linear bounded operators from $\mathcal{X}$ and $\mathcal{Y}$, and by $\mathcal{X} \oplus \mathcal{Y}$ the Banach space of all ordered pairs (x,y), $x \in \mathcal{X}$, $y \in \mathcal{Y}$, with the norm $\|(x,y)\|^2 = \|x\|^2 + \|y\|^2$ (so, if both $\mathcal{X}$ and $\mathcal{Y}$ are Hilbert spaces, $\mathcal{X} \oplus \mathcal{Y}$ is a Hilbert space as well). $I_{\mathcal{X}}$ (often abbreviated to I) is the identity operator of $\mathcal{X}$. Given an operator polynomial (1.1.1), and given an open set $\Omega \subset \mathbb{C}$, an operator $A \in L(\mathcal{Y})$, where $\mathcal{Y}$ is some Banach space, is called a *linearization* of $L(\lambda)$ with respect to Ω if the following conditions hold:

(i) $\sigma(A) \subset \Omega$;

(ii) for some Banach spaces $\mathcal{Z}_1$ and $\mathcal{Z}_2$, and for some operator-valued functions $E: \Omega \to L(\mathcal{Y} \oplus \mathcal{Z}_2, \mathcal{X} \oplus \mathcal{Z}_1)$, $F: \Omega \to L(\mathcal{Y} \oplus \mathcal{Z}_2, \mathcal{X} \oplus \mathcal{Z}_1)$ which are analytic and invertible on Ω, we have

$$L(\lambda) \oplus I_{\mathcal{Z}_1} = E(\lambda)((\lambda I - A) \oplus I_{\mathcal{Z}_2})F(\lambda), \quad \lambda \in \Omega. \tag{1.1.2}$$

Recall that an operator function $M: \Omega \to L(\mathcal{U}, \mathcal{V})$, where $\mathcal{U}, \mathcal{V}$ are Banach spaces, is called *analytic* on Ω if in a neighborhood N_{λ_0} of every $\lambda_0 \in \Omega$ it is represented by a convergent power series:

$$M(\lambda) = \sum_{j=0}^{\infty} (\lambda - \lambda_0)^j M_j, \quad \lambda \in N_{\lambda_0}, \tag{1.1.3}$$

where $M_j \in L(\mathcal{U}, \mathcal{V})$. If M is in addition invertible on Ω (which means that each operator $M(\lambda) \in L(\mathcal{U}, \mathcal{V})$ has (two-sided) inverse), then the inverse $M^{-1}: \Omega \to L(\mathcal{U}, \mathcal{V})$ is also analytic. Indeed, assuming $M(\lambda)$ is given by (1.1.3), for $\lambda \in N_{\lambda_0}$ which are sufficiently close to λ_0 we have $\|M(\lambda_0)^{-1}(M(\lambda_0) - M(\lambda))\| < 1$, and hence

$$M(\lambda)^{-1} = \left[\sum_{j=0}^{\infty} [M(\lambda_0)^{-1}(M(\lambda_0) - M(\lambda))]^j\right] M(\lambda_0)^{-1}$$

is easily seen to be representable by a convergent power series

$$\sum_{j=0}^{\infty} (\lambda - \lambda_0)^j \tilde{M}_j$$

for some $\tilde{M}_j \in L(\mathcal{V}, \mathcal{U})$.

So, if (i) and (ii) hold, then $E(\lambda)^{-1}$ and $F(\lambda)^{-1}$ are analytic functions on Ω as well. Note also that the spaces $\mathcal{Y} \oplus \mathcal{Z}_2$ and $\mathcal{X} \oplus \mathcal{Z}_1$ are necessarily isomorphic.

A linearization of $L(\lambda)$ with respect to $\mathbb{C}$ will be called *global linearization*.

The equality (1.1.2) shows that (taking into account (i)) that

$$\sigma(A) = \Sigma(L) \cap \Omega.$$

Since $\sigma(A)$ is compact, the obvious necessary condition for existence of a linearization of $L(\lambda)$ with respect to Ω is that $\Sigma(L) \cap \Omega$ be a compact subset of $\mathbb{C}$. We shall see later that this condition is also sufficient.

Let us give some examples of linearizations.

EXAMPLE 1.1.2. Let $\mathcal{X} = \mathbb{C}$ and

$$L(\lambda) = (\lambda-\alpha_1)\cdots(\lambda-\alpha_n),$$

where $\alpha_1,\ldots,\alpha_n$ are distinct complex numbers. The number α_j is a linearization of $L(\lambda)$ with respect to a sufficiently small disc containing α_j. In this case, $\mathcal{Z}_1 = \mathcal{Z}_2 = \{0\}$, $E(\lambda) = L(\lambda)/(\lambda-\alpha_j)$ and $F(\lambda) \equiv 1$. ∎

EXAMPLE 1.1.3. Let $\mathcal{X} = \mathbb{C}^n$, and

$$L(\lambda) = \begin{bmatrix} \lambda-\alpha_1 & & & 0 \\ & \lambda-\alpha_2 & & \\ & & \ddots & \\ 0 & & & \lambda-\alpha_n \end{bmatrix}$$

where $\alpha_n,\ldots,\alpha_1$ are distinct complex numbers. Again the number α_j is a linearization of $L(\lambda)$ with respect to a sufficiently small disc containing α_j. In this case $\mathcal{Z}_1 = \{0\}$, $\mathcal{Z}_2 = \mathbb{C}^{n-1}$,

$$E(\lambda) = P \operatorname{diag}[1,\lambda-\alpha_2,\ldots,\lambda-\alpha_{j-1},\lambda-\alpha_{j+1},\ldots,\lambda-\alpha_n],\ F(\lambda) \equiv P,$$

where P is the $n \times n$ matrix obtained from I by permuting the first and the jth rows. ∎

The next example will play a very important role in the sequel.

EXAMPLE 1.1.4. Let

$$L(\lambda) = \lambda^{\ell} I + \sum_{j=0}^{\ell-1} \lambda^{j} A_j$$

be a *monic* operator polynomial, i.e., with leading coefficient I. The operator

$$C_L = \begin{bmatrix} 0 & I & 0 & \cdot & \cdot & \cdot & 0 \\ 0 & 0 & I & \cdot & \cdot & \cdot & 0 \\ \cdot & \cdot & \cdot & & & & \cdot \\ \cdot & \cdot & \cdot & & & & \cdot \\ \cdot & \cdot & \cdot & & & & \cdot \\ 0 & 0 & 0 & \cdot & \cdot & \cdot & I \\ -A_0 & -A_1 & & \cdot & \cdot & \cdot & -A_{\ell-1} \end{bmatrix} : \mathcal{X}^{\ell} \to \mathcal{X}^{\ell}$$

is called the *companion operator* of $L(\lambda)$ (here and elsewhere $\mathcal{X}^{\ell} = \mathcal{X} \oplus \cdots \oplus \mathcal{X}$ ℓ times). By straightforward multiplication one verifies that

$$(L(\lambda) \oplus I_{\mathcal{X}^{\ell-1}})F(\lambda) = E(\lambda)(\lambda I - C_L), \tag{1.1.5}$$

where

$$F(\lambda) = \begin{bmatrix} I & 0 & \cdot & \cdot & \cdot & & 0 \\ -\lambda I & I & \cdot & \cdot & \cdot & & 0 \\ 0 & -\lambda I & & & & & \cdot \\ \cdot & \cdot & & & & & \cdot \\ \cdot & \cdot & & & & & \cdot \\ \cdot & \cdot & & & & & \cdot \\ 0 & 0 & \cdot & \cdot & \cdot & -\lambda I & I \end{bmatrix}, \tag{1.1.6}$$

$$(1.1.7)\qquad E(\lambda) = \begin{bmatrix} B_{\ell-1}(\lambda) & B_{\ell-2}(\lambda) & \cdot & \cdot & \cdot & B_0(\lambda) \\ -I & 0 & \cdot & \cdot & \cdot & 0 \\ 0 & -I & & & & \cdot \\ \cdot & \cdot & & & & \cdot \\ \cdot & \cdot & & & & \cdot \\ \cdot & \cdot & & & & \cdot \\ 0 & 0 & \cdot & \cdot & -I & 0 \end{bmatrix}$$

with $B_0(\lambda) \equiv I$ and $B_n(\lambda) = \lambda^n I + \lambda^{n-1} A_{\ell-1} + \cdots + A_{\ell-n}$ for $1 \leq n \leq \ell-1$. From (1.1.6) and (1.1.7), it is easily seen that $F(\lambda)$ and $E(\lambda)$ are operator polynomials invertible for all $\lambda \in \mathbb{C}$. Moreover, the inverses $F(\lambda)^{-1}$ and $E(\lambda)^{-1}$ are operator polynomials as well:

$$F(\lambda)^{-1} = \begin{bmatrix} I & 0 & \cdot & \cdot & \cdot & 0 \\ \lambda I & I & \cdot & \cdot & \cdot & 0 \\ \lambda^2 I & \lambda I & \cdot & \cdot & \cdot & 0 \\ \cdot & \cdot & & & & \cdot \\ \cdot & \cdot & & & & \cdot \\ \cdot & \cdot & & & & \cdot \\ \lambda^{\ell-1} I & \lambda^{\ell-2} I & \cdot & \cdot & \lambda I & I \end{bmatrix};$$

$$E(\lambda)^{-1} = \begin{bmatrix} 0 & -I & 0 & \cdot & \cdot & \cdot & 0 \\ 0 & 0 & -I & \cdot & \cdot & \cdot & 0 \\ \cdot & \cdot & \cdot & & & & \cdot \\ \cdot & \cdot & \cdot & & & & \cdot \\ \cdot & \cdot & \cdot & & & & \cdot \\ 0 & 0 & 0 & \cdot & \cdot & \cdot & -I \\ I & B_{\ell-1}(\lambda) & B_{\ell-2}(\lambda) & \cdot & \cdot & \cdot & B_1(\lambda) \end{bmatrix}.$$

In view of (1.1.5), C_L is a global linearization of $L(\lambda)$. ∎

Example 1.1.4 makes it natural to introduce the following definition. A global linearization A of an operator polynomial $L(\lambda)$ will be called *polynomially induced* if for some Banach spaces $\mathcal{Z}_1$ and $\mathcal{Z}_2$

$$L(\lambda) \oplus I_{\mathcal{Z}_1} = E(\lambda)((\lambda I-A)+I_{\mathcal{Z}_2})F(\lambda), \quad \lambda \in \mathbb{C}$$

where $E(\lambda)$, $F(\lambda)$, $E(\lambda)^{-1}$ and $F(\lambda)^{-1}$ are operator polynomials (in particular, $E(\lambda)$ and $F(\lambda)$ are analytic and invertible on $\mathbb{C}$). Formulas (1.1.4)-(1.1.7) show that the companion operator is a polynomially induced global linearization. In finite dimensional case ($\dim \mathcal{X} < \infty$) every global linearization of an operator polynomial $L(\lambda)$ with $\det L(\lambda) \not\equiv 0$ is polynomially induced. This follows, for instance, from the Smith form of an $n\times n$ matrix $L(\lambda)$ whose entries are polynomials:

$$L(\lambda) = E(\lambda) \operatorname{diag}[d_1(\lambda),d_2(\lambda),\dots,d_n(\lambda)]F(\lambda),$$

where $E(\lambda)$, $F(\lambda)$ are $n\times n$ matrix polynomials with constant non-zero determinants (so $E(\lambda)^{-1}$ and $F(\lambda)^{-1}$ are matrix polynomials as well) and $d_1(\lambda),\dots,d_n(\lambda)$ are scalar polynomials with leading coefficients 1. The Smith form is a well-known result and its proof can be found, for instance, in Gantmacher [1], Gohberg-Lancaster-Rodman [2,3].

In the infinite dimensional case, not every operator polynomial with compact spectrum admits a polynomially induced global linearization.

EXAMPLE 1.1.5. Let $A \in L(\mathcal{X})$ be a quasinilpotent with $A^n \neq 0$ for $n = 0,1,\dots$. The operator polynomial $I+\lambda A$ has no polynomially induced global linearization (otherwise, its inverse $(I+\lambda A)^{-1}$ would be a polynomial, which is false). The global linearization of $I+\lambda A$ is the operator acting on the trivial Banach space which consists of zero only. ■

1.2 Uniqueness of linearizations

It is evident from the definition of a linearization of the operator polynomial $L(\lambda)$ that the linearization is generally not unique. For instance, any operator similar to a linearization is again a linearization. It turns out, however, that this is the only freedom, as the next result shows. The problem of existence of linearizations (which exists always provided the obvious necessary condition that $\Sigma(L) \cap \Omega$ be compact is satisfied) will be dealt with in the next section.

THEOREM 1.2.1. *Let*

$$L(\lambda) = \sum_{j=0}^{\ell} \lambda^j A_j, \quad A_j \in L(\mathcal{X})$$

be an operator polynomial, and let $\Omega \subset \mathbb{C}$ *be an open set. If* $A \in L(\mathcal{Y})$ *and* $\tilde{A} \in L(\tilde{\mathcal{Y}})$ *are two linearizations of* $L(\lambda)$ *with respect to* Ω, *then* $\tilde{A} = S^{-1}AS$ *for some invertible* $S \in L(\tilde{\mathcal{Y}},\mathcal{Y})$ *(in particular, the Banach spaces* $\mathcal{Y}$ *and* $\tilde{\mathcal{Y}}$ *are isomorphic).*

PROOF. Let $\Omega_1 \subset \Omega_2$ be bounded Cauchy domains such that $\sigma(A) \cup \sigma(\tilde{A}) \subset \Omega_1 \subset \overline{\Omega}_1 \subset \Omega_2 \subset \overline{\Omega}_2 \subset \Omega$. Recall that a bounded open set is called a *Cauchy domain* if its boundary consists of finite number of disjoint closed rectifiable curves. The existence of the bounded Cauchy domain Ω_1 and Ω_2 with the required properties is intuitively clear. A proof of the fact that for any open set $V \subset \mathbb{C}$ and any compact $U \subset V$, there is a bounded Cauchy domain W such that $U \subset W \subset \overline{W} \subset V$ can be done by considering the set $Q = \bigcup_{i=1}^{k} Q_i$, where $Q_1,\ldots,Q_k$ are open discs such that $Q \supset U$ and $\overline{Q} \subset V$.

From the hypotheses of the theorem, we have

$$(\lambda-A) \oplus I_{\mathcal{Z}} = E(\lambda)[(\lambda-\tilde{A}) \oplus I_{\tilde{\mathcal{Z}}}]F(\lambda), \quad \lambda \in \Omega,$$

where $E(\lambda)$ and $F(\lambda)$ are analytic and invertible operator-valued functions on Ω, and $\mathcal{Z}$ and $\tilde{\mathcal{Z}}$ are some Banach spaces. Let $\pi: \mathcal{Y} \oplus \mathcal{Z} \to \mathcal{Y}$ and $\tilde{\pi}: \tilde{\mathcal{Y}} \oplus \tilde{\mathcal{Z}} \to \tilde{\mathcal{Y}}$ be the projections on the first component (so $\pi(y,z) = y$, $y \in \mathcal{Y}$, $z \in \mathcal{Z}$, and analogously for $\tilde{\pi}$), and let

$$\tau: \mathcal{Y} \to \mathcal{Y} \oplus \mathcal{Z}, \quad \tilde{\tau}: \tilde{\mathcal{Y}} \to \tilde{\mathcal{Y}} \oplus \tilde{\mathcal{Z}}$$

be natural imbeddings (so $\tau(y) = (y,0)$, $y \in \mathcal{Y}$; $\tilde{\tau}(\tilde{y}) = (\tilde{y},0)$, $\tilde{y} \in \tilde{\mathcal{Y}}$). Define

$$(1.2.1) \qquad S = \frac{1}{2\pi i} \int_{\partial\Omega_2} (\lambda-\tilde{A})^{-1}\tilde{\pi}E(\lambda)^{-1}\tau d\lambda \in L(\mathcal{Y},\tilde{\mathcal{Y}})$$

and

$$(1.2.2) \qquad T = \frac{1}{2\pi i} \int_{\partial\Omega_1} \pi F(\lambda)^{-1}\tilde{\tau}(\lambda-\tilde{A})^{-1}d\lambda \in L(\tilde{\mathcal{Y}},\mathcal{Y}).$$

It will suffice to show that S and T are inverses to each other and that

$$(1.2.3) \qquad SA = \tilde{A}S.$$

We have:

$$SA = \frac{1}{2\pi i} \int_{\partial\Omega_2} (\lambda-\tilde{A})^{-1}\tilde{\pi}E(\lambda)^{-1}\tau A d\lambda.$$

In view of the equalities

$$F(\lambda)[(\lambda-A)^{-1} \oplus I_{\mathcal{Z}}] = [(\lambda-\tilde{A})^{-1} \oplus I_{\tilde{\mathcal{Z}}}]E(\lambda)^{-1}, \quad \lambda \in \Omega$$

and

$$\tilde{\pi}[(\lambda-\tilde{A})^{-1} \oplus I_{\tilde{\mathcal{Z}}}] = (\lambda-\tilde{A})^{-1}\tilde{\pi}, \quad [(\lambda-A)^{-1} \oplus I_{\mathcal{Z}}]\tau = \tau(\lambda-A)^{-1},$$

it follows that

$$SA = \frac{1}{2\pi i} \int_{\partial\Omega} \tilde{\pi}F(\lambda)\tau(\lambda-A)^{-1}A d\lambda$$

$$= \frac{1}{2\pi i} \int_{\partial\Omega_2} [-\tilde{\pi}F(\lambda)]\tau d\lambda + \frac{1}{2\pi i} \int_{\partial\Omega_2} \lambda\tilde{\pi}F(\lambda)\tau(\lambda-A)^{-1}d\lambda.$$

The first integral here is zero since $F(\lambda)$ is analytic inside and on Ω_2. The second integral is rewritten as follows:

$$\frac{1}{2\pi i}\int_{\partial\Omega_2} \lambda\tilde{\pi}F(\lambda)\tau(\lambda-A)^{-1}d\lambda = \frac{1}{2\pi i}\int_{\partial\Omega_2} \lambda(\lambda-\tilde{A})^{-1}\tilde{\pi}E(\lambda)^{-1}\tau d\lambda$$

$$= \frac{1}{2\pi i}\int_{\partial\Omega_2} \tilde{\pi}E(\lambda)^{-1}\tau d\lambda + \frac{1}{2\pi i}\int_{\partial\Omega_2} \tilde{A}(\lambda-\tilde{A})^{-1}\tilde{\pi}E(\lambda)^{-1}\tau d\lambda$$

$$= \frac{1}{2\pi i}\int_{\partial\Omega_2} \tilde{A}(\lambda-\tilde{A})^{-1}\tilde{\pi}E(\lambda)^{-1}\tau d\lambda = \tilde{A}S,$$

and (1.2.3) is verified.

Further,

$$TS = \frac{1}{4\pi^2}\{\int_{\partial\Omega_1} \pi F(\lambda)^{-1}\tilde{\tau}(\lambda-\tilde{A})^{-1}d\lambda\}\{\int_{\partial\Omega_2} (\mu-\tilde{A})^{-1}\tilde{\pi}E(\mu)^{-1}\tau d\mu\}$$

$$= -\frac{1}{4\pi^2}\int_{\partial\Omega_1}\int_{\partial\Omega_2} \pi F(\lambda)^{-1}\tilde{\tau}(\lambda-\tilde{A})^{-1}(\mu-\tilde{A})^{-1}\tilde{\pi}E(\mu)^{-1}\tau d\mu d\lambda.$$

Using the resolvent equality

$$(\lambda-\mu)(\lambda-\tilde{A})^{-1}(\mu-\tilde{A})^{-1} = (\mu-\tilde{A})^{-1}-(\lambda-\tilde{A})^{-1}$$

rewrite the product TS in the form

$$TS = \frac{1}{4\pi^2}\int_{\partial\Omega_1}\int_{\partial\Omega_2} (\lambda-\mu)^{-1}\pi F(\lambda)^{-1}\tilde{\tau}(\mu-\tilde{A})^{-1}\tilde{\pi}E(\mu)^{-1}\tau d\mu d\lambda$$

$$+ \frac{1}{4\pi^2}\int_{\partial\Omega_1}\int_{\partial\Omega_2} (\lambda-\mu)^{-1}\pi F(\lambda)^{-1}\tilde{\tau}(\lambda-\tilde{A})^{-1}\tilde{\pi}E(\mu)^{-1}\tau d\mu d\lambda$$

$$= -\frac{1}{4\pi^2}\int_{\partial\Omega_2}\{\int_{\partial\Omega_1} (\lambda-\mu)^{-1}\pi F(\lambda)^{-1}d\lambda\}\tilde{\tau}(\mu-\tilde{A})^{-1}\tilde{\pi}E(\mu)^{-1}\tau d\mu$$

$$+ \frac{1}{4\pi^2}\int_{\partial\Omega_1} \pi F(\lambda)^{-1}\tilde{\tau}(\lambda-\tilde{A})^{-1}\tilde{\pi}\{\int_{\partial\Omega_2} (\lambda-\mu)^{-1}E(\mu)^{-1}d\mu\}\tau d\lambda.$$

As the function $(\lambda-\mu)^{-1}\pi F(\lambda)^{-1}$ is analytic (as a function on λ) inside and on Ω_1 for every fixed $\mu \in \partial\Omega_2$, the first summand is zero. To compute the second summand, observe that

$$\frac{1}{2\pi i}\int_{\partial\Omega_2} (\lambda-\mu)^{-1}E(\mu)^{-1}d\mu = -E(\lambda)^{-1}, \quad \lambda \in \partial\Omega_1$$

by the Cauchy's formula, and so

$$TS = \frac{1}{2\pi i}\int_{\partial\Omega_1} \pi F(\lambda)^{-1}\tilde{\tau}(\lambda-\tilde{A})^{-1}\tilde{\pi}E(\lambda)^{-1}\tau d\lambda$$

$$= \frac{1}{2\pi i}\int_{\partial\Omega_1} \pi F(\lambda)^{-1}[(\lambda-\tilde{A})^{-1} \oplus 0_{\tilde{\mathcal{Z}}}]\ E(\lambda)^{-1}\tau d\lambda$$

$$= \frac{1}{2\pi i}\int_{\partial\Omega_1} \pi F(\lambda)^{-1}[(\lambda-\tilde{A})^{-1} \oplus I_{\tilde{\mathcal{Z}}}]\ E(\lambda)^{-1}\tau d\lambda,$$

where the last equality follows in view of the analyticity of the function

$$\pi F(\lambda)^{-1}[0_{\tilde{\mathcal{Y}}} \oplus I_{\tilde{\mathcal{Z}}}]\ E(\lambda)^{-1}\tau$$

inside and on Ω_1. So

$$TS = \frac{1}{2\pi i}\int_{\partial\Omega_1} \pi F(\lambda)^{-1}F(\lambda)[(\lambda-A)^{-1} \oplus I_{\mathcal{Z}}]\tau d\lambda$$

$$= \frac{1}{2\pi i}\int_{\partial\Omega_1} (\lambda-A)^{-1}d\lambda = I.$$

It remains to prove that $ST = I$. To this end observe that

$$(\lambda-\tilde{A})^{-1}\tilde{\pi}E(\lambda)^{-1}\tau = \tilde{\pi}F(\lambda)\tau(\lambda-A)^{-1} \tag{1.2.4}$$

$$\pi F(\lambda)^{-1}\tilde{\tau}(\lambda-\tilde{A})^{-1} = (\lambda-A)^{-1}\pi E(\lambda)\tilde{\tau}. \tag{1.2.5}$$

Indeed, (1.2.4) easily follows from

$$[(\lambda-\tilde{A})^{-1} \oplus I_{\tilde{\mathcal{Z}}}]E(\lambda)^{-1} = F(\lambda)[(\lambda-A)^{-1} \oplus I_{\mathcal{Z}}]$$

after premultiplication by $\tilde{\pi}$ and postmultiplication by τ. The equality (1.2.5) is proved analogously. So

$$S = \frac{1}{2\pi i}\int_{\partial\Omega_1} \tilde{\pi}F(\lambda)\tau(\lambda-A)^{-1}d\lambda, \qquad T = \frac{1}{2\pi i}\int_{\partial\Omega_2} (\lambda-A)^{-1}\pi E(\lambda)\tilde{\tau}.$$

Now the equality $ST = I$ can be proved in the same way as $TS = I$, with the roles of S and T interchanged. ■

Some comments concerning Theorem 1.2.1 are in order. First, a similarity S between two linearizations A and $\tilde{A}$ of $L(\lambda)$ with respect to Ω is given, together with its inverse, by explicit formulas (1.2.1) and (1.2.2). Secondly, the assumption that $\Sigma(L) \cap \Omega$ is compact, is implicit in the hypothesis of Theorem 1.2.1 (this assumption is necessary for the existence of a linearization of $L(\lambda)$ with respect to Ω). Finally, let us indicate the following important corollary from Theorem 1.2.1.

COROLLARY 1.2.2. *Let* $A,B \in L(\mathcal{X})$, *and assume that there exist analytic and invertible* $L(\mathcal{X})$*-valued functions* $E(\lambda)$ *and* $F(\lambda)$ *defined on an open set* Ω *which contains* $\sigma(A) \cup \sigma(B)$, *such that*

$$E(\lambda)(\lambda-A) = (\lambda-B)F(\lambda), \quad \lambda \in \Omega.$$

Then the operators A *and* B *are similar.*

Indeed, both A and B are linearizations of $\lambda-A$ with respect to Ω.

1.3 Existence of linearization

The next theorem shows that a linearization for an operator polynomial $L(\lambda)$ with respect to a *bounded* open set Ω exists provided the necessary condition that $\Sigma(L) \cap \Omega$ is compact is satisfied. We shall see later that the theorem is true also for unbounded Ω. However, here the boundedness of Ω will be used essentially in the proof.

THEOREM 1.3.1. *Let*

$$L(\lambda) = \sum_{j=0}^{\ell} \lambda^j A_j, \quad A_j \in L(\mathcal{X})$$

be an operator polynomial, and let $\Omega_0 \subset \mathbf{C}$ *be a bounded Cauchy domain such that* $\Sigma(L) \cap \Omega_0$ *is compact. Then there is a linearization* $A \in L(\mathcal{Y})$ *of* $L(\lambda)$ *with respect to* Ω*. If* $\mathcal{X}$ *is a Hilbert space (resp. a separable Hilbert space), then* $\mathcal{Y}$ *can be chosen a Hilbert space (resp. a separable Hilbert space) as well.*

PROOF. Without loss of generality, we shall assume that $0 \in \Omega_0$ (otherwise replace $L(\lambda)$ by $L(\lambda-\lambda_0)$ and Ω_0 by $\{\lambda-\lambda_0 \mid \lambda \in \Omega_0\}$.

Denote by $C(\partial\Omega_0,\mathcal{X})$ the Banach space of all $\mathcal{X}$-valued continuous functions on $\partial\Omega_0$ endowed with the supremum norm.

Let $P \in L(C(\partial\Omega_0,\mathcal{X}))$ be defined by

$$(Pf)(z) = \frac{1}{2\pi i}\int_{\partial\Omega_0} \zeta^{-1}f(\zeta)d\zeta, \qquad z \in \partial\Omega_0.$$

Observe that Pf is a constant function and (by the Cauchy's integral formula) Pf = f for every constant function $f \in C(\partial\Omega_0,\mathcal{X})$. So P is a projection of $C(\partial\Omega_0,\mathcal{X})$ whose range may be identified with $\mathcal{X}$. We have the direct sum decomposition

$$C(\partial\Omega_0,\mathcal{X}) = \mathcal{X} \dotplus \operatorname{Ker} P.$$

Next define $V \in L(C(\partial\Omega_0,\mathcal{X}))$ by

$$(Vf)(z) = zf(z), \quad z \in \partial\Omega_0.$$

Note that Ω_0 belongs to the resolvent set of V. In fact, for each $\lambda \in \Omega_0$ we have

$$(V-\lambda I)^{-1}f(z) = (z-\lambda)^{-1}f(z), \qquad z \in \partial\Omega_0.$$

Further, let $M \in L(C(\partial\Omega_0,\mathcal{X}))$ be defined by

$$(Mf)(z) = L(z)f(z), \qquad z \in \partial\Omega_0.$$

Using the definitions and the Cauchy integral formula one sees easily that

$$PV(V-\lambda I)^{-1}P = P, \qquad \lambda \in \Omega_0, \tag{1.3.1}$$

and

$$PV(V-\lambda I)^{-1}MP = L(\lambda)P, \quad \lambda \in \Omega_0. \tag{1.3.2}$$

Observe also that M commutes with V and hence with $(V-\lambda I)^{-1}$ for every $\lambda \in \Omega_0$.

Finally, put

$$T = V-PV+PVM$$

and

$$B(\lambda) = (T-\lambda I)(V-\lambda I)^{-1}, \quad \lambda \in \Omega_0.$$

Then

$$B(\lambda) = I-PV(V-\lambda I)^{-1}+PV(V-\lambda I)^{-1}M$$

for all $\lambda \in \Omega_0$. But then one can use the identities (1.3.1) and (1.3.2) to show that for each $\lambda \in \Omega_0$

$$B(\lambda)P = PL(\lambda)P, \quad (I-P)B(\lambda) = I-P. \tag{1.3.3}$$

Formula (1.3.3) implies that for each $\lambda \in \Omega_0$ the matrix of $B(\lambda)$ with respect to the direct sum decomposition $C(\partial\Omega_0,\mathcal{X}) = \mathcal{X} \dot{+} \mathcal{Z}$ (here $\mathcal{Z}$ = Ker P) is given by

$$B(\lambda) = \begin{bmatrix} L(\lambda) & C(\lambda) \\ 0 & I_{\mathcal{Z}} \end{bmatrix}, \tag{1.3.4}$$

where $C: \Omega_0 \to L(\mathcal{Z},\mathcal{X})$ is analytic operator function. But then

$$T-\lambda I = \begin{bmatrix} I_{\mathcal{X}} & C(\lambda) \\ 0 & I_{\mathcal{Z}} \end{bmatrix} \begin{bmatrix} L(\lambda) & 0 \\ 0 & I_{\mathcal{Z}} \end{bmatrix} (V-\lambda I) \tag{1.3.5}$$

for each $\lambda \in \Omega$. In the right hand side of (1.3.5), the first and third factor are invertible operators on $\mathcal{X} \dot{+} \mathcal{Z}$. Formula (1.3.5) shows in particular that $\sigma(T) \cap \Omega_0 = \Sigma(L) \cap \Omega_0$ which is compact. So $\sigma(T)$ is a union of two disjoint compact sets $\sigma(T) \cap \Omega_0$ and $\sigma(T) \cap (\mathbb{C}\setminus\Omega_0)$, and consequently there is a direct sum decomposition

$$C(\partial\Omega_0,\mathcal{X}) = \mathcal{Y}_1 \dot{+} \mathcal{Y}_2,$$

where $\mathcal{Y}_1$ and $\mathcal{Y}_2$ are T-invariant subspaces such that $\sigma(T|\mathcal{Y}_1) = \sigma(T) \cap \Omega_0$ and $\sigma(T|\mathcal{Y}_2) = \sigma(T) \cap (\mathbf{C}\backslash\Omega_0)$. With respect to this decomposition, rewrite (1.3.5) in the form

$$\begin{bmatrix} T|\mathcal{Y}_1-\lambda I_{\mathcal{Y}_1} & 0 \\ 0 & I_{\mathcal{Y}_2} \end{bmatrix} = \begin{bmatrix} I_{\mathcal{X}} & C(\lambda) \\ 0 & I_{\mathcal{Z}} \end{bmatrix} \begin{bmatrix} L(\lambda) & 0 \\ 0 & I_{\mathcal{Z}} \end{bmatrix} (V-\lambda I) \cdot \begin{bmatrix} T|\mathcal{Y}_1-\lambda I_{\mathcal{Y}_1} & 0 \\ 0 & I_{\mathcal{Y}_2} \end{bmatrix}$$

This equality shows that $T|\mathcal{Y}_1$ is a linearization of $L(\lambda)$ with respect to Ω.

In case $\mathcal{X}$ is a Hilbert space or a separable Hilbert space, use $L_2(\partial\Omega_0,\mathcal{X})$ in the above arguments in place of $C(\partial\Omega_0,\mathcal{X})$. For the properties of L_2 spaces of functions with values in a Hilbert space, we refer the reader to Mikusinski [1]. ∎

Note that the proof of Theorem 1.3.1 furnishes also a formula for a linearization of $L(\lambda)$. Namely, let $T \in L(C(\partial\Omega_0,\mathcal{X}))$ be defined by

$$(Tf)(z) = zf(z) - \frac{1}{2\pi i}\int_{\partial\Omega_0} [I-L(\varsigma)]f(\varsigma)d\varsigma$$

(it is assumed that $0 \in \Omega_0$). Then

$$A = T|\operatorname{Im} Q,$$

where

$$Q = \frac{1}{2\pi i}\int_{\partial\Omega_1} (\lambda I-T)^{-1}d\lambda,$$

and Ω_1 is another Cauchy domain such that $\overline{\Omega}_1 \subset \Omega_0$ and $\Sigma(L) \cap \Omega_0 \subset \Omega_1$, is a linearization of $L(\lambda)$ with respect to Ω_0.

1.4 Operator polynomials that are multiples of identity modulo compacts

Let $\mathcal{X}$ be a Banach space, and consider the following algebra of operators:

$$\mathcal{A} = \{X \in L(\mathcal{X}) \mid X = \alpha I+K \text{ for some } \alpha \in \mathbf{C} \text{ and some compact operator } K\}.$$

Thus, $X \in \mathcal{A}$ if and only if X is a scalar multiple of I modulo the ideal of compact operators. In this section, we develop some basic spectral properties of operator polynomials with values in $\mathcal{A}$. More generally, we consider also the algebras

$$\mathcal{A}_m = \{X \in L(\mathcal{X}^m) \mid X = [X_{ij}]_{i,j=1}^m \text{ where } X_{ij} \in \mathcal{A}\}$$

for m = 1,2,... . In particular, $\mathcal{A}_1 = \mathcal{A}$.

We start with a simple proposition, the proof of which is left to the reader.

PROPOSITION 1.4.1. *An operator polynomial* $L(\lambda) = \sum_{j=0}^{\ell} \lambda^j L_j$ *has the property that* $L(\lambda) \in \mathcal{A}_m$ *for all* $\lambda \in \mathbb{C}$ *(here* m *is fixed) if and only if all its coefficients* L_j *belong to* $\mathcal{A}_m$.

Thus, we study operator polynomials with coefficients in $\mathcal{A}_m$.

Let $L(\lambda)$ be an operator polynomial. We say that a point $\lambda_0 \in \mathbb{C}$ is a *normal eigenvalue* of $L(\lambda)$ if $L(\lambda_0)$ is not invertible, $L(\lambda)$ is invertible for all λ with $0 < |\lambda-\lambda_0| < \varepsilon$ (where $\varepsilon > 0$ is sufficiently small), and if $L(\lambda_0)$ is a Fredholm operator. Recall that the index of a Fredholm operator $A \in L(\mathcal{X})$ is defined by

$$\text{ind } A = \dim \text{Ker } A - \dim(\mathcal{X}/\text{Im } A).$$

It follows from the perturbation theory for Fredholm operators (see, e.g., Schechter [1]) that given a Fredholm operator $A \in L(\mathcal{X})$ there is $\varepsilon > 0$ such that any operator $B \in L(\mathcal{X})$ with $\|B-A\| < \varepsilon$ is Fredholm and ind B = ind A. As a consequence, we obtain that ind $L(\lambda_0) = 0$ for a normal eigenvalue λ_0 of an operator polynomial $L(\lambda)$. In particular, λ_0 is indeed an eigenvalue of $L(\lambda)$, i.e., Ker $L(\lambda_0) \neq \{0\}$.

The condition of being a normal eigenvalue can be expressed in terms of linearizations:

PROPOSITION 1.4.2. *Let* $L(\lambda)$ *be an operator polynomial such that* λ_0 *is an isolated point of* $\Sigma(L)$. *Then* λ_0 *is a normal eigenvalue of* $L(\lambda)$ *if and only if a linearization* C *of* $L(\lambda)$ *with respect to the disc* $\{\lambda \mid |\lambda-\lambda_0| < \varepsilon\}$ *for* $\varepsilon > 0$ *sufficiently small acts in a finite dimensional space*: $C \in L(\mathcal{Y})$ *with* $\dim \mathcal{Y} < \infty$.

The existence of such linearization C is ensured by Theorem 1.3.1, and its uniqueness (Theorem 1.2.1) implies that Proposition 1.4.2 does not depend on the choice of C. The dimension of the finite dimensional space on which C acts is called the *algebraic multiplicity* of λ_0 (as a normal eigenvalue of $L(\lambda)$).

PROOF. By definition of a linearization, we have

$$L(\lambda) \oplus I_{\mathcal{Z}_1} = E(\lambda)((\lambda I-C) \oplus I_{\mathcal{Z}_2})F(\lambda), \qquad \lambda \in \Omega, \tag{1.4.1}$$

where $E(\lambda)$ and $F(\lambda)$ are analytic and invertible in the disc $\Omega = \{\lambda \in \mathbf{C} \mid |\lambda-\lambda_0| < \varepsilon\}$. If C acts in a finite dimensional space, then obviously $(\lambda_0 I-C) \oplus I_{\mathcal{Z}_2}$ is Fredholm, and hence, so is $L(\lambda_0)$. Conversely, assume $L(\lambda_0)$ is Fredholm. Then so is $\lambda_0 I-C$, and the index of $\lambda_0 I-C$ is zero because $\operatorname{ind} L(\lambda_0) = 0$. Let $\mathcal{Y}$ be the Banach space on which C acts, and consider the closed ideal $\mathcal{C}(\mathcal{Y})$ of compact operators acting in $\mathcal{Y}$. There is a natural algebra homomorphism ϕ: $L(\mathcal{Y}) \to L(\mathcal{Y})/\mathcal{C}(\mathcal{Y})$ defined by $\phi(X) = X+\mathcal{C}(\mathcal{Y})$ for any $X \in L(\mathcal{Y})$. Observe that the quotient $L(\mathcal{Y})/\mathcal{C}(\mathcal{Y})$ is a Banach algebra with the norm

$$\|X+\mathcal{C}(\mathcal{Y})\| = \inf_{K\in\mathcal{C}(\mathcal{Y})} \|X+K\|.$$

The next observation is that

$$\lambda I-\phi(C)$$

is invertible in $L(\mathcal{Y})/\mathcal{C}(\mathcal{Y})$ for all $\lambda \in \mathbf{C}$ (here I stands for the identify in the Banach algebra $L(\mathcal{Y})/\mathcal{C}(\mathcal{Y})$). Indeed, if $\lambda \neq \lambda_0$, then $\lambda I-C$ is invertible (in $L(\mathcal{Y})$), and hence $\phi(\lambda I-C) = \lambda I-\phi(C)$ is invertible as well. In the case $\lambda = \lambda_0$, there is a compact (even finite rank) operator $K \in L(\mathcal{Y})$ such that $\lambda_0 I-C+K$ is invertible (because $\lambda_0 I-C$ is Fredholm with index zero). So $\phi(\lambda_0 I-C+K) = \lambda_0 I-\phi(C)$ is also invertible.

We have found taht the spectrum of the element $\phi(C)$ in the Banach algebra $L(\mathcal{Y})/\mathcal{C}(\mathcal{Y})$ is empty. This implies that the

Banach algebra is the zero algebra, i.e., $\mathcal{C}(\mathcal{Y}) = L(\mathcal{Y})$. In particular, $I_{\mathcal{Y}}$ is a compact operator. It remains to use the well-known fact that the identity operator on a Banach space is compact if and only if this Banach space is finite dimensional. (A transparent proof of this fact can be found, for instance, in Kreyszig [1], Theorem 2.5-5.) ■

We return now to the operator polynomials with coefficients in A_m. Here is the main result of this section:

THEOREM 1.4.3. *Let* $L(\lambda)$ *be an operator polynomial with coefficients in* A_m:

$$L(\lambda) = \sum_{j=0}^{\ell} \lambda^j L_j,$$

where

$$L_j = \left[\alpha_{pq}^{(j)} I + K_{pq}^{(j)}\right]_{p,q=1}^{m}$$

with complex numbers $\alpha_{pq}^{(j)}$ *and compact operators* $K_{pq}^{(j)}$. *Let*

$$\hat{L}(\lambda) = \sum_{j=0}^{\ell} \lambda^j \hat{L}_j,$$

where

$$\hat{L}_j = [\alpha_{pq}^{(j)}]_{p,q=1}^{m}$$

is an $m \times m$ *matrix. Then the spectrum of* $L(\lambda)$ *contains all the zeros of* $\det \hat{L}(\lambda)$, *and under the additional assumption that* $\Sigma(L) \neq \mathbb{C}$, *every point in* $\Sigma(L)$ *which is not a zero of* $\det \hat{L}(\lambda)$ *is a normal eigenvalue of* $L(\lambda)$.

PROOF. Let $\lambda_0 \in \mathbb{C}$ be such that $\det \hat{L}(\lambda_0) \neq 0$. Define the operator polynomial

$$\tilde{L}(\lambda) = \sum_{j=0}^{\ell} \lambda^j \left[\alpha_{pq}^{(j)} I_{\mathcal{X}}\right]_{p,q=1}^{m}.$$

Then obviously $\tilde{L}(\lambda_0)$ is invertible, and since the difference $L(\lambda_0) - \tilde{L}(\lambda_0)$ is compact, the operator $L(\lambda_0)$ is Fredholm with index zero.

It remains to prove that every point $\lambda_0 \in \Sigma(L)$ with $\det \hat{L}(\lambda_0) \neq 0$ is an isolated point of spectrum of $L(\lambda)$.

Let $\Omega = \{\lambda_0 \in \mathbb{C} \mid \det \hat{L}(\lambda_0) = 0\}$ and consider the connected open set $\mathbb{C}\backslash\Omega$. We now prove the following auxilliary statement:

Given $\lambda_0 \in \mathbb{C}\backslash\Omega$, *there is* $\varepsilon > 0$ *such that* $\dim \operatorname{Ker} L(\lambda)$ *is constant for all* λ *in the punctured disc* $\{\lambda \in \mathbb{C} \mid 0 < |\lambda-\lambda_0| < \varepsilon\}$.

For the proof, use the fact that $L(\lambda_0)$ is Fredholm with index 0, so let $\mathcal{N}$ be a direct complement to $\operatorname{Im} L(\lambda_0)$, with a basis $g_1,\dots,g_n$. Further, let $f_1,\dots,f_n$ be a linearly independent set in the dual space $\mathcal{X}^*$ such that $f_j|_{\mathcal{M}} = 0$, $j = 1,\dots,n$ for some direct complement $\mathcal{M}$ to $\operatorname{Ker} L(\lambda_0)$, and $f_j(e_k) = \delta_{jk}$ $(1 \le j,k \le n)$, where $e_1,\dots,e_n$ is a fixed basis in $\operatorname{Ker} L(\lambda_0)$. Put

$$A(\lambda) = L(\lambda) - \sum_{j=1}^{n} f_j(\cdot)g_j.$$

Clearly, $A(\lambda)$ is an operator polynomial, and $\operatorname{Ker} A(\lambda_0) = \{0\}$, $\operatorname{Im} A(\lambda_0) = \mathcal{X}$. Thus, $A(\lambda_0)$ is invertible, and consequently $A(\lambda)$ is invertible for all λ sufficiently close to λ_0. Further, $x \in \operatorname{Ker} L(\lambda)$ if and only if

$$A(\lambda)x = \sum_{j=1}^{n} f_j(x)g_j,$$

or

$$x = \sum_{j=1}^{n} \alpha_j A(\lambda)^{-1}g_j, \tag{1.4.2}$$

where

$$\alpha_j = f_j(x). \tag{1.4.3}$$

Substituting x from (1.4.2) into (1.4.3), we obtain the following system of linear equations to determine α_j:

$$\alpha_j = \sum_{k=1}^{n} \alpha_k f_j(A(\lambda)^{-1}g_k), \qquad j = 1,\dots,n. \tag{1.4.4}$$

Clearly, the number of linearly independent solutions $\{\alpha_j\}_{j=1}^n$ of (1.4.4) coincides with the dimension of Ker $L(\lambda)$, for $|\lambda-\lambda_0| < \varepsilon_1$, where $\varepsilon_1 > 0$ is sufficiently small.

The matrix $D(\lambda)$ of the system (1.4.4) is an analytic function on λ for $|\lambda-\lambda_0| < \varepsilon_1$. Let p be the largest size of a not identically zero minor of $D(\lambda)$, and let $M(\lambda)$ be some minor of $D(\lambda)$ of size p with $M(\lambda) \not\equiv 0$. As $M(\lambda)$ is an analytic function of λ, for some ε, $0 < \varepsilon \leq \varepsilon_1$, the punctured disc

$$\{\lambda \in \mathbb{C} \mid 0 < |\lambda-\lambda_0| < \varepsilon\}$$

does not contain zeros of $M(\lambda)$. Then clearly

$$\dim \operatorname{Ker} L(\lambda) = n-p$$

for all λ in this punctured disc, and the auxilliary statement is proved.

Now we can easily finish the proof of Theorem 1.4.3. Let $\lambda_0 \in \mathbb{C}\backslash\Omega$ be the spectrum of $L(\lambda)$, and let $\lambda_1 \in \mathbb{C}\backslash\Omega$ be such that $L(\lambda_1)$ is invertible. Pick a connected compact set $K \subset \mathbb{C}\backslash\Omega$ that contains both points λ_0 and λ_1. For every $\lambda' \in K$ let $\varepsilon(\lambda') > 0$ be such that dim Ker $L(\lambda)$ is constant for $0 < |\lambda-\lambda'| < \varepsilon(\lambda')$. Using compactness of K, choose a finite set $\lambda^{(1)},\ldots,\lambda^{(m)} \in K$ such that

$$K \subset \bigcup_{j=1}^{m} \{\lambda \in \mathbb{C} \mid |\lambda-\lambda^{(j)}| < \varepsilon(\lambda^{(j)})\}.$$

It is now clear that dim Ker $L(\lambda)$ is constant for all $\lambda \in K\backslash\{\lambda^{(1)},\ldots,\lambda^{(m)}\}$. Since dim Ker $L(\lambda) = 0$ in a neighborhood of λ_1, we have dim Ker $L(\lambda) = 0$ in a punctured neighborhood of λ_0. So λ_0 is an isolated point in $\Sigma(L)$. ∎

1.5 Inverse linearizations of operator polynomials

We consider here operator polynomials of the form

$$L(\lambda) = I - \sum_{j=0}^{m} \lambda^j H_j, \tag{1.5.1}$$

where $H_j \in L(\mathcal{X})$. For such operators, it is useful to introduce the following definition: An operator $T \in L(\mathcal{Y})$, where $\mathcal{Y}$ is some Banach space, is called an *inverse linearization* of $L(\lambda)$ if

$$L(\lambda) \oplus I_{\mathcal{Z}} = E(\lambda)(I_{\mathcal{Y}}-\lambda T)F(\lambda), \quad \lambda \in \mathbf{C},$$

where $\mathcal{Z}$ is some Banach space and $E(\lambda)$ and $F(\lambda)$ are some operator-valued functions $E(\lambda)$ and $F(\lambda)$ that are analytic and invertible on $\mathbf{C}$. If both $E(\lambda)$ and $F(\lambda)$, as well as their inverses, are operator polynomials we say that the inverse linearization T is *polynomially induced*. One could also define inverse linearizations with respect to open sets in $\mathbf{C}$, but this notion will not be used in the sequel.

In contrast with the linearization (Theorem 1.2.1), the inverse linearization is not unique up to similarity (see Exercise 1.6).

In general, the operator polynomial $L(\lambda)$ need not have compact spectrum. However, assume this happens, and A be a linearization of $L(\lambda)$ with respect to $\mathbf{C}$ (the existence of such linearizations will be proved later; see Section 6.3):

$$L(\lambda) \oplus I_{\mathcal{Z}_1} = E(\lambda)((\lambda I-A) \oplus I_{\mathcal{Z}_2})F(\lambda), \quad \lambda \in \mathbf{C},$$

where $E(\lambda)$ and $F(\lambda)$ are analytic and invertible on $\mathbf{C}$. Then it is easy to see that A is invertible and the operator $A^{-1} \oplus 0_{\mathcal{Z}_2}$ is an inverse linearization for $L(\lambda)$.

We now construct a particular inverse linearization for $L(\lambda)$. Let $L(\lambda)$ be given by (1.5.1). For $n = 1,\dots,m$, let A_{jn} $(j = 1,\dots,n)$ be operators acting on $\mathcal{X}$ such that

$$H_n = A_{1n}A_{2n}\cdots A_{nn}, \quad (n = 1,\dots,m). \tag{1.5.2}$$

Form the operator $T_L = [T_L]_{i,j=1}^{2^{m-1}}$ acting on $\mathcal{X} \oplus\cdots\oplus \mathcal{X}$ (2^{m-1} times) by setting

(1.5.3)

$$[T_L]_{ij} = \begin{cases} A_{1\alpha} & ; \ i = 1, \ j = 2^{\alpha-1}, \ 1 \le \alpha \le m; \\ A_{\alpha\beta} & ; \ i = 2^{\beta-1}-2^{\alpha-2}+1, \ j = 2^{\beta-1}-2^{\alpha-1}+1, \ 2 \le \alpha \le \beta \le m; \\ 0 & ; \ \text{otherwise}. \end{cases}$$

For example, for m = 4 we have

$$T_L = \begin{bmatrix} A_{11} & A_{12} & 0 & A_{13} & 0 & 0 & 0 & A_{14} \\ A_{22} & 0 & 0 & 0 & 0 & 0 & 0 & 0 \\ A_{33} & 0 & 0 & 0 & 0 & 0 & 0 & 0 \\ 0 & 0 & A_{23} & 0 & 0 & 0 & 0 & 0 \\ A_{44} & 0 & 0 & 0 & 0 & 0 & 0 & 0 \\ 0 & 0 & 0 & 0 & 0 & 0 & 0 & 0 \\ 0 & 0 & 0 & 0 & A_{34} & 0 & 0 & 0 \\ 0 & 0 & 0 & 0 & 0 & 0 & A_{24} & 0 \end{bmatrix}.$$

It is easy to see that every row and column of T_L, except the first, contains at most one entry different from zero. Indeed, if

$$i = 2^{\beta_1-1}-2^{\alpha_1-2}+1 = 2^{\beta_2-1}-2^{\alpha_2-2}+1,$$

for $\beta_2 > \beta_1$, then

$$2^{\beta_2-1}-2^{\beta_1-1} = 2^{\alpha_2-2}-2^{\alpha_1-2}.$$

Thus

$$2^{\beta_1-1}(2^{\beta_2-\beta_1}-1) = 2^{\alpha_1-2}(2^{\alpha_2-\alpha_1}-1),$$

hence

$$2^{\beta_1+1-\alpha_1}(2^{\beta_2-\beta_1}-1) = 2^{\alpha_2-\alpha_1}-1.$$

Since $\alpha_1 < \beta_1+1$, the left hand side of the last equality is even. So, $\alpha_2 = \alpha_1$. Similarly, one can see that the only column which may contain more than one entry different from zero is the first one. The possible non-zero entries in the first column are: $[T_L]_{11}(= A_{11})$ and $[T_L]_{i1}(= A_{\alpha+1,\alpha+1})$ for $i = 2^{\alpha-1}+1$, $\alpha \ge 1$.

Next, introduce the operator

$$S = T_L - \tau\pi T_L,$$

where $\tau: \mathcal{X} \to \mathcal{X}^{2^{m-1}}$ denotes the canonical embedding from $\mathcal{X}$ into the first coordinate space of $\mathcal{X}^{2^{m-1}}$, and $\pi: \mathcal{X}^{2^{m-1}} \to \mathcal{X}$ is the canonical projection onto the first coordinate space of $\mathcal{X}^{2^{m-1}}$. It is not difficult to see that S is block lower triangular with zeros on the main diagonal: $S = [S_{ij}]_{i,j=1}^{2^{m-1}}$ where $S_{ij} = 0$ for $i \geq j$. Consequently, S is nilpotent and $I-\lambda S$ is an invertible operator polynomial in $\mathbb{C}$ whose inverse is again a polynomial.

THEOREM 1.5.1. *The operator* T_L *is a polynomially induced inverse linearization of* $L(\lambda)$. *In more detail,*

$$L(\lambda) \oplus I_{\mathcal{X}^p} = E(\lambda)(I-\lambda T_L)(I-\lambda S)^{-1}, \qquad \lambda \in \mathbb{C}, \tag{1.5.4}$$

where $p = 2^{m-1}-1$, *the operator* S *is defined above, and the invertible operator polynomial* $E(\lambda)$ *with polynomial inverse is given by*

$$E(\lambda) = I + \sum_{\nu=2}^{m} \lambda^{\nu-1} E_\nu$$

with $E_\nu = [E_{\nu_{ij}}]_{i,j=1}^{2^{m-1}}$ *and*

$$E_{\nu_{ij}} = \begin{cases} A_{1\beta}A_{2\beta}\cdots A_{\nu-1,\beta} & \text{if } i = 1,\ j = 2^{\beta-1}-2^{\nu-2}+1;\ 2 \leq \nu \leq \beta \leq m; \\ 0 & \text{otherwise.} \end{cases}$$

PROOF. Let us verify first that $E(\lambda)$ indeed has all the required properties. All the entries in the difference $E(\lambda)-I$ are zeros with the possible exception of the off-diagonal entries in the first row. Hence, $E(\lambda)$ is invertible for all $\lambda \in \mathbb{C}$ and

$$E(\lambda)^{-1} = I - \sum_{\nu=2}^{m} \lambda^{\nu-1} E_\nu .$$

Now consider the formula (1.5.4). The block rows (except for the first block row) of $I-\lambda T_L$ and $I-\lambda S$ are the same; so the second, third, etc., block rows of $L(\lambda) \oplus I_{\mathfrak{X}^p}$ and $E(\lambda)(I-\lambda T_L)(I-\lambda S)^{-1}$ are the same. The first block row of $(L(\lambda) \oplus I)(I-\lambda S)$ is $[L(\lambda)0\cdots]$; we shall see that the same is true for the first block row of the operator function $E(\lambda)(I-\lambda T_L)$. Indeed, the $(n,1)$ entry of $I-\lambda T_L$ is

$$(I-\lambda T_L)_{n1} = \begin{cases} I-\lambda A_{11} & \text{for } n = 1; \\ -\lambda A_{\alpha+1,\alpha+1} & \text{for } n = 2^{\alpha-1}+1; \\ 0 & \text{otherwise.} \end{cases}$$

The $(1,1)$ entry of $E(\lambda)$ is $(E(\lambda))_{11} = I$, and its $(1,2^{\alpha-1}+1)$ entry is

$$(E(\lambda))_{1,2^{\alpha-1}+1} = \lambda^{\alpha} \prod_{j=1}^{\alpha} A_{j,\alpha+1}, \quad 1 \leq \alpha \leq m-1.$$

Therefore

$$\sum_{n=1}^{2^{m-1}} (E(\lambda))_{1n}(I-\lambda T_L)_{n1} = I-\lambda A_{11}-\lambda \sum_{\alpha=1}^{m-1} \lambda^{\alpha} A_{1,\alpha+1}\cdots A_{\alpha+1,\alpha+1} = L(\lambda).$$

Next we prove that

$$\sum_{n=1}^{2^{m-1}} (E(\lambda))_{1n}(I-\lambda T_L)_{nj} = 0 \tag{1.5.5}$$

for $j > 1$. Indeed, for $j = 2^{\alpha}$ $(\alpha \geq 1)$ we have

$$(I-\lambda T_L)_{n,2^{\alpha}} = \begin{cases} -\lambda A_{1,\alpha+1} & \text{if } n = 1 \\ I & \text{if } n = 2^{\alpha} \\ 0 & \text{otherwise} \end{cases}$$

and

$$(E(\lambda))_{11} = I; \ (E(\lambda))_{1,2^{\alpha}} = \lambda A_{1,\alpha+1}.$$

So, (1.5.5) holds for $j = 2^{\alpha}$. For $j = 2^{\beta-1}-2^{\alpha-1}+1$, $2 \le \alpha \le \beta \le m$, we have

$$(I-\lambda T_L)_{nj} = \begin{cases} I & \text{for } n = j \\ -\lambda A_{\alpha\beta} & \text{for } n = 2^{\beta-1}-2^{\alpha-2}+1 \\ 0 & \text{otherwise,} \end{cases}$$

and

$$(E(\lambda))_{1j} = \lambda^{\alpha} A_{1\beta} \cdots A_{\alpha\beta}; \quad (E(\lambda))_{1i} = \lambda^{\alpha-1} A_{1\beta} \cdots A_{\alpha-1,\beta},$$

where $i = 2^{\beta-1}-2^{\alpha-1}+1$. For all other values of $j > 1$, (1.5.5) is evident. This completes the proof of (1.5.4). ∎

We now specialize to a particular case when the operator H_j in (1.5.1) are compact. *From now until the end of this section, it will be assumed that $\mathfrak{X}$ is an infinite-dimensional separable Hilbert space.*

Let $H_n = U_n D_n$ $(n = 1,\dots,m)$ be a polar decomposition of H_n (so $D_n = (H_n^* H_n)^{1/2}$ and U_n is a partial isometry that maps $\operatorname{Im} H_n^*$ onto $\operatorname{Im} H_n$). Put $A_{1n} = U_n D_n^{1/n}$ and $A_{jn} = D_n^{1/n}$, $j = 2,\dots,n$. Then the equalities (1.5.2) are satisfied. The operator T_L by (1.5.3) with this special choice of A_{jn} will be called the *special inverse linearization* of $L(\lambda)$. Since H_j are compact operators, so are A_{jn} $(j = 1,\dots,n)$, and consequently the special inverse linearization is compact.

We need at this point to introduce ideals of compact operators. (For extensive treatment of this topic, see the books Gohberg-Kreĭn [1]; Dunford-Schwartz [2]; Simon [1].) Let $A \in L(\mathfrak{X})$ be a compact operator. Then the positive semidefinite operator $(A^*A)^{1/2}$ is compact as well. So the non-zero eigenvalues of $(A^*A)^{1/2}$ form a non-increasing sequence of positive numbers

$$s_1(A) \ge s_2(A) \ge s_3(A) \ge \cdots . \tag{1.5.6}$$

We go by the conventions that a non-zero eigenvalue of $(A^*A)^{1/2}$ is repeated in (1.5.6) the number of times equal to its algebraic multiplicity (recall that all non-zero eigenvalues are normal), and that in case $(A^*A)^{1/2}$ has only finite number of non-zero eigenvalues, the sequence (1.5.6) is augmented by infinite number of zeros. So, in any case, the sequence (1.5.6) is infinite.

A compact operator A is said to belong to the *class* S_p (where $p > 0$) if the sequence (1.5.6) has the property that

$$\sum_{j=1}^{\infty} (s_j(A))^p < \infty.$$

The class S_1 which is of special importance is called the *trace class*. We note that if $A \in S_p$ and $B \in L(\mathcal{X})$, then both products AB and BA belong to S_p (this follows from the inequalities

$$s_j(BA) \le \|B\| s_j(A); \; s_j(AB) \le \|B\| s_j(A); \quad j = 1,2,\ldots$$

(see Section II.2 in Gohberg-Kreĭn [1])). Also note that for $p \ge 1$, the class S_p is closed under addition: if $A,B \in S_p$, then also $A+B \in S_p$. Indeed, we have (see, e.g., Section II.4 in Gohberg-Kreĭn [1])

$$\sum_{j=1}^{k} (s_j(A+B))^p \le \sum_{j=1}^{k} (s_j(A)+s_j(B))^p, \quad k = 1,2,\ldots,$$

and an application of the Minkowski's inequality proves our claim. It follows from these two observations that S_p is an *ideal* in $L(\mathcal{X})$ for every $p \ge 1$. The ideal S_p (for $p \ge 1$) is a Banach space with the norm

$$\|A\|_p = \left(\sum_{j=1}^{\infty} (s_j(A))^p\right)^{1/p}.$$

It will be convenient to use the notation S_∞ to designate the ideal of all compact operators in $L(\mathcal{X})$.

THEOREM 1.5.2. *The special inverse linearization of* $L(\lambda)$ *belongs to the class* S_p *for some fixed* p, $1 \le p \le \infty$ *if and only if*

$$H_n \in S_{p/n}, \quad n = 1,2,\ldots,m.$$

The proof of Theorem 1.5.2 is based on the following fact.

PROPOSITION 1.5.3. *An operator* $X = [X_{ij}]_{i,j=1}^{k} \in L(\mathfrak{X}^k)$ *belongs to* S_p *(where* $1 \leq p \leq \infty$ *and p is fixed) if and only if all operators* X_{ij} *belong to* S_p.

PROOF. We use the notation $[Y^{(ij)}]$ to designate the operator in $L(\mathfrak{X}^k)$ whose (i,j) operator entry is $Y \in L(\mathfrak{X})$ and all other entries are zeros.

Assume all X_{ij} belong to S_p. It is easy to see that

$$s_q(X_{ij}) = s_q([X_{ij}^{(ij)}]), \quad q = 1,2,\dots .$$

So $[X_{ij}^{(ij)}] \in S_p$ for all $1 \leq i,j \leq k$. As

$$X = \sum_{i,j=1}^{k} [X_{ij}^{(ij)}]$$

and S_p is closed under addition, the operator X is in S_p.

Assume now that $X \in S_p$. We use the following property of the numbers $s_q(A)$ (see, e.g., Theorem II.2.1 in Gohberg-Kreĭn [1]):

$$s_q(A) = \min \|A-K\| \tag{1.5.7}$$

where the minimum is taken over all finite rank operators K with $\dim \operatorname{Im} K \leq q-1$. Applying (1.5.7) with $A = X$, let $K = [K_{ij}]_{i,j=1}^{k}$ be finite rank operator with $s_q(X) = \|X-K\|$ and $\dim \operatorname{Im} K \leq q-1$. Then for fixed indices i,j we have

$$s_q(X) = \|X-K\| \geq \|X_{ij}-K_{ij}\| \geq s_q(X_{ij}), \tag{1.5.8}$$

where the last inequality follows from (1.5.7) applied with $A = X_{ij}$. The inequality (1.5.8) obviously implies that $X_{ij} \in S_p$ for all i,j. ∎

PROOF OF THEOREM 1.5.2. If the special inverse linearization T_L is in S_p, then by Proposition 1.5.3, all its operator entries are in S_p. In particular, $A_{jn} \in S_p$ $(1 \leq j \leq n \leq m)$, which implies that $H_n \in S_{p/n}$. (Here we use the fact that $s_q(A_{jn}) = (s_q(H_n))^{1/n}$, $q = 1,2,\dots$.) Conversely, let $H_n \in S_{p/n}$, $n = 1,\dots,m$. Then $A_{jn} \in S_p$, and hence by Proposition 1.5.3, $T_L \in S_p$. ∎

1.6 Exercises

<u>Ex. 1.1</u>. Let $\mathcal{X}$ be a Banach algebra with identity e. Let $a,b \in \mathcal{X}$ be such that

$$f(\lambda)(\lambda e-a)g(\lambda) = \lambda e-b, \quad \lambda \in \Omega,$$

where Ω is an open set containing $\sigma(a) \cup \sigma(b)$, and $f(\lambda)$ and $g(\lambda)$ are invertible analytic $\mathcal{X}$-valued functions defined on Ω. Prove that $a = c^{-1}bc$ for some invertible $c \in \mathcal{X}$.

<u>Ex. 1.2</u>. Let $L(\lambda)$ be an operator polynomial with coefficients in the algebra

$$\mathcal{K} \overset{\text{def}}{=} \{X \in L(\mathcal{X}) \mid X = \alpha I+K \text{ for some } \alpha \in \mathbb{C} \text{ and some compact } K \in L(\mathcal{X})\}.$$

Prove that if $L(\lambda_0)$ is invertible for some $\lambda_0 \in \mathbb{C}$, then $\Sigma(L)$ is at most countable and has not more than degree L accumulation points in the finite complex plane.

<u>Ex. 1.3</u>. Prove Proposition 1.4.1.

<u>Ex. 1.4</u>. Let $L(\lambda) = I + \sum_{j=0}^{\ell} \lambda^j K_j$, where $K_j \in L(\mathcal{X})$ are finite rank operators (i.e., $\dim \operatorname{Im} K_j < \infty$). Prove that $\Sigma(L)$ consists of finite number of points (unless $\Sigma(L) = \mathbb{C}$). Hint: Prove that there exists a finite dimensional subspace $\mathcal{M} \subset \mathcal{X}$ and a direct complement $\mathcal{M}'$ to $\mathcal{M}$ in $\mathcal{X}$ such that $K_j\mathcal{M} \subset \mathcal{M}$ and $K_j\mathcal{M}' = \{0\}$ for $j = 0,\dots,\ell$.

<u>Ex. 1.5</u>. Generalize the result of the previous exercise as follows: Let $L(\lambda) = \sum_{j=0}^{\ell} \lambda^j(\alpha_j I+K_j)$, where $K_0,\dots,K_\ell$ are finite rank operators, and assume $\Sigma(L) \neq \mathbb{C}$. Prove that the set $\Sigma(L)\backslash\{\lambda_0 \mid \sum_{j=0}^{\ell} \alpha_j\lambda_0^j = 0\}$ is finite (this set consists of normal eigenvalues of $L(\lambda)$ by Theorem 1.4.3).

Ex. 1.6. Prove that an inverse linearization is generally not unique up to similarity, even assuming that $\mathcal{X}$ is finite dimensional.

Ex. 1.7. Let $L(\lambda) = \lambda^2 I+\lambda B+C$, where $B,C \in L(\mathcal{X})$ and C is an invertible operator such that $C = D^2$ for some $D \in L(\mathcal{X})$ (the operator D is necessarily invertible). Prove that

$$\begin{bmatrix} 0 & D \\ -D & -B \end{bmatrix}$$

is a global linearization of $L(\lambda)$.

Ex. 1.8. Given a linearization of $L(\lambda)$ with respect to Ω, find a linearization for the polynomial $L(\lambda+\alpha)$ with respect to the open set $\{\lambda+\alpha \mid \lambda \in \Omega\}$ (here α is a fixed complex number).

Ex. 1.9. Given a global linearization of an operator polynomial $L(\lambda)$, find a global linearization of an operator polynomial $L(p(\lambda))$, where $p(\lambda) = \lambda^n+\alpha_{n-1}\lambda^{n-1}+\cdots+\lambda_0$ is a given scalar polynomial.

Ex. 1.10. Show that the direct sum of linearizations of operator polynomials $L_1(\lambda),\ldots,L_r(\lambda)$ (with respect to the same open set Ω) is a linearization for $L_1(\lambda) \oplus\cdots\oplus L_r(\lambda)$.

Ex. 1.11. Let (λ) be an operator polynomial in the upper triangular form

$$L(\lambda) = \begin{bmatrix} L_{11}(\lambda) & L_{12}(\lambda) & \cdots & L_{1k}(\lambda) \\ 0 & L_{22}(\lambda) & \cdots & L_{2k}(\lambda) \\ \vdots & \vdots & & \vdots \\ 0 & 0 & \cdots & L_{kk}(\lambda) \end{bmatrix} : \mathcal{X}^k \to \mathcal{X}^k.$$

Assume that the spectra $\Sigma(L_{11}),\ldots,\Sigma(L_{kk})$ are disjoint. Prove that if $T_1,\ldots,T_k$ are linearizations of $L_{11},\ldots,L_{kk}$ respectively (with respect to Ω), then $T_1 \oplus\cdots\oplus T_k$ is a linearization of $L(\lambda)$ with respect to Ω.

Ex. 1.12. Assume $\mathcal{X}$ is a Hilbert space, and let $L(\lambda)$ be an operator polynomial with coefficients in $L(\mathcal{X})$ and with linearization T with respect to Ω. Prove that T^* is a linearization for the polynomial $(L(\bar{\lambda}))^*$ with respect to $\bar{\Omega} = \{\bar{\lambda} \mid \lambda \infty \Omega\}$.

Ex. 1.13. State and prove the fact analogous to Ex. 1.12 for the case when $\mathcal{X}$ is a Banach space.

Ex. 1.14. Verify that the main results of Sections 1.2-1.5 (namely, Theorems 1.2.1, 1.3.1, 1.4.3 and Proposition 1.4.2) remain valid also in the framework of operator-valued functions $W: \Omega \rightarrow L(\mathcal{X})$ which are analytic in the open set Ω and with the property that the spectrum of W:

$$\Sigma(W) = \{\lambda \in \Omega \mid W(\lambda) \text{ is not invertible}\}.$$

is a compact set.

Ex. 1.15. Prove that

$$L(\lambda) = I-\lambda\pi T_L(I-\lambda S)^{-1}\tau$$

(in the notation introduced before and in Theorem 1.5.1).

1.7 Notes

The material in Section 1.1 is standard. The companion operator is a well-known linearization for operator polynomials used by many authors. The uniqueness of linearizations (in the more general framework of analytic operator functions with compact spectrum) was proved in Kaashoek-van der Mee-Rodman [1]. Corollary 1.2.2 was proved in Rodman [2]. The results of Section 1.3 are taken from Gohberg-Kaashoek-Lay [1]. Proof of Theorem 1.4.3 is an adaptation of the proof (given in Gohberg-Kreĭn [1]) of a theorem due to Gohberg [1]. Theorem 1.5.1 is particular case of results proved in Kaashoek-van de Ven [1]. This paper contains results on regularized determinants of $I-\lambda T$, where T is the special inverse linearization of an operator polynomial with coefficients in certain operator ideals. Inverse linearizations with their applications to traces and determinants for entire operator functions were studied in Perelson [1] (see also Perelson [3]). Traces and determinants for operator polynomials with coefficients in certain Banach algebras of $m \times m$ matrices with operator entries were studied in Perelson [2,3]. Some of the exercises are adapted from Gohberg-Lancaster-Rodman [3].

Linearizations of various classes of operator-valued functions were extensively studied, see den Boer [1], Gohberg-Kaashoek-Lay [1], Mityagin [1,2], van der Mee [1], Heinig [1].

It should be noted that the construction used to establish the existence of linearizations is analogous to the universal models of operators (see Rota [1], Herrero [3]).

For connections between linearizations of operator polynomials and the Sz.-Nagy-Foias characteristic function (Sz.-Nagy-Foias [1]), see Kaashoek-van der Mee-Rodman [3], van der Mee [1].

CHAPTER 2: REPRESENTATIONS AND DIVISORS OF MONIC OPERATOR POLYNOMIALS

We develop here the notions and properties of spectral pairs and spectral triples for monic operator polynomials. These notions are used throughout the book, and their usefulness is based on the possibility to express the divisibility properties of operator polynomials in terms of spectral pairs and triples.

2.1 Spectral pairs

Let $L(\lambda) = \sum_{j=0}^{\ell} \lambda^j A_j$ be an operator polynomial, with $A_j \in L(\mathcal{X})$, where $\mathcal{X}$ is a Banach space, and assume $L(\lambda)$ is *monic*, i.e., $A_\ell = I$. As we have seen in Chapter 1, the companion operator

$$C_L = \begin{bmatrix} 0 & I & 0 & \cdots & 0 \\ 0 & 0 & I & \cdots & 0 \\ \vdots & \vdots & & & \vdots \\ -A_0 & -A_1 & & \cdots & -A_{\ell-1} \end{bmatrix} : \mathcal{X}^\ell \to \mathcal{X}^\ell$$

or, more generally, any operator similar to it is a global linearization of $L(\lambda)$. In particular, the spectrum of C_L coincides with that of $L(\lambda)$. However, C_L alone cannot adequately describe the spectral structure of $L(\lambda)$. For instance, the case when $\dim \mathcal{X} < \infty$ and $\ell = 1$ represents a very familiar situation. Here one is often interested not only in the Jordan form of $L(\lambda) = \lambda I + A_0$ (which is fully described by the class of operators similar to $C_L = -A_0$), but also in the eigenvectors and generalized eigenvectors for $L(\lambda)$. More generally, in the case of finite dimensional $\mathcal{X}$ which we identify with $\mathbb{C}^n$, the following construction of eigenvectors and generalized eigenvectors for a monic operator polynomial $L(\lambda)$ is used.

Write $L(\lambda) = \lambda^{\ell} I + \sum_{j=0}^{\ell-1} \lambda^j A_j$, where $A_0,\ldots,A_{\ell-1} \in L(\mathbb{C}^n)$, and we interpret A_j as $n\times n$ matrices with respect to the standard orthonormal basis in $\mathbb{C}^n$. A chain of vectors $x_0,\ldots,x_p$ from $\mathbb{C}^n$ is called a *Jordan chain* of $L(\lambda)$ corresponding to $\lambda_0 \in \mathbb{C}^n$ if

$$L(\lambda_0)x_0 = 0, \; x_0 \neq 0; \; \sum_{j=0}^{k} \frac{1}{j!} L^{(j)}(\lambda_0)x_{k-j} = 0$$

for $k = 1,\ldots,p$ (here $L^{(j)}(\lambda_0)$ stands for the jth derivative of $L(\lambda)$ evaluated at λ_0). Clearly a necessary and sufficient condition for existence of a Jordan chain of $L(\lambda)$ corresponding to λ_0 is that Ker $L(\lambda_0) \neq \{0\}$. Such λ_0 are called the *eigenvalues* of $L(\lambda)$. Because $L(\lambda)$ is monic, det $L(\lambda)$ is a polynomial of λ of degree $n\ell$, and hence the number of eigenvalues of $L(\lambda)$ counted with multiplicities is precisely $n\ell$. Let $\lambda_1,\ldots,\lambda_p$ be all the *distinct* eigenvalues of $L(\lambda)$. For each j, we construct a pair of matrices X_j (of size $n\times m_j$, where m_j is the multiplicity of λ_j as a zero of det $L(\lambda)$) and T_j (of size $m_j\times m_j$) as follows:

$$X_j = [x_{11}^{(j)}\cdots x_{1r_1}^{(j)} \; x_{21}^{(j)}\cdots x_{2r_2}^{(j)}\cdots x_{k1}^{(j)}\cdots x_{kr_k}^{(j)}];$$

$$T_j = \begin{bmatrix} J_{r_1}(\lambda_j) & & 0 \\ & J_{r_2}(\lambda_j) & \\ 0 & & \ddots \\ & & & J_{r_k}(\lambda_j) \end{bmatrix}.$$

Here $x_{s1}^{(j)},\ldots,x_{sr_s}^{(j)}$ for $s = 1,\ldots,k$ are Jordan chains of $L(\lambda)$ corresponding to λ_j such that the vectors $x_{11}^{(j)}, x_{21}^{(j)},\ldots,x_{k1}^{(j)}$ are linearly independent and $r_1+\cdots+r_k = m_j$, and $J_{r_s}(\lambda_j)$ is the $r_s\times r_s$ upper triangular Jordan block with eigenvalue λ_j. Finally, put

$$X = [X_1 X_2 \cdots X_p]; \quad T = \begin{bmatrix} T_1 & 0 & \cdots & 0 \\ 0 & T_2 & \cdots & 0 \\ \vdots & & & \vdots \\ 0 & 0 & \cdots & T_p \end{bmatrix}.$$

It can be shown that X and T are correctly defined (i.e., the Jordan chains $x_{s1}^{(j)}, \ldots, x_{sr_s}^{(j)}$ with the required properties indeed exist), the matrix

$$\begin{bmatrix} X \\ XT \\ \vdots \\ XT^{\ell-1} \end{bmatrix}$$

is square and invertible, and

$$XT^{\ell} + \sum_{j=0}^{\ell-1} A_j XT^j = 0.$$

Since we do not need these properties in the sequel, they will not be proved here. For the proof we refer the reader to Gohberg-Lancaster-Rodman [1,2].

In case of infinite dimensional $\mathcal{X}$, we take the properties of the pair of matrices (X,T) described above as definition. So, a pair (X,T), where $X \in L(\mathcal{Y},\mathcal{X})$, $T \in L(\mathcal{Y})$ for some Banach space $\mathcal{Y}$, will be called a *right spectral pair* for the monic operator polynomial $L(\lambda) = \lambda^{\ell} I + \sum_{j=0}^{\ell-1} \lambda^j A_j$ if the operator

$$\begin{bmatrix} X \\ XT \\ \vdots \\ XT^{\ell-1} \end{bmatrix} : \mathcal{Y} \to \mathcal{X}^{\ell} \tag{2.1.1}$$

is invertible and

$$A_0 X + A_1 XT + \cdots + A_{\ell-1} XT^{\ell-1} + XT^{\ell} = 0. \tag{2.1.2}$$

Throughout the book we shall use the notation $\text{col}[Z_i]_{i=p}^{q}$ to designate the operator column

$$\begin{bmatrix} z_p \\ z_{p+1} \\ \vdots \\ z_q \end{bmatrix}.$$

Thus, the operator (2.1.1) will be written as $\mathrm{col}[XT^i]_{i=0}^{\ell-1}$.

THEOREM 2.1.1. *A right spectral pair of* $L(\lambda)$ *exists and is unique up to similarity: If* (X,T) *and* (X_1,T_1) *are two right spectral pairs of* $L(\lambda)$*, then*

$$X_1 = XS, \quad T_1 = S^{-1}TS \tag{2.1.3}$$

for a uniquely determined invertible operator $S \in L(\mathcal{Y}',\mathcal{Y})$*, where* $T \in L(\mathcal{Y})$, $T' \in L(\mathcal{Y}')$. *Moreover, in every right spectral pair* (X,T) *of* $L(\lambda)$*, the operator* T *is a global linearization of* $L(\lambda)$.

PROOF. Let $X_0 = [I0\cdots0] \in L(\mathcal{X}^\ell,\mathcal{X})$ and $T_0 = C_L$. Using the structure of C_L, one easily verifies that for $j = 0,\ldots,\ell-1$

$$X_0T_0^j = [0\cdots0I0\cdots0]$$

with I on the $(j+1)^{\text{th}}$ place, and

$$X_0T_0^\ell = [-A_0,-A_1,\ldots,-A_{\ell-1}].$$

The equality (2.1.2) now follows for $X = X_0$, $T = T_0$. Also,

$$\mathrm{col}[XT^i]_{i=0}^{\ell-1} = I \in L(\mathcal{X}^\ell)$$

in this case. So (X_0,T_0) is a right spectral pair of $L(\lambda)$. To prove the uniqueness of a right spectral pair, observe that (2.1.3) holds with

$$S = \{\mathrm{col}[XT^i]_{i=0}^{\ell-1}\}^{-1}\mathrm{col}[X_1T_1^i]_{i=0}^{\ell-1}. \tag{2.1.4}$$

Indeed,

$$X_1\{\mathrm{col}[X_1T_1^i]_{i=0}^{\ell-1}\}^{-1} = X\{\mathrm{col}[XT^i]_{i=0}^{\ell-1}\}^{-1} = [I0\cdots0]$$

and

$$\{\operatorname{col}[X_1T_1^i]_{i=0}^{\ell-1}\}T_1\{\operatorname{col}[X_1T_1^i]_{i=0}^{\ell-1}\}^{-1} = \{\operatorname{col}[XT^i]_{i=0}^{\ell-1}\}T\{\operatorname{col}[XT^i]_{i=0}^{\ell-1}\}^{-1} = C_L \tag{2.1.5}$$

in view of the property (2.1.2) of right spectral pairs. Further, using the invertibility of $\operatorname{col}[XT^i]_{i=0}^{\ell-1}$ and $\operatorname{col}[X_1T_1^i]_{i=0}^{\ell-1}$, it is easily seen that any S which satisfies (2.1.3) must be of the form (2.1.4). Finally, since $([I0\cdots0],C_L)$ is a right spectral pair for $L(\lambda)$ and any other right spectral pair (X,T) is similar to it, clearly T is a global linearization for $L(\lambda)$. ∎

Analogously a *left spectral pair* (T,Y) for $L(\lambda) = \lambda^\ell I + \sum_{j=0}^{\ell-1}\lambda^j A_j$ is defined: here $T \in L(\mathcal{Y})$, $Y \in L(\mathcal{X},\mathcal{Y})$ for some Banach space $\mathcal{Y}$ are such that

$$[Y, TY, \ldots, T^{\ell-1}]: \mathcal{X}^\ell \to \mathcal{Y} \tag{2.1.6}$$

is invertible and

$$YA_0+TYA_1+\cdots+T^{\ell-1}YA_{\ell-1}+T^\ell Y = 0. \tag{2.1.7}$$

For instance, an easy calculation verifies that

(2.1.8)

$$T = \begin{bmatrix} 0 & 0 & \cdots & 0 & -A_0 \\ I & 0 & \cdots & 0 & -A_1 \\ 0 & I & & \vdots & \vdots \\ \vdots & \vdots & \cdots & 0 & -A_{\ell-2} \\ 0 & 0 & & I & -A_{\ell-1} \end{bmatrix}: \mathcal{X}^\ell \to \mathcal{X}^\ell; \quad Y = \begin{bmatrix} I \\ 0 \\ \vdots \\ 0 \end{bmatrix}: \mathcal{X} \to \mathcal{X}^\ell$$

is a left spectral pair for $L(\lambda)$. Indeed, here

$$[Y, TY, \ldots, T^{\ell-1}Y] = I_{\mathcal{X}^\ell}.$$

As in Theorem 2.1.1, one proves that a left spectral pair is unique up to similarity: if (T_1,Y_1) and (T_2,Y_2) are left spectral pairs for $L(\lambda)$, then

$$T_1 = S^{-1}T_2S, \quad Y_1 = S^{-1}Y_2$$

for some invertible S which is uniquely determined by (T_1,Y_1) and (T_2,Y_2).

2.2 Representations in terms of spectral pairs

The following theorem describes the monic operator polynomial in terms of its spectral pairs.

THEOREM 2.2.1. *Let* (X,T_1), $T_1 \in L(\mathcal{Y})$ *be a right spectral pair and* (T_2,Y), $T_2 \in L(\mathcal{Y})$ *be a left spectral pair of a monic operator polynomial* $L(\lambda)$. *Then*

$$L(\lambda) = \lambda^\ell I - XT_1^\ell(V_1+V_2\lambda+\cdots+V_\ell\lambda^{\ell-1}), \tag{2.2.1}$$

where $V_j \in L(\mathcal{X},\mathcal{Y})$ *are defined by*

$$[V_1V_2\cdots V_\ell] = \{\mathrm{col}[XT^i]_{i=0}^{\ell-1}\}^{-1}$$

$$L(\lambda) = \lambda^\ell I-(W_1+\lambda W_2+\cdots+\lambda^{\ell-1}W_\ell)T_2^\ell Y, \tag{2.2.2}$$

where $W_j \in L(\mathcal{Y},\mathcal{X})$ *are defined by*

$$\mathrm{col}[W_j]_{j=1}^\ell = [Y, TY, \ldots, T^{\ell-1}Y]^{-1}.$$

The representation (2.2.1) and (2.2.2) are called *right* and *left canonical forms* of $L(\lambda)$, respectively.

PROOF. We shall prove (2.2.1) only (the proof of (2.2.2) is analogous). In view of the uniqueness of a right spectral pair up to similarity (Theorem 2.1.1), it suffices to prove (2.2.1) for

$$X = [I0\cdots0] \in L(\mathcal{X}^\ell,\mathcal{X}),$$

$$T_1 = C_L,$$

the companion operator of $L(\lambda)$, (cf. the proof of Theorem 2.1.1). As in this case,

$$XT_1^{\ell} = [-A_0, -A_1, \dots, -A_{\ell-1}],$$

where A_j is the jth coefficient of $L(\lambda)$, and $V_j = \mathrm{col}[\delta_{ji} I_{\mathcal{X}}]_{i=1}^{\ell}$ (here and elsewhere in this book, δ_{ji} stands for the Kronecker symbol: $\delta_{ji} = 1$ if $j = i$ and $\delta_{ji} = 0$ if $j \neq i$), the formula (2.2.1) follows immediately. ∎

In particular, a right (or left) spectral pair defines uniquely the monic operator polynomial.

Let us state explicitly the following important consequence of Theorem 2.2.1.

COROLLARY 2.2.2. *A pair of operators* (X,T) *where* $X \in L(\mathcal{Y},\mathcal{X})$, $T \in L(\mathcal{Y})$ *is a right spectral pair for a monic operator polynomial* $L(\lambda)$ *of degree* ℓ *if and only if*

$$\mathrm{col}[XT^i]_{i=0}^{\ell-1} \in L(\mathcal{Y},\mathcal{X}^{\ell})$$

is invertible. A pair (T,Y), *where* $T \in L(\mathcal{Y})$, $Y \in L(\mathcal{X},\mathcal{Y})$ *is a left spectral pair for a monic operator polynomial of degree* ℓ *if and only if*

$$[Y, TY, \dots, T^{\ell-1}Y] \in L(\mathcal{X}^{\ell},\mathcal{Y})$$

is invertible.

Indeed, the part "only if" follows from the definitions of spectral pairs. The part "if" follows by defining $L(\lambda)$ as in (2.2.1) or (2.2.2).

Observe that if (X,T_1) is a right spectral pair of a monic operator polynomial $L(\lambda)$, then T_1 is a global linearization of $L(\lambda)$. Indeed, as the proof of Theorem 2.1.1 shows, the pair

$$([I0\cdots 0], C_L) \tag{2.2.3}$$

is a right spectral pair of $L(\lambda)$ and C_L is a global linearization of $L(\lambda)$. Since any two right spectral pairs are similar, our observation follows. Analogously, the operator T_2 taken from a left spectral pair (T_2,Y) of $L(\lambda)$ is a global linearization for $L(\lambda)$. By Theorem 1.2.1, T_1 and T_2 are similar. In fact, for the right and left spectral pairs given by (2.2.3) and (2.1.8), one can write down this similarity explicitly:

(2.2.4)
$$\begin{bmatrix} 0 & 0 & \cdots & & 0 & -A_0 \\ I & 0 & & & 0 & -A_1 \\ 0 & I & & & \vdots & \vdots \\ \vdots & & & & & \\ & & & I & 0 & -A_{\ell-2} \\ 0 & 0 & \cdots & 0 & I & -A_{\ell-1} \end{bmatrix} \begin{bmatrix} A_1 & A_2 & \cdots & A_{\ell-1} & I \\ A_2 & & & \cdot & \\ \vdots & & \cdot & & \\ & \cdot & & & \\ A_{\ell-1} & I & & & \\ I & & & & 0 \end{bmatrix}$$
$$= \begin{bmatrix} A_1 & A_2 & \cdots & A_{\ell-1} & I \\ A_2 & & & \cdot & \\ \vdots & & \cdot & & \\ & \cdot & & & \\ A_{\ell-1} & I & & & \\ I & & & & 0 \end{bmatrix} \begin{bmatrix} 0 & 0 & \cdots & & 0 & -A_0 \\ I & 0 & & & 0 & -A_1 \\ 0 & I & & & \vdots & \vdots \\ \vdots & & & & & \\ & & & I & 0 & -A_{\ell-2} \\ 0 & 0 & \cdots & 0 & I & -A_{\ell-1} \end{bmatrix}.$$

This formula can be verified by a straightforward multiplication.

We have seen in Section 2.1 that the motivation to introduce and study right spectral pairs comes from the Jordan chains in the finite-dimensional case. One can show that, conversely, if some "part" of a monic operator polynomial $L(\lambda)$ is finite dimensional, then one can recover Jordan chain (as introduced above) from the restriction of a right spectral pair to that part:

THEOREM 2.2.3. *Let* (X,T), *where* $X \in L(\mathcal{Y},\mathcal{X})$, $T \in L(\mathcal{Y})$, *be a right spectral pair for the monic operator polynomial* $L(\lambda)$, *and suppose that* $\mathcal{L}$ *is a finite dimensional* T*-invariant subspace in* $\mathcal{Y}$.

Let $e_1, e_2, \ldots, e_k$ *be a basis for* $\mathcal{L}$ *and assume that the representation of* $T|_{\mathcal{L}}$ *in this basis is a Jordan normal form. If* $e_i, e_{i+1}, \ldots, e_{i+r}$ *is a basis for a Jordan block of* $T|_{\mathcal{L}}$ *corresponding to eigenvalue* λ_0, *then* $\{Xe_i, Xe_{i+1}, \ldots, Xe_{i+r}\}$ *is a Jordan chain for* L *corresponding to eigenvalue* λ_0.

For the proof, we refer the reader to Gohberg-Lancaster-Rodman [4].

2.3 Linearizations

We study here global linearizations of monic operator polynomials. Given a Banach space $\mathcal{X}$, consider the set $\mathcal{YL}_\ell(\mathcal{X})$ of all global linearizations of monic operator polynomials of degree ℓ with coefficients in $L(\mathcal{X})$. As we have seen above, the set of $\mathcal{YL}_\ell(\mathcal{X})$ consists of all operators which are similar to operators of type

$$\begin{bmatrix} 0 & I & 0 & \cdots & 0 \\ 0 & 0 & I & \cdots & 0 \\ \vdots & & & & \vdots \\ & & & & I \\ -A_0 & -A_1 & & \cdots & -A_{\ell-1} \end{bmatrix} \in L(\mathcal{X}^\ell), \tag{2.3.1}$$

where $A_0,\ldots,A_{\ell-1}$ belong to $L(\mathcal{X})$. Hence we shall assume that $\mathcal{YL}_\ell(\mathcal{X}) \subset L(\mathcal{X}^\ell)$.

THEOREM 2.3.1. *The set* $\mathcal{YL}_\ell(\mathcal{X})$ *is open in* $L(\mathcal{X}^\ell)$ *in the operator norm topology.*

This theorem follows immediately from the more informative

PROPOSITION 2.3.2. *For every monic operator polynomial* $L(\lambda)$ *of degree* ℓ *with coefficients in* $L(\mathcal{X})$ *there exist positive constants* ε *and* K *such that any operator* $B \in L(\mathcal{X}^\ell)$ *with* $\|B-C_L\| < \varepsilon$, *where* C_L *is the companion operator of* $L(\lambda)$, *is similar to the companion operator* C_M *of some monic operator polynomial* $M(\lambda)$ *of degree* ℓ; *moreover,*

$$\|C_M-C_L\| \le K\|B-C_L\|, \tag{2.3.2}$$

and an invertible operator $S \in L(\mathcal{X}^\ell)$ *exists such that* $B = S^{-1}C_MS$ *and* $\|I-S\| \le K\|B-C_L\|$.

PROOF. Write B in the block matrix form $B = [B_{ij}]_{i,j=1}^{\ell}$, where $B_{ij} \in L(\mathcal{X})$. Letting $P_1 = [I0\cdots 0]: \mathcal{X}^{\ell} \to \mathcal{X}$, observe that $\mathrm{col}[P_1 C_L^i]_{i=0}^{\ell-1} = I$. Hence there exists $\varepsilon > 0$ such that for every $B \in L(\mathcal{X}^{\ell})$ with $\|B-C_L\| < \varepsilon$, the operator $Q(B) \overset{\mathrm{def}}{=} \mathrm{col}[P_1 B^i]_{i=0}^{\ell-1}: \mathcal{X}^{\ell} \to \mathcal{X}^{\ell}$ is invertible. Put $M(\lambda) = \lambda^{\ell} I - P_1 B(V_1 + V_2\lambda + \cdots + V_{\ell}\lambda^{\ell-1})$, where $[V_1 V_2 \cdots V_{\ell}] = Q(B)^{-1}$, $V_i: \mathcal{X} \to \mathcal{X}^{\ell}$. Then one easily verifies the equality $Q(B)B = C_M Q(B)$. Further,

$$\|Q(B)^{-1} - I\| \leq K_1\|B-C_L\|, \quad \|B^{\ell} - C_L^{\ell}\| \leq K_2\|B-C_L\|,$$

for some positive constants K_1 and K_2 (where $B \in L(\mathcal{X}^{\ell})$ is such that $\|B-C_L\| < \varepsilon$). Taking into account the equality

$$L(\lambda) = \lambda^{\ell} I - P_1 C_L^{\ell}(U_1 + U_2\lambda + \cdots + U_{\ell}\lambda^{\ell-1}),$$

where $[U_1 U_2 \cdots U_{\ell}] = I$ (see Theorem 2.2.1), we obtain (2.3.2). ∎

We propose the following open

PROBLEM 2.3.1. *Characterize intrinsically (i.e., in terms of the operator itself) the global linearizations of monic operator polynomials.*

In the finite-dimensional case ($\dim \mathcal{X} < \infty$), this problem is solved.

THEOREM 2.3.3. *Let* $\mathcal{X} = \mathbb{C}^n$, *and let* $X \in L(\mathcal{X}^{\ell})$. *The following statements are equivalent:* (i) X *is a global linearization of a monic operator polynomial of degree* ℓ *with coefficients in* $L(\lambda)$; (ii) $\dim \mathrm{Ker}(\lambda_0 I - X) \leq n$ *for every* $\lambda_0 \in \mathbb{C}$; (iii) *there exists* $n(\ell-1)$*-dimensional subspace* $\mathcal{M} \subset \mathcal{X}^{\ell}$ *such that*

$$\mathrm{Ker}(\lambda_0 I - X) \cap \mathcal{M} = \{0\}$$

for every $\lambda_0 \in \mathbb{C}$.

For the proof of equivalence (i) $\longleftrightarrow$ (ii) we refer the reader to Gohberg-Lancaster-Rodman [2], Section 1.3. Clearly, (iii) $\to$ (ii). To verify that (i) $\to$ (iii), we may assume that X is of the form (2.3.1). It is easily seen from (2.3.1) that any $x \in \mathrm{Ker}(\lambda_0 I-X)$ has the form $\mathrm{col}[\lambda_0^i x_0]_{i=0}^{\ell-1}$ for some $x_0 \in \mathcal{X}$. We use the notation $\mathrm{col}[z_j]_{j=p}^{q}$ to designate the vector column

$$\begin{bmatrix} z_p \\ z_{p+1} \\ \vdots \\ z_q \end{bmatrix} .$$

So we can take

$$\mathcal{M} = 0 \oplus \mathcal{X} \oplus \cdots \oplus \mathcal{X}$$

to satisfy (iii).

In the infinite-dimensional case, the condition (iii), properly interpreted, is necessary.

PROPOSITION 2.3.4. *Let* $X \in \mathcal{X}^\ell$ *be a global linearization of a monic operator polynomial* $L(\lambda)$ *of degree* ℓ *with coefficients in* $L(\mathcal{X})$. *Then there is a (closed) subspace* $\mathcal{M} \subset \mathcal{X}^\ell$ *isomorphic to* $\mathcal{X}^{\ell-1}$ *such that for every* $\lambda_0 \in \mathbb{C}$ *the subspaces* $\mathrm{Ker}(\lambda_0 I-X)$ *and* $\mathcal{M}$ *form a direct sum, i.e.,* $\mathrm{Ker}(\lambda_0 I-X)+\mathcal{M}$ *is closed and* $(\mathrm{Ker}(\lambda_0 I-X)) \cap \mathcal{M} = \{0\}$.

PROOF. Without loss of generality we can assume that $X = C_L$. Take $\mathcal{M} \subset \mathcal{X}^\ell$ be the set of elements from $\mathcal{X}^\ell$ whose first coordinate (which belongs to $\mathcal{X}$) is zero. As in the finite-dimensional case above, one verifies that $(\mathrm{Ker}(\lambda_0 I-X)) \cap \mathcal{M} = \{0\}$. To prove the closedness of $\mathrm{Ker}(\lambda_0 I-X)+\mathcal{M}$, let $x^{(m)} = x_1^{(m)}+x_2^{(m)}$, $m = 1,2,\ldots$ be a sequence from $\mathrm{Ker}(\lambda_0 I-X)+\mathcal{M}$, with $x_1^{(m)} \in \mathrm{Ker}(\lambda_0 I-X)$ and $x_2^{(m)} \in \mathcal{M}$ such that $\lim_{m\to\infty} x^{(m)} = y$.

We have to prove that $y \in \mathrm{Ker}(\lambda_0 I-X)+\mathcal{M}$. Write $x_1^{(m)} = \mathrm{col}[\lambda_0^i x_0^{(m)}]_{i=0}^{\ell-1}$; $y = \mathrm{col}[y_i]_{i=1}^{\ell}$; $x_2^{(m)} = \mathrm{col}[x_{2i}^{(m)}]_{i=1}^{\ell}$. Then $x_{21}^{(m)} = 0$, so $\lim_{m\to\infty} x_0^{(m)} = y_1$, and

$$y = \mathrm{col}[\lambda_0^i y_1]_{i=0}^{\ell-1} + \mathrm{col}[y_i-\lambda_0^{i-1} y_1]_{i=1}^{\ell} \in \mathrm{Ker}(\lambda_0 I-X)+\mathcal{M}. \qquad \blacksquare$$

2.4 Generalizations of canonical forms

We have seen in Corollary 2.2.2 that a pair of operators (X,T), where $X \in L(\mathcal{Y},\mathcal{X})$, $T \in L(\mathcal{Y})$ is a right spectral pair for a monic operator polynomial $L(\lambda)$ of degree ℓ if and only if $Q \overset{\text{def}}{=} \text{col}[XT^i]_{i=0}^{\ell-1}$ is invertible, and then $L(\lambda)$ is uniquely determined by (2.2.1). However, the formula (2.2.1) makes sense also if Q^{-1} is replaced by a one-sided inverse (if such exists). The relationships between the resulting $L(\lambda)$ and the pair (X,T) will be explored in this section.

THEOREM 2.4.1. *Let $X \in L(\mathcal{Y},\mathcal{X})$ and $T \in L(\mathcal{Y})$ be such that the operator $Q \in L(\mathcal{Y},\mathcal{X}^\ell)$ has a left inverse Q_L^{-1}. Define operators $A_i \in L(\mathcal{X})$ by $A_i = -XT^\ell G_{i+1}$, $i = 0,1,\ldots,\ell-1$ where $[G_1\cdots G_\ell] = Q_L^{-1} \in L(\mathcal{X}^\ell,\mathcal{Y})$. Then*

$$A_0X+A_1XT+\cdots+A_{\ell-1}XT^{\ell-1}+XT^\ell = 0. \tag{2.4.1}$$

Let L be the monic operator polynomial on $\mathcal{X}$ defined by $L(\lambda) = \Sigma_{i=0}^{\ell}A_i\lambda^i$, $A_\ell = I$, and let C_L be the companion operator of L. Then

$$T = Q_L^{-1}|_{\mathcal{M}}\, C_L|_{\mathcal{M}}\, Q \tag{2.4.2}$$

where $\mathcal{M} = \text{Im } Q$.

PROOF. Using the definition of the A_i, we have

$$A_0X+\cdots+A_{\ell-1}XT^{\ell-1} = [A_0A_1\cdots A_{\ell-1}]Q = -XT^\ell[G_1\cdots G_\ell]Q = -XT^\ell,$$

which gives (2.4.1). With L and C_L as defined, it is easily verified that, as a consequence of (2.4.1), $QT = C_LQ$. It is apparent from this relation that $\mathcal{M}$ is invariant under C_L. Using the decomposition $\mathcal{X}^\ell = (\text{Ker } Q_L^{-1}) \dot{+} \mathcal{M}$, the representation (2.4.2) follows from the equality $QT = C_LQ$. ∎

Observe that obviously we also have

$$X = X_0Q,$$

where $X_0 = [I0\cdots 0] \in L(\mathcal{X}^\ell,\mathcal{X})$. Thus, in a certain sense (which will be made precise later on), the pair (X,T) is a restriction to Im Q of the right spectral pair (X_0,C_L) of the monic operator polynomial $L(\lambda)$ defined in Theorem 2.4.1.

THEOREM 2.4.2. *Let* $X \in L(\mathcal{Y},\mathcal{X})$ *and* $T \in L(\mathcal{Y})$ *be such that the operator* $Q(X,T) \in L(\mathcal{Y},\mathcal{X}^\ell)$ *has a right inverse* Q_R^{-1}. *Define operators* $V_i \in L(\mathcal{X})$ *by* $A_i = -XT^\ell V_{i+1}$, $i = 0,1,\ldots,\ell-1$ *where* $[V_1\cdots V_\ell] = Q_R^{-1} \in L(\mathcal{X}^\ell,\mathcal{Y})$, *and hence the monic operator polynomial* $L(\lambda) = \Sigma_{i=0}^{\ell} A_i\lambda^i$, $A_\ell = I$, *is defined. Then, if* $\mathcal{M} = \operatorname{Im} Q_R^{-1}$ *and* P *is the projection in* $L(\mathcal{Y})$ *on* $\mathcal{M}$ *along* Ker Q, *we have*

$$C_L = Q(PT)\Big|_{Q_R^{-1}} . \tag{2.4.3}$$

PROOF. Using the definition of the A_i and the fact that $QQ_R^{-1} = I$, it can be verified that

$$QTQ_R^{-1} = \operatorname{col}[XT^i]_{i=1}^{\ell}\,[V_1\cdots V_\ell] = C_L.$$

In view of the decomposition $\mathcal{Y} = (\operatorname{Ker} Q) \dot{+} \mathcal{M}$, the relation (2.4.3) follows immediately. ∎

Here, under the hypotheses of Theorem 2.4.2, the right spectral pair $([I0\cdots 0],C_L)$ of $L(\lambda)$ can be seen as a restriction of the given pair (X,T).

A sequence of parallel remarks can be made concerning operator pairs $T \in L(\mathcal{Y})$, $Y \in L(\mathcal{X},\mathcal{Y})$ via the study of operator

$$R = [Y,TY,\ldots,T^{\ell-1}Y]$$

from $\mathcal{X}^\ell$ to $\mathcal{Y}$. For example, the dual of Theorem 2.4.1 is:

THEOREM 2.4.3. *Let* $T \in L(\mathcal{Y})$ *and* $Y \in L(\mathcal{X},\mathcal{Y})$ *be such that the operator* $R \in L(\mathcal{X}^\ell,\mathcal{Y})$ *has a right inverse* R_R^{-1}. *Define operators* $A_i \in L(\mathcal{X})$ *by* $A_i = -W_{i+1}T^\ell Y$, $i = 0,1,\ldots,\ell-1$ *where*

$$\text{col}[W_i]_{i=1}^{\ell} = R_R^{-1},$$

and hence the monic operator polynomial $L(\lambda) = \Sigma_{i=0}^{\ell} A_i\lambda^i$, $A_\ell = I$ *is defined. Then*

$$YA_0 + TYA_1 + \cdots + T^{\ell-1}YA_{\ell-1} + T^{\ell}Y = 0$$

and letting

$$C_2 = \begin{bmatrix} 0 & 0 & \cdots & 0 & -A_0 \\ I & 0 & \cdots & 0 & -A_1 \\ 0 & I & \cdots & 0 & -A_2 \\ \vdots & & & & \vdots \\ 0 & 0 & \cdots & I & -A_{\ell-1} \end{bmatrix}$$

we have also

$$T = R(PC_2|_{\mathcal{M}})R_R^{-1}$$

where $\mathcal{M} = \text{Im } R_R^{-1}$ *and* $P \in L(\mathcal{X}^{\ell})$ *is the projection on* $\mathcal{M}$ *along* Ker R.

Observe that C_2 is a global linearization of $L(\lambda)$ (see the formula (2.1.8)).

2.5 Spectral triples

Here we introduce a basic notion of spectral triples which, together with right and left spectral pairs, form the main tools in our investigation of operator polynomials.

Let $L(\lambda)$ be a monic operator polynomial on of degree ℓ, and let (X,T) where $X \in L(\mathcal{Y},\mathcal{X})$, $T \in L(\mathcal{Y})$ be a right spectral pair of $L(\lambda)$. Define the operator $Y \in L(\mathcal{X},\mathcal{Y})$ by

$$XT^jY = \begin{cases} 0, & \text{for } j = 0,\dots,\ell-2 \\ I, & \text{for } j = \ell-1. \end{cases} \tag{2.5.1}$$

These properties define Y correctly and uniquely in view of the invertibility of $\text{col}[XT^i]_{i=0}^{\ell-1}$. The triple of operators (X,T,Y) will be called *spectral triple* of $L(\lambda)$. Basic properties of spectral triples are summarized in the following proposition.

PROPOSITION 2.5.1. (a) *A spectral triple of a monic operator polynomial* $L(\lambda)$ *is unique up to similarity: If* (X_1, T_1, Y_1) *and* (X_2, T_2, Y_2) *are two spectral triples of* $L(\lambda)$, *then there is a unique invertible operator* S *such that* $X_2 = X_1S$, $T_2 = S^{-1}T_1S$, $Y_2 = S^{-1}Y_1$; (b) *if* (X, T, Y) *is a spectral triple of* $L(\lambda)$, *then* (T, Y) *is a left spectral pair of* $L(\lambda)$; (c) *for a left spectral pair* (T, Y) *of* $L(\lambda)$ *define uniquely* X *by*

$$X[Y, TY, \ldots, T^{\ell-1}Y] = [0 \cdots 0I].$$

(This definition is correct in view of invertibility of $[Y, TY, \ldots, T^{\ell-1}Y]$.) *Then* (X, T, Y) *is a spectral triple of* $L(\lambda)$.

PROOF. The part (a) follows from Theorem 2.1.1. To prove the part (b), it is sufficient to consider the case when

$$X = [I0 \cdots 0] \in L(\mathcal{X}^{\ell}, \mathcal{X}),$$
$$T = C_L,$$

and the corresponding Y turns out to be $Y = \operatorname{col}[\delta_{\ell i} I]_{i=1}^{\ell} \in L(\mathcal{X}, \mathcal{X}^{\ell})$. Now use the fact that (2.1.8) is a left spectral pair of $L(\lambda)$. Hence, so is

$$\left(B^{-1}\begin{bmatrix} 0 & 0 & \cdots & & -A_0 \\ I & 0 & \cdots & & -A_1 \\ 0 & I & \cdots & & \\ \vdots & \vdots & & & \vdots \\ 0 & 0 & \cdots & I & -A_{\ell-1} \end{bmatrix} B, \quad B^{-1}\begin{bmatrix} I \\ 0 \\ \vdots \\ 0 \end{bmatrix}\right), \tag{2.5.2}$$

where

$$B = \begin{bmatrix} A_1 & A_2 & \cdots & A_{\ell-1} & I \\ A_2 & & & I & 0 \\ \vdots & & \cdot^{\cdot^{\cdot}} & \vdots & \vdots \\ A_{\ell-1} & I & & & 0 \\ I & 0 & \cdots & 0 & 0 \end{bmatrix}.$$

It remains to observe that in view of the formula (2.2.4), the pair (2.5.2) is just $(C_L, \mathrm{col}[\delta_{\ell i}I]_{i=1}^{\ell})$.

Finally, for the proof of (c), one can assume without loss of generality that T and Y are given by (2.1.8). The details are left to the reader. ∎

The importance of spectral triples stems from the following representation theorem for the inverses of monic operator polynomials.

THEOREM 2.5.2. *Let* $L(\lambda)$ *be a monic operator polynomial with spectral triple* (X,T,Y). *Then*

$$L(\lambda)^{-1} = X(\lambda I-T)^{-1}Y, \quad \lambda \notin \sigma(L). \tag{2.5.3}$$

The representation (2.5.3) will be called the *resolvent form* of $L(\lambda)$.

PROOF. Because of Proposition 2.5.1(a), it is enough to check (2.5.3) for

$$X = [I0\cdots 0]; \; T = C_L; \; Y = \mathrm{col}[\delta_{i\ell}I_{\mathcal{X}}]_{i=1}^{\ell}.$$

Rewrite formula (1.1.5) in the form

$$L(\lambda)^{-1} \oplus I_{\mathcal{X}^{\ell-1}} = F(\lambda)\,(\lambda I-C_L)E(\lambda)^{-1},$$

premultiply by $[I0\cdots 0]$ and postmultiply by $\mathrm{col}[\delta_{i1}I_{\mathcal{X}}]_{i=1}^{\ell}$. Since the first row of $F(\lambda)$ is X, and the first column of $E(\lambda)^{-1}$ is Y, the equality (2.5.3) follows. ∎

Note that the converse statement to Theorem 2.5.2 is false in general. Namely, if $X \in L(\mathcal{Y},\mathcal{X})$, $T \in L(\mathcal{Y})$, $Y \in L(\mathcal{X},\mathcal{Y})$ are such that (2.5.2) holds, then (X,T,Y) is not necessarily a spectral triple for $L(\lambda)$, as the following example shows. Take $L(\lambda) = \lambda I - I$ and let $X, Y \in L(\mathcal{X})$ be any operators for which $XY = I$, $YX \neq I$; then $L^{-1}(\lambda) = X(\lambda I - I)^{-1}Y$ but (X, I, Y) is not a spectral triple for $L(\lambda)$. However, if $\mathcal{X}$ is finite dimensional and (2.5.3) holds for $X \in L(\mathcal{X}^\ell, \mathcal{X})$, $T \in L(\mathcal{X}^\ell)$, $Y \in L(\mathcal{X}, \mathcal{X}^\ell)$ (as usual, ℓ is the degree of $L(\lambda)$), then (X, T, Y) is indeed a spectral triple for $L(\lambda)$ (see Gohberg-Lancaster-Rodman [2]).

In connection with the observation made in the preceding paragraph, note the following fact.

THEOREM 2.5.3. *Let* $L(\lambda)$ *be a monic operator polynomial of degree* ℓ, *and let* $X \in L(\mathcal{Y}, \mathcal{X})$, $T \in L(\mathcal{Y})$ *be operators such that*

$$L(\lambda)^{-1} = X(\lambda I - T)^{-1}Y, \quad \lambda \notin \sigma(L). \tag{2.5.4}$$

Then the operator $Q \overset{\text{def}}{=} \text{col}[XT^i]_{i=0}^{\ell-1}$ *is right invertible and the operator*

$$R \overset{\text{def}}{=} [Y, TY, \ldots, T^{\ell-1}Y]$$

is left invertible. Moreover, (X, T, Y) *is a spectral triple for* $L(\lambda)$ *if and only if* Q *is invertible, or, equivalently, if and only if* R *is invertible.*

PROOF. Observe that

$$\lambda^{-\ell}L(\lambda) = I + \sum_{j=0}^{\ell-1} \lambda^{j-\ell}A_j$$

is invertible in a neighborhood of infinity, and hence, for $|\lambda|$ sufficiently large, $L(\lambda)$ can be developed into a power series

$$L(\lambda)^{-1} = \sum_{j=-\ell}^{-\infty} \lambda^j Z_j, \quad Z_{-\ell} = I \tag{2.5.5}$$

for some operators Z_j. Let $r > 0$ be so large that $\sigma(T) \subset \{\lambda \in \mathbb{C} \mid |\lambda| < r\}$ and (2.5.5) holds for $|\lambda| \geq \frac{r}{2}$. Then

$$\frac{1}{2\pi i}\int_{|\lambda|=r} \lambda^j L(\lambda)^{-1} d\lambda = \begin{cases} 0, & \text{if } j = 0, \ldots, \ell-2 \\ I, & \text{if } j = \ell-1, \end{cases}$$

but also

$$\frac{1}{2\pi i}\int_{|\lambda|=r} \lambda^j L(\lambda)^{-1}d\lambda = \frac{1}{2\pi i}\int_{|\lambda|=r} \lambda^j X(\lambda I-T)^{-1}Yd\lambda = XT^jY,$$

$$j = 0,1,\dots .$$

Comparing these formulas we see that

$$QR = \begin{bmatrix} 0 & & I \\ & I^{\cdot^{\cdot^{\cdot}}} & \\ I & & * \end{bmatrix} \tag{2.5.6}$$

is invertible, and the left (resp. right) invertibility of R (resp. Q) follows.

If (X,T,Y) is a spectral triple of $L(\lambda)$, then (X,T) is a right spectral pair and (T,Y) is a left spectral pair of $L(\lambda)$, so the invertibility of Q and R follows. Conversely, assume that one of Q and R, say Q, is invertible. Formula (2.5.6) shows that R is invertible as well. Further, for $j = 0,\dots,\ell-1$ we have

$$0 = \frac{1}{2\pi i}\int_{|\lambda|=r} \lambda^j L(\lambda)L(\lambda)^{-1}d\lambda = \frac{1}{2\pi i}\int_{|\lambda|=r} \lambda^j\left(\sum_{k=0}^{\ell-1}\lambda^k A_k+\lambda^\ell I\right)$$

$$\cdot X(\lambda I-T)^{-1}Yd\lambda = (A_0X+\cdots+A_{\ell-1}XT^{\ell-1}+XT^\ell)T^jY,$$

so

$$(A_0X+\cdots+A_{\ell-1}XT^{\ell-1}+XT^\ell)R = 0.$$

By the invertibility of R,

$$A_0X+\cdots+A_{\ell-1}XT^{\ell-1}+XT^\ell = 0,$$

and hence (X,T) is a right spectral pair of $L(\lambda)$. ∎

2.6 Multiplication and division theorems

Using the right and left canonical forms and the resolvent form developed in the preceding section, we derive here formulas for multiplication and division of monic operator polynomials in terms of these forms.

THEOREM 2.6.1. *Let* L_1, L_2 *be monic operator polynomials on* $\mathcal{X}$ *with spectral triples,* X_1, T_1, Y_1 *and* X_2, T_2, Y_2, *respectively, and let* $L(\lambda) = L_2(\lambda)L_1(\lambda)$. *Then*

(a) $L^{-1}(\lambda) = X(I\lambda - T)^{-1}Y$

where

$$X = [X_1, 0], \quad T = \begin{bmatrix} T_1 & Y_1X_2 \\ 0 & T_2 \end{bmatrix}, \quad Y = \begin{bmatrix} 0 \\ Y_2 \end{bmatrix},$$

(b) (X, T, Y) *is a spectral triple for* L.

PROOF. It is easily verified that

$$(\lambda I - T)^{-1} = \begin{bmatrix} (\lambda I - T_1)^{-1} & (\lambda I - T_1)^{-1}Y_1X_2(\lambda I - T_2)^{-1} \\ 0 & (\lambda I - T_2)^{-1} \end{bmatrix},$$

$$X(\lambda I - T)^{-1}Y = X_1(\lambda I - T_1)^{-1}Y_1X_2(\lambda I - T_2)^{-1}Y_2$$

$$= L_1(\lambda)^{-1}L_2(\lambda)^{-1} = L(\lambda)^{-1}.$$

For the part (b) we have to prove (by Theorem 2.5.3) that $Q = \mathrm{col}[XT^i]_{i=0}^{\ell-1}$ is invertible, where ℓ is the degree of $L(\lambda)$. By induction on k one easily proves that

$$XT^k = [X_1T_1^k, \sum_{i=0}^{k-1}(X_1T_1^iY_1)X_2T_2^{k-1-i}], \quad k = 1,2,\ldots . \tag{2.6.1}$$

Using the relations

$$X_1T_1^kY_1 = \begin{cases} 0 & \text{if } k = 0,\ldots,k_1-2 \\ I & \text{if } k = k_1-1, \end{cases}$$

(here k_1 is the degree of L_1), rewrite (2.6.1) in the form

$$XT^k = \begin{cases} [X_1T_1^k, \quad 0] & \text{if } k = 1,\ldots,k_1-1 \\ [X_1T_1^k, \quad \sum_{i=k_1}^{k-1} X_1T_1^iY_1X_2T_2^{k-1-i} + X_2T_2^{k-k_1}] & \text{if } k \geq k_1. \end{cases}$$

So Q has the following block form

$$Q = \begin{bmatrix} Q_{11} & 0 \\ Q_{21} & Q_{22} \end{bmatrix},$$

where $Q_{11} = \mathrm{col}[X_1 T_1^i]_{i=0}^{k_1-1}$ is invertible, and

$$Q_{22} = \begin{bmatrix} I & 0 & \cdots & 0 \\ X_1 T_1^{k_1} Y_1 & I & \cdots & 0 \\ \vdots & \vdots & & \\ X_1 T_1^{\ell-2} Y_1 & X_1 T_1^{\ell-3} Y_1 & \cdots X_1 T_1^{k_1} Y_1 & I \end{bmatrix} \begin{bmatrix} X_2 \\ X_2 T_2 \\ \vdots \\ X_2 T_2^{k_2-1} \end{bmatrix}$$

is invertible as well (here $k_2 = \ell - k_1$ is the degree of $L_2(\lambda)$). Therefore, Q is invertible. ∎

We pass now to the division of monic operator polynomials. Let $L(\lambda)$ be an operator polynomial (not necessarily monic) and let $L_1(\lambda)$ be a *monic* operator polynomial (both on the same Banach space $\mathcal{X}$). Applying long division of polynomials (which is possible because the leading coefficient of $L_1(\lambda)$ is invertible), write

$$L(\lambda) = Q_1(\lambda)L_1(\lambda)+R_1(\lambda) = L_1(\lambda)Q_2(\lambda)+R_2(\lambda),$$

where $Q_1(\lambda)$, $R_1(\lambda)$, $Q_2(\lambda)$, and $R_2(\lambda)$ are operator polynomials such that the degrees of $R_1(\lambda)$ and $R_2(\lambda)$ are smaller than the degree of $L_1(\lambda)$. Moreover, these operator polynomials are uniquely determined by $L(\lambda)$ and $L_1(\lambda)$.

The following theorem describes division in terms of the right and left canonical forms of $L_1(\lambda)$.

THEOREM 2.6.2. *Let* $L(\lambda) = \sum_{j=0}^{\ell} \lambda^j A_j$ *be an operator polynomial (not necessarily monic), and let*

$$L_1(\lambda) = \lambda^k I - X_1 T_1^k (V_1 + \cdots + V_k \lambda^{k-1})$$

be a monic operator polynomial of degree $k \leq \ell$ *in the right canonical form where* (X_1, T_1) *is a right spectral pair of* $L_1(\lambda)$ *and*

$$[V_1 V_2 \cdots V_k] = \{\mathrm{col}[X_1 T_1^i]_{i=0}^{k-1}\}^{-1}.$$

Then

$$L(\lambda) = Q_1(\lambda) L_1(\lambda) + R_1(\lambda), \tag{2.6.2}$$

where

$$Q_1(\lambda) = \sum_{j=0}^{\ell-k} \lambda^j \Big(\sum_{i=j+1}^{\ell} A_i X_1 T_1^{i-j-1} \Big) V_k$$

and

$$R_1(\lambda) = \sum_{j=1}^{k} \lambda^{j-1} \Big(\sum_{i=0}^{\ell} A_i X_1 T_1^i \Big) V_j .$$

If

$$L_1(\lambda) = \lambda^k I - (W_1 + \cdots + W_k \lambda^{k-1}) T_1^k Y_1$$

is a left canonical form of $L_1(\lambda)$ *(so* (T_1, Y_1) *is a left spectral pair of* $L_1(\lambda)$ *and*

$$\mathrm{col}[W_i]_{i=1}^{k} = [Y_1, T_1 Y_1, \ldots, T_1^{k-1} Y_1]^{-1}),$$

then

$$L(\lambda) = L_1(\lambda) Q_2(\lambda) + R_2(\lambda), \tag{2.6.3}$$

where

$$Q_2(\lambda) = \sum_{j=0}^{\ell-k} \lambda^j \Big(\sum_{i=j+1}^{\ell} W_k T_1^{i-j-1} Y_1 A_1 \Big)$$

and

$$R_2(\lambda) = \sum_{j=1}^{k} \lambda^{j-1} \Big(W_j \sum_{i=0}^{\ell} T_1^i Y_1 A_i \Big).$$

PROOF. As the proof of (2.6.2) is the same as in the finite-dimensional case (see Gohberg-Lancaster-Rodman [2]), and the proof of (2.6.3) is analogous, we indicate only the main steps in the proof of (2.6.2).

Define

$$G_{\alpha\beta} = X_1 T_1^{\alpha} V_{\beta}, \quad 1 \leq \beta \leq k, \quad \alpha = 0, 1, 2, \ldots . \tag{2.6.4}$$

Then for each i, $1 \leq i \leq k$, we have

$$(2.6.5)\qquad G_{p+1,i} = G_{pk}G_{ki}+G_{p,i-1}, \quad p = 0,1,2,\ldots$$

(and we set $G_{p0} = 0$, $p = 0,1,\ldots$).

Next, we check the following equalities:

$$(2.6.6)\qquad V_j = \sum_{m=0}^{k-j} T_1^m V_k B_{j+m}, \quad j = 1,\ldots,k,$$

where B_p are the coefficients of $L_1(\lambda)$: $L_1(\lambda) = \Sigma_{j=0}^{k} B_j\lambda^j$ with $B_k = I$.

Define also the following operator polynomials:

$L_{1,j}(\lambda) = B_j+B_{j+1}\lambda+\cdots+B_k\lambda^{k-j}$, $j = 0,1,\ldots,k$. In particular, $L_{1,0}(\lambda) = L_1(\lambda)$ and $L_{1,k}(\lambda) = I$. We need the following property of the polynomials $L_{1,j}(\lambda)$:

$$(2.6.7)\qquad V_k L_1(\lambda) = (I\lambda-T_1)\Big(\sum_{j=0}^{k-1} T_1^j V_k L_{1,j+1}(\lambda)\Big).$$

We are now ready to prove that the difference $L(\lambda)-R_1(\lambda)$ is divisible by $L_1(\lambda)$. Indeed, using (2.6.6) and then (2.6.7)

$$R_1(\lambda)(L_1(\lambda))^{-1} = \Big(\sum_{i=0}^{\ell} A_i X_1 T_1^i\Big) \cdot \Big(\sum_{j=0}^{k-1} T_1^j V_k L_{1,j+1}(\lambda)\Big) \cdot L_1^{-1}(\lambda)$$

$$= \sum_{i=0}^{\ell} A_i X_1 T_1^i (I\lambda-T_1)^{-1} V_k.$$

Also,

$$L(\lambda)(L_1(\lambda))^{-1} = \sum_{i=0}^{\ell} A_j\lambda^j \cdot X_1(I\lambda-T_1)^{-1}V_k$$

(and here we use the resolvent form of $L_1(\lambda)$). So

$$L(\lambda)(L_1(\lambda))^{-1}-R_1(\lambda)(L_1(\lambda))^{-1} = \sum_{i=0}^{\ell} A_i X_1(I\lambda^i-T_1^i)(I\lambda-T_1)^{-1}V_k.$$

Using the equality $I\lambda^i-T_1^i = (\Sigma_{p=0}^{i-1}\lambda^p T_1^{i-1-p}) \cdot (\lambda I-T_1)$ $(i > 0)$, we obtain

(2.6.8) $$L(\lambda)(L_1(\lambda))^{-1}-R(\lambda)(L_1(\lambda))^{-1} = \sum_{p=0}^{\ell-1} \lambda^p \sum_{i=p+1}^{\ell} A_i X_1 T_1^{i-1-p} V_k .$$

Because of the relations $X_1 T_1^j V_k = 0$ for $j = 0,\ldots,k-1$, all the terms in (2.6.8) with $p = \ell-k+1,\ldots,p = \ell-1$ vanish, and formula (2.6.2) is proved. ∎

The following important corollary follows at once from Theorem 2.6.2. We say that an operator polynomial $L_1(\lambda)$ is a *right* (resp. *left*) divisor of an operator polynomial $L(\lambda)$ if $L(\lambda) = Q(\lambda)L_1(\lambda)$ (resp. $L(\lambda) = L_1(\lambda)Q(\lambda)$) for some operator polynomial $Q(\lambda)$.

COROLLARY 2.6.3. *Let* $L(\lambda) = \sum_{j=0}^{\ell} \lambda^j A_j$ *be an operator polynomial, and let* $L_1(\lambda)$ *be a monic operator polynomial with right spectral pair* (X_1,T_1) *and left spectral pair* (T_1,Y_1). *Then* $L_1(\lambda)$ *is a right divisor of* $L(\lambda)$ *if and only if*

$$\sum_{j=0}^{\ell} A_j X_1 T_1^j = 0.$$

$L_1(\lambda)$ *is a left divisor of* $L(\lambda)$ *if and only if*

$$\sum_{j=0}^{\ell} T_1^j Y_1 A_j = 0.$$

2.7 Characterization of divisors in terms of subspaces

We describe here right divisors of monic operator polynomials in geometric terms of invariant subspaces of their linearizations.

THEOREM 2.7.1. *Let* $L(\lambda)$ *be a monic operator polynomial of degree* ℓ *with right spectral pair* (X,T), *where* $X \in L(\mathcal{Y},\mathcal{X})$, $T \in L(\mathcal{Y})$. *Then for every* T*-invariant subspace* $\mathcal{M} \subset \mathcal{Y}$ *such that the operator*

(2.7.1) $$\operatorname{col}[XT^i]_{i=0}^{k-1}\Big|_{\mathcal{M}} \in L(\mathcal{M},\mathcal{X}^k)$$

is invertible, there exists a unique monic operator polynomial $L_1(\lambda)$ *of degree* k *which is a right divisor of* $L(\lambda)$ *and whose right spectral pair is similar to* $(X|_{\mathcal{M}}, T|_{\mathcal{M}})$.

Conversely, for every monic operator polynomial $L_1(\lambda)$ *of degree* k *which is a right divisor of* $L(\lambda)$ *and has right spectral pair* (X_1, T_1), *the (closed) subspace*

$$\mathcal{M} = \operatorname{Im}\{\{\operatorname{col}[XT^i]_{i=0}^{\ell-1}\}^{-1}\operatorname{col}[X_1T_1^i]_{i=0}^{\ell-1}\} \tag{2.7.2}$$

is T*-invariant, the restriction*

$$\{\operatorname{col}[XT^i]_{i=0}^{k-1}\}|_{\mathcal{M}} \in L(\mathcal{M}, \mathcal{X}^k)$$

is invertible and $(X|_{\mathcal{M}}, T|_{\mathcal{M}})$ *is similar to* (X_1, T_1).

PROOF. Let $\mathcal{M} \subset \mathcal{Y}$ be T-invariant such that (2.7.1) is invertible. Construct the monic operator polynomial $L_1(\lambda)$ with right spectral pair $(X|_{\mathcal{M}}, T|_{\mathcal{M}})$ (cf. (2.2.1)):

$$L_1(\lambda) = I\lambda^k - X|_{\mathcal{M}}(T|_{\mathcal{M}})^k(V_1+V_2\lambda+\cdots V_k\lambda^{k-1}),$$

where $[V_1\cdots V_k] = [\operatorname{col}[X|_{\mathcal{M}}(T|_{\mathcal{M}})^i]_{i=0}^{k-1}]^{-1}$. Appeal to Corollary 2.6.3 (bear in mind the equality

$$A_0X+A_1XT+\cdots+A_{\ell-1}XT^{\ell-1}+XT^{\ell} = 0,$$

where A_j are the coefficients of $L(\lambda)$) to deduce that $L_1(\lambda)$ is a right divisor of $L(\lambda)$.

For the converse statement observe first that the subspace $\mathcal{M}$ defined by (2.7.2) is indeed closed because obviously $k \le \ell$ and hence $\operatorname{col}[X_1T_1^i]_{i=0}^{\ell-1}$ is left invertible. Further, since $L_1(\lambda)$ is a right divisor of $L(\lambda)$, Corollary 2.6.3 implies

$$C_L \operatorname{col}[X_1T_1^i]_{i=0}^{\ell-1} = \operatorname{col}[X_1T_1^i]_{i=0}^{\ell-1}T_1, \tag{2.7.3}$$

where C_L is the companion operator for $L(\lambda)$. Also

$$C_L \operatorname{col}[XT^i]_{i=0}^{\ell-1} = \operatorname{col}[XT^i]_{i=0}^{\ell-1}T. \tag{2.7.4}$$

Eliminating C_L from (2.7.3) and (2.7.4), we obtain

(2.7.5) $$T[\mathrm{col}[XT^i]_{i=0}^{\ell-1}]^{-1}[\mathrm{col}[X_1T_1^i]_{i=0}^{\ell-1}]$$

$$= [\mathrm{col}[XT^i]_{i=0}^{\ell-1}]^{-1}[\mathrm{col}[X_1T_1^i]_{i=0}^{\ell-1}]T_1.$$

This equality readily implies that the subspace $\mathcal{M}$ given by (2.7.2) is T-invariant.

Further,

(2.7.6) $$X[\mathrm{col}[XT^i]_{i=0}^{\ell-1}]^{-1}[\mathrm{col}[X_1T_1^i]_{i=0}^{\ell-1}] = X_1.$$

Since the operator

$$S = [\mathrm{col}[XT^i]_{i=0}^{\ell-1}]^{-1}[\mathrm{col}[X_1T_1^i]_{i=0}^{\ell-1}]$$

is left invertible with Im $S = \mathcal{M}$, the equalities (2.7.5) and (2.7.6) show that (X_1,T_1) is similar to $(X|_{\mathcal{M}},T|_{\mathcal{M}})$ (indeed,

$$X_1 = X|_{\mathcal{M}}\cdot\hat{S}, \quad T_1 = \hat{S}^{-1}T\hat{S},$$

where $\hat{S} \in L(\mathcal{Y}_1,\mathcal{M})$ (here $\mathcal{Y}_1$ is the Banach space on which T_1 acts) is the invertible operator defined by $\hat{S}x = Sx$, $x \in \mathcal{Y}_1$). ∎

Note that the subspace $\mathcal{M}$ defined by (2.7.2) does not depend on the choice of the right spectral pair (X_1,T_1) of $L_1(\lambda)$, because

$$\mathrm{Im}\ \mathrm{col}[X_1T_1^i]_{i=0}^{\ell-1} = \mathrm{Im}\ \mathrm{col}[X_1S\cdot(S^{-1}T_1S)^i]_{i=0}^{\ell-1}$$

for any invertible operator S. Thus, for every monic right divisor $L_1(\lambda)$ of $L(\lambda)$ of degree k we have constructed a subspace $\mathcal{M}$, which will be called the *supporting subspace* of $L_1(\lambda)$. As (2.7.2) shows, the supporting subspace does depend on the right spectral pair (X,T); but once the pair (X,T) is fixed, the supporting subspace depends only on the divisor $L_1(\lambda)$. If we wish to stress the dependence of $\mathcal{M}$ on (X,T) also (not only on $L_1(\lambda)$), we shall speak in terms of a supporting subspace relative to the right spectral pair (X,T).

So Theorem 2.7.1 gives a one-to-one corresondence between the right monic divisors of $L(\lambda)$ of degree k and T-invariant subspaces $\mathcal{M} \subset \mathcal{Y}$, such that $\mathrm{col}[X|_{\mathcal{M}}(T|_{\mathcal{M}})^i]_{i=0}^{k-1}$ is invertible, which are in fact the supporting subspaces of the right divisors. Thus, we have a description of the algebraic relation (divisibility of monic polynomials) in a geometric language of supporting subspaces.

For two divisors of $L(\lambda)$, it may happen that one of them is in turn a divisor of the other. In terms of supporting subspaces, such a relationship means nothing more than inclusion, as the following corollary shows.

COROLLARY 2.7.2. *Let* $L_{11}(\lambda)$ *and* $L_{12}(\lambda)$ *be monic right divisors of* $L(\lambda)$. *Then* $L_{11}(\lambda)$ *is a right divisor of* $L_{12}(\lambda)$ *if and only if for the supporting subspaces* $\mathcal{M}_1$ *and* $\mathcal{M}_2$ *of* $L_{11}(\lambda)$ *and* $L_{12}(\lambda)$, *respectively, the relation* $\mathcal{M}_1 \subset \mathcal{M}_2$ *holds.*

PROOF. Let (X,T) be the right spectral pair of $L(\lambda)$ relative to which the supporting subspaces $\mathcal{M}_1$ and $\mathcal{M}_2$ are defined. Then, by Theorem 2.7.1 $(X|_{\mathcal{M}_i}, T|_{\mathcal{M}_i})$ $(i = 1,2)$ is a right spectral pair of $L_{1i}(\lambda)$. If $\mathcal{M}_1 \subset \mathcal{M}_2$, then, by Theorem 2.7.1 (when applied to $L_{12}(\lambda)$ in place of $L(\lambda)$), $L_{11}(\lambda)$ is a right divisor of $L_{12}(\lambda)$. Suppose now $L_{11}(\lambda)$ is a right divisor of $L_{12}(\lambda)$. Then, by Theorem 2.7.1, there exists a supporting subspace $\mathcal{M}_{12} \subset \mathcal{M}_2$ of $L_{11}(\lambda)$ as a right divisor of $L_{12}(\lambda)$, so that $(X|_{\mathcal{M}_{12}}, T|_{\mathcal{M}_{12}})$ is a right spectral pair of $L_{11}(\lambda)$ as a divisor of $L(\lambda)$. Since the supporting subspace is unique, it follows that $\mathcal{M}_1 = \mathcal{M}_{12} \subset \mathcal{M}_2$. ∎

It is possible to deduce results analogous to Theorem 2.7.1 and Corollary 2.7.2 for left divisors by using left spectral pairs. However, we will describe left divisors in Section 2.6.9 in terms of the description of quotients.

For future reference let us record the following result which is obtain by successive applications of Corollary 2.7.2.

THEOREM 2.7.3. *Let* $L(\lambda)$ *and* (X,T) *be as in Theorem* 2.7.1, *and let* $\mathcal{M}_1 \subset \cdots \subset \mathcal{M}_r \subset \mathcal{Y}$ *be a finite chain of* T-*invariant subspaces such that the operators*

$$(2.7.7) \qquad [\mathrm{col}[XT^i]_{i=0}^{k_j-1}]\big|_{\mathcal{M}_j} \in L(\mathcal{M}_j, \mathcal{X}^{k_j}), \quad j = 1,\ldots,r$$

are invertible for some integers $k_1 < \cdots < k_r < \ell$. *Then there is a unique factorization*

$$(2.7.8) \qquad L(\lambda) = L_{r+1}(\lambda)L_r(\lambda)\cdots L_1(\lambda)$$

such that $L_j(\lambda)$ *are monic operator polynomials, the degree of the product* $M_j(\lambda) \stackrel{\mathrm{def}}{=} L_j(\lambda)L_{j-1}(\lambda)\cdots L_1(\lambda)$ is k_j for $j = 1,\ldots,r$, *and right spectral pair of* $M_j(\lambda)$ *is similar to* $(X|_{\mathcal{M}_j}, T|_{\mathcal{M}_j})$. *Conversely, for every factorization* (2.7.8) *there is a unique chain* $\mathcal{M}_1 \subset \cdots \subset \mathcal{M}_r$ *of T-invariant subspaces such that the operators* (2.7.7) *are invertible, where* k_j *is the degree of* $M_j(\lambda) \stackrel{\mathrm{def}}{=} L_j(\lambda)L_{j-1}(\lambda)\cdots L_1(\lambda)$, *and* $(X|_{\mathcal{M}_j}, T|_{\mathcal{M}_j})$ *is a right spectral pair of* $M_j(\lambda)$ *for* $j = 1,\ldots,r$. *In fact*

$$\mathcal{M}_j = \mathrm{Im}\{\{\mathrm{col}[XT^i]_{i=0}^{\ell-1}\}^{-1}\mathrm{col}[X_jT_j^i]_{i=0}^{\ell-1}\},$$

where (X_j, T_j) *is a right spectral pair of* $M_j(\lambda)$.

We conclude this section with an important particular case: divisors of the first degree. It follows from Corollary 2.6.3 and also easily verified directly, that $\lambda I - Z$ is a right divisor of an operator polynomial $L(\lambda) = \sum_{j=0}^{\ell} \lambda^j A_j$ if and only if

$$(2.7.9) \qquad \sum_{j=0}^{\ell} A_j Z^j = 0.$$

In this case, it is natural to consider (2.7.9) as an operator equation with an operator unknown Z, and we say that $Z_0 \in L(\mathcal{X})$ is a *right operator root* of the operator equation (2.7.9) (or of the polynomial $L(\lambda)$) if

$$\sum_{j=0}^{\ell} A_j Z_0^j = 0.$$

Analogously, $Z_0 \in L(\mathcal{X})$ is a *left operator root* of the equation

$$\sum_{j=0}^{\ell} Z^j A_j = 0 \tag{2.7.10}$$

if (2.7.10) is true with Z replaced by Z_0. Theorem 2.7.1 specialized to the case of right divisors of first degree can be stated as follows.

COROLLARY 2.7.4. *Let* (X,T) *be a right spectral pair of the monic operator polynomial* $L(\lambda) = \lambda^\ell I + \sum_{j=0}^{\ell-1} \lambda^j A_j$, *where* $A_j \in L(\mathcal{X})$. *Then there exists one-to-one correspondence between operators* $Z \in L(\mathcal{X})$ *satisfying*

$$Z^\ell + \sum_{j=0}^{\ell-1} A_j Z^j = 0$$

and T-*invariant subspaces* $\mathcal{M}$ *for which*

$$X|_{\mathcal{M}}: \mathcal{M} \to \mathcal{X}$$

is invertible. This correspondence is given by the formula

$$\mathcal{M} = \operatorname{Im}\{\{\operatorname{col}[XT^i]_{i=0}^{\ell-1}\}^{-1}\operatorname{col}[Z^i]_{i=0}^{\ell-1}\}.$$

The analogous characterization of left operator roots will be given in Section 2.9.

2.8 Factorable indexless polynomials

As first application of the characterization of divisors given in the preceding section, we consider here factorable polynomials. A monic operator polynomial $L(\lambda)$ on a Banach space $\mathcal{X}$ is called *factorable* if it admits factorization of type

$$L(\lambda) = (\lambda I+X_1)\cdots(\lambda I+X_\ell)$$

for some $X_1,\ldots,X_\ell \in L(\mathcal{X})$.

Not all monic operator polynomials are factorable, even for finite-dimensional $\mathcal{X}$ (unless, of course, $\dim \mathcal{X} = 1$). The following example shows that for the case of two-dimensional $\mathcal{X}$.

EXAMPLE 2.8.1. Let $\mathcal{X} = \mathbb{C}^2$, and

$$L(\lambda) = \begin{bmatrix} \lambda^2 & -1 \\ 0 & \lambda^2 \end{bmatrix}.$$

If there were a factorization $L(\lambda) = (\lambda I+X_1)(\lambda I+X_2)$, then necessarily $X_1+X_2 = 0$ and

$$X_1^2 = \begin{bmatrix} 0 & 1 \\ 0 & 0 \end{bmatrix}.$$

However, there is no 2×2 complex matrix whose square is $\begin{bmatrix} 0 & 1 \\ 0 & 0 \end{bmatrix}$. ∎

In the infinite-dimensional case there is also an index obstruction to factorability:

EXAMPLE 2.8.2. Let $A \in L(\mathcal{X})$ be Fredholm with an odd index. Then the operator polynomial $L(\lambda) = \lambda^2 I+A$ is not factorable. If it were, say

$$L(\lambda) = \lambda^2 I+A = (\lambda I+X_1)(\lambda I+X_2),$$

then $X_1^2 = -A$. It follows that Ker $X_1 \subset$ Ker A and Im $X_1 \supset$ Im A. Hence, X_1 is Fredholm, and by the logarithmic property of the index (ind (B_1B_2) = ind B_1 + ind B_2 for Fredholm B_1 and B_2) we have 2 ind X_1 = ind A. However, this contradicts the choice of A to have an *odd* index. ∎

We consider operator polynomials which avoid the index obstruction. To this end introduce the following definition. An operator polynomial $L(\lambda)$ on $\mathcal{X}$ will be called *indexless* if for every $\lambda_0 \in \mathcal{X}$ such that the operator $L(\lambda_0)$ is semifredholm, the index of $L(\lambda_0)$ is zero. (Recall that an operator $X \in L(\mathcal{X})$ is called *semifredholm* if Im X is closed and at least one of the numbers codim Im X and dim Ker X is finite; then the index of X is defined as dim Ker X - codim Im X, and its value can be integer or $+\infty$ or $-\infty$.) Introduce also the norm in the set of all operator polynomials on $\mathcal{X}$ of degree ℓ in a natural way:

$$|||\sum_{j=0}^{\ell} \lambda^j A_j||| = \sum_{j=0}^{\ell} ||A_j||.$$

The following theorem is the main result of this section.

THEOREM 2.8.1. *Let $\mathcal{X}$ be a separable Hilbert space. Then the set of factorable indexless monic operator polynomials of degree ℓ on $\mathcal{X}$ is dense in the set of all indexless monic operator polynomials of degree ℓ on $\mathcal{X}$, in the sense of the norm* $|||\cdot|||$.

The proof of Theorem 2.8.1 requires substantial preparation.

An operator $A \in L(\mathcal{X})$ will be called *indexless* if for every $\lambda_0 \in \mathbb{C}$ such that $\lambda_0 I-A$ is semifredholm the index of $\lambda_0 I-A$ is zero. (In the literature the term "biquasitriangular" is often used to designate indexless operators; however, here the property of having zero index is crucial, so "indexless" seems to be more appropriate to this context.) We need the following important approximation result. An operator $A \in L(\mathcal{X})$ will be called *simple* if there is a direct sum decomposition $\mathcal{X} = \mathcal{X}_1 \dot{+} \cdots \dot{+} \mathcal{X}_k$ with A-invariant subspaces $\mathcal{X}_1, \ldots, \mathcal{X}_k$, and there are numbers $\lambda_1, \ldots, \lambda_k \in \mathbb{C}$ such that $A|_{\mathcal{X}_i} = \lambda_i I_{\mathcal{X}_i}$ for $i = 1, \ldots, k$.

THEOREM 2.8.2. *Let $\mathcal{X}$ be a separable Hilbert space. Then the closure (in the operator norm) of the set of all simple operators on $\mathcal{X}$ coincides with the set of all indexless operators acting in $\mathcal{X}$.*

The proof of Theorem 2.8.2 is difficult and will not be presented here; it can be found in Apostol-Voiculescu [1], or Apostol-Foias [1].

It is not difficult to describe all invariant subspaces for a simple operator.

PROPOSITION 2.8.3. *Let $A \in L(\mathcal{X})$ be a simple operator defined by the direct sum decomposition $\mathcal{X} = \mathcal{X}_1 \dot{+} \cdots \dot{+} \mathcal{X}_k$ and by $A|_{\mathcal{X}_i} = \lambda_i I_{\mathcal{X}_i}$ for different complex numbers $\lambda_1, \ldots, \lambda_k$. Then a subspace $\mathcal{M} \subset \mathcal{X}$ is A-invariant precisely when*

$$\mathcal{M} = \mathcal{N}_1 \dot{+} \cdots \dot{+} \mathcal{N}_k \tag{2.8.1}$$

for some subspaces $\mathcal{N}_1 \subset \mathcal{X}_1, \ldots, \mathcal{N}_k \subset \mathcal{X}_k$.

PROOF. Clearly, every subspace of the form (2.8.1) is A-invariant. Conversely, for an A-invariant subspace $\mathcal{M}$, define $\mathcal{N}_i = \mathcal{M} \cap \mathcal{X}_i$, $i = 1, \ldots, k$. To verify that (2.8.1) holds, we have only to check that for $x \in \mathcal{M}$ written in the form $x = x_1 + \cdots + x_k$, where $x_1 \in \mathcal{X}_1, \ldots, x_k \in \mathcal{X}_k$, we have $x_1, \ldots, x_k \in \mathcal{M}$. Indeed,

$$x_j = \Big(\prod_{i \neq j} (\lambda_i - \lambda_j) \Big)^{-1} \prod_{i \neq j} (\lambda_i I - A) x$$

belongs to $\mathcal{M}$ because $\mathcal{M}$ is A-invariant. ∎

Next consider the companion operators of indexless operator polynomials.

PROPOSITION 2.8.4. *A monic operator polynomial* $L(\lambda)$ *is indexless if and only if its companion operator is such.*

PROOF. Apply formula (1.1.5). ∎

We need also a characterization of simple operators in terms of their set of all invariant subspaces. An operator $A \in L(\mathcal{X})$ is said to have the *subspace complementedness* property if for every subspace $\mathcal{N} \subset \mathcal{X}$ there is an A-invariant subspace $\mathcal{M}$ which is a direct complement to $\mathcal{N}$ in $\mathcal{X}$: $\mathcal{M} \dot{+} \mathcal{N} = \mathcal{X}$. The *chain complementedness* property of $A \in L(\mathcal{X})$ means that for every finite chain of subspaces

$$\mathcal{X} \supset \mathcal{N}_1 \supset \cdots \supset \mathcal{N}_\ell \tag{2.8.1}$$

there is a chain of A-invariant subspaces

$$\mathcal{M}_1 \subset \cdots \subset \mathcal{M}_\ell \subset \mathcal{X} \tag{2.8.2}$$

such that $\mathcal{M}_i \dot{+} \mathcal{N}_i = \mathcal{X}$ for $i = 1, \ldots, \ell$.

THEOREM 2.8.5. *Let* $\mathcal{X}$ *be a Hilbert space. The following statements are equivalent for an operator* $A \in L(\mathcal{X})$:

(i) A *is a simple operator;*

(ii) A *has the subspace complementedness property;*

(iii) A *has the chain complementedness property.*

PROOF. We prove here only that (i) → (iii) (this is the part of Theorem 2.8.5 which will be used in the sequel). For the proof of the implication (iii) → (i) the reader is referred to Fong-Herrero-Rodman [1]. Note that the implication (iii) → (ii) is trivial.

First, we prove that (i) implies (ii).

Let A be a simple operator, so $A|_{\mathcal{X}_j} = \lambda_j I_{\mathcal{X}_j}$, $j = 1,\dots,m$ for some direct sum decomposition $\mathcal{X} = \mathcal{X}_1 \dot{+} \cdots \dot{+} \mathcal{X}_m$ and distinct complex numbers $\lambda_1,\dots,\lambda_m$. Without loss of generality we can assume that $\mathcal{X}_1,\dots,\mathcal{X}_m$ are orthogonal to each other (otherwise introduce in $\mathcal{X}$ a topologically equivalent inner product in which $\mathcal{X}_1,\dots,\mathcal{X}_m$ become orthogonal to each other). Proceed by induction on m. For m = 1 the assertion (ii) is trivial. We assume that (ii) is proved for simple operators with at most m-1 points in the spectrum. Let $\mathcal{M} \subset \mathcal{X}$ be a subspace. Denote by R_j the orthogonal projection on $\mathcal{X}_j$ (so Im $R_j = \mathcal{X}_j$, and Ker $R_j = \mathcal{X}_1 \oplus \cdots \oplus \mathcal{X}_{j-1} \oplus \mathcal{X}_{j+1} \oplus \cdots \oplus \mathcal{X}_m$). Put $\mathcal{R}_1 = \overline{R_1\mathcal{M}}$; $\mathcal{R}_2 = \mathcal{M} \cap (\mathcal{X}_2 \oplus \cdots \oplus \mathcal{X}_m)$; $\mathcal{R}_3 = \mathcal{M} \ominus \mathcal{R}_2$ (here and elsewhere in this proof, $\ominus$ stands for the orthogonal complement). It follows immediately from the definitions that R_1 maps $\mathcal{R}_3$ onto $R_1\mathcal{M}$ in a one-to-one manner. Consequently, dim $\mathcal{R}_1$ = dim $\mathcal{R}_3$ (the dimensions are understood here as the cardinalities of orthonormal bases). Hence there exists an isometry W which maps $\mathcal{R}_1$ onto $\mathcal{R}_3$. Introduce the operator $T = WR_1|_{\mathcal{R}_3} \in L(\mathcal{R}_3)$, and let T = UB be the polar decomposition of T (so B is positive semidefinite, and U is an isometry from $\overline{\text{Im } T^*}$ onto $\overline{\text{Im } T}$). In our case, U is actually unitary, because $\overline{\text{Im } T} = \mathcal{R}_3$ and Ker T = Ker$(R_1|_{\mathcal{R}_3}) = \{0\}$, i.e., $\overline{\text{Im } T^*} = \mathcal{R}_3$. Obviously, $\|B\| = \|T\| \le 1$. As $B \in L(\mathcal{R}_3)$ is a positive semidefinite contraction, its spectral measure $E(\lambda)$ is supported by the interval [0,1], and we can write

$$B = \int_0^1 \lambda \, dE(\lambda).$$

Put

$$\mathcal{Y}_1 = E(\tfrac{1}{2})\mathcal{R}_3, \ \mathcal{Y}_2 = \mathcal{R}_3 \ominus \mathcal{Y}_1.$$

Obviously, $B\mathcal{Y}_j \in \mathcal{Y}_j$ for $j = 1,2$. As $R_1 = W^{-1}UB$, and the operator $W^{-1}U$ is an isometry, the linear sets $R_1\mathcal{Y}_1$ and $R_1\mathcal{Y}_2$ are orthogonal. Hence, $R_1\mathcal{Y}_1$ and $\mathcal{Y}_2$ are orthogonal as well.

Next, we verify that $R_1\mathcal{Y}_2$ is closed, i.e., is a subspace. Indeed, if $f \in \mathcal{Y}_2$, then $\|R_1 f\| = \|Bf\| \geq \frac{1}{2}\|f\|$, and our assertion follows. Also, the sum of the subspaces $\mathcal{Y}_1$ and $\overline{R_1\mathcal{Y}_1}$ is closed. To verify this assertion, observe that if $f \in \mathcal{Y}_1$, then $\|R_1 f\| = \|Bf\| \leq \frac{1}{2}\|f\|$, and consequently, for $g \in \overline{R_1\mathcal{Y}_1}$, we have

$$\|f-g\| \geq \|f-R_1 f\| \geq \|f\|-\|R_1 f\| \geq \tfrac{1}{2}\|f\|.$$

Consider the subspace

$$\mathcal{Z}_1 = (\mathcal{Y}_1 \dot{+} \overline{R_1\mathcal{Y}_1}) \oplus \mathcal{Y}_2 \oplus \mathcal{R}_2.$$

Our next step is to find an A-invariant direct complement $\mathcal{Z}_2$ for $\mathcal{Z}_1$. To do that, use the induction hypothesis and find a subspace $\mathcal{Z}_4$ of type $\mathcal{Z}_4 = \mathcal{X}_2' \oplus \cdots \oplus \mathcal{X}_m'$, $\mathcal{X}_j' \subset \mathcal{X}_j$ for $j = 1,2,\ldots,m$ such that

$$\mathcal{Z}_3 \dot{+} \mathcal{Z}_4 = \mathcal{X}_2 \oplus \cdots \oplus \mathcal{X}_m.$$

Let

$$\mathcal{Z}_2 = \mathcal{Z}_4 \oplus (\mathcal{X}_1 \ominus (\overline{R_1\mathcal{Y}_1} \oplus R_1\mathcal{Y}_2)).$$

The subspace $\mathcal{Z}_2$ is obviously A-invariant, and it turns out to be a direct complement to $\mathcal{Z}_1$. Let us verify that; in other words, we have to check that $\mathcal{Z}_1 + \mathcal{Z}_2 = \mathcal{X}$ and $\mathcal{Z}_1 \cap \mathcal{Z}_2 = \{0\}$. Let $h \in \mathcal{Z}_1 \cap \mathcal{Z}_2$. By the definition of $\mathcal{Z}_2$, we have $h = x+y$, where $x \in \mathcal{Z}_4$, $y \in \mathcal{X}_1 \ominus (\overline{R_1\mathcal{Y}_1} \oplus R_1\mathcal{Y}_2)$. The vector y is orthogonal to $\mathcal{Z}_1$ because $R_1\mathcal{Z}_1 = \overline{R_1\mathcal{Y}_1} \oplus R_1\mathcal{Y}_2$. Also, y is orthogonal to $\mathcal{Z}_4$. So

$$(y,y) = (y,h)-(y,x) = 0,$$

and hence $y = 0$, i.e., $h \in \mathcal{Z}_4$. However, $\mathcal{Z}_1 \cap \mathcal{Z}_4 = \{0\}$, so $h = 0$. This verifies the equality $\mathcal{Z}_1 \cap \mathcal{Z}_2 = \{0\}$. Next, we prove that $\mathcal{Z}_1+\mathcal{Z}_2 = \mathcal{X}$. Obviously,

$$\mathcal{X}_2\oplus\cdots\oplus\mathcal{X}_m = \mathcal{Z}_3\dot{+}\mathcal{Z}_4 \subset \mathcal{Z}_1+\mathcal{Z}_2,$$

and since $\overline{R_1\mathcal{Y}_1} \subset \mathcal{Z}_1$, it is enough to show that $R_1\mathcal{Y}_2 \subset \mathcal{Z}_1+\mathcal{Z}_2$. Given $f \in \mathcal{R}_1\mathcal{Y}_2$, let $g \in \mathcal{Y}_2$ be such that $R_1g = f$. As $g \in \mathcal{Z}_1$, and

$$(I-R_1)g \in \mathcal{X}_2\oplus\cdots\oplus\mathcal{X}_m \subset \mathcal{Z}_1+\mathcal{Z}_2,$$

we have

$$f = g-(I-R_1)g \in \mathcal{Z}_1+\mathcal{Z}_2.$$

Having found an A-invariant direct complement $\mathcal{Z}_2$ for $\mathcal{Z}_1$, the proof of the implication (i) → (ii) is easily completed. Namely, the equality

$$\mathcal{Y}_1+\mathcal{Y}_2+\overline{R_1\mathcal{Y}_1}+\mathcal{R}_2+\mathcal{Z}_2 = \mathcal{X}$$

implies that $\overline{R_1\mathcal{Y}_1}\dot{+}\mathcal{Z}_2$ is an A-invariant direct complement to $\mathcal{M}$.

Secondly, assume that A is a simple operator. We prove by induction on ℓ that for every chain (2.8.1) there is a chain (2.8.2) with $\mathcal{M}_i \dot{+} \mathcal{N}_i = \mathcal{X}$ for $i = 1,\ldots,\ell$. For $\ell = 1$ this is just the hypothesis that (ii) holds, which is true by the first part of the proof. Assume this statement is proved already for all chains (2.8.1) with ℓ replaced by $\ell-1$. Let

$$\mathcal{X} = \mathcal{X}_1 \dot{+}\cdots\dot{+} \mathcal{X}_k$$

be a direct sum decomposition, where $\mathcal{X}_1,\ldots,\mathcal{X}_k$ are A-invariant subspaces and $A|_{\mathcal{X}_i} = \lambda_i I_{\mathcal{X}_i}$, $i = 1,\ldots,k$ for some complex numbers $\lambda_1,\ldots,\lambda_k$. Let

$$\mathcal{X} \supset \mathcal{N}_1 \supset\cdots\supset \mathcal{N}_\ell$$

be a chain of subspaces. Let $\mathcal{M}_{1\ell} \subset \mathcal{X}_1,\ldots,\mathcal{M}_{k\ell} \subset \mathcal{X}_k$ be subspaces with the property that $\mathcal{M}_\ell = \mathcal{M}_{1\ell} \dot{+}\cdots\dot{+} \mathcal{M}_{k\ell}$ is a direct complement to $\mathcal{N}_\ell$ in $\mathcal{X}$ (here we use property (ii) and Proposition 2.8.3). By the induction hypothesis, and using again Proposition 2.8.3, there exist chains of subspaces $\mathcal{M}_{i1} \subset \mathcal{M}_{i2} \subset \cdots \subset \mathcal{M}_{i,\ell-1}$ in $\mathcal{M}_{i\ell}$, $i = 1,\dots,k$ such that $\mathcal{M}_{1j}+\mathcal{M}_{2j}+\cdots+\mathcal{M}_{kj}$ is a direct complement to $\mathcal{M}_\ell \cap \mathcal{N}_j$ in $\mathcal{M}_\ell$, for $j = 2,\dots,\ell-1$. Put $\mathcal{M}_j = \mathcal{M}_{1j}\dot{+}\cdots\dot{+}\mathcal{M}_{kj}$, $j = 1,\dots,\ell-1$. As $\mathcal{N}_j = \mathcal{N}_\ell\dot{+}(\mathcal{M}_\ell \cap \mathcal{N}_j)$, it follows that $\mathcal{M}_j$ is also a direct complement to $\mathcal{N}_j$ in $\mathcal{X}$. ■

PROOF OF THEOREM 2.8.1. Let $L(\lambda)$ be an indexless monic operator polynomial of degree ℓ, with the companion operator C_L. By Proposition 2.8.4, C_L is indexless as well.

Using Theorem 2.8.2 for given $\varepsilon > 0$, find $B \in L(\mathcal{X}^\ell)$ such that $\|B-C_L\| < \varepsilon$ and $B|_{\mathcal{H}_i} = \lambda_i I$, $i = 1,\dots,k$, for some decomposition into the direct sum $\mathcal{X}^\ell = \mathcal{H}_1\dot{+}\cdots\dot{+}\mathcal{H}_k$. Taking ε small enough, in view of Theorem 2.3.2, we can ensure that B is similar to the companion operator $C_{\tilde{L}}$ of a monic operator polynomial $\tilde{L}(\lambda)$ of degree ℓ; moreover,

$$\|C_{\tilde{L}}-C_L\| \le K\|B-C_L\|, \tag{2.8.2}$$

where the positive constant K depends on C_L only. We have

$$C_{\tilde{L}}|_{\mathcal{Y}_i} = \lambda_i I, \quad i = 1,\dots,k,$$

where $\mathcal{Y}_i$, $i = 1,\dots,k$ are subspaces such that $\mathcal{X}^\ell = \mathcal{Y}_1\dot{+}\cdots\dot{+}\mathcal{Y}_k$. Let

$$P_j = \begin{bmatrix} I & 0 & \cdots & 0 & 0 & \cdots & 0 \\ 0 & I & \cdots & 0 & 0 & \cdots & 0 \\ \vdots & & & \vdots & \vdots & & \vdots \\ 0 & 0 & \cdots & I & 0 & \cdots & 0 \end{bmatrix} \in L(\mathcal{X}^\ell, \mathcal{X}^j),$$

$j = 1,\dots,\ell-1$.

By Theorem 2.8.5, for $j = 1,\dots,\ell-1$, there is a direct complement $\mathcal{M}_j$ to Ker P_j in $\mathcal{X}^\ell$ of the form $\mathcal{M}_j = \mathcal{M}_{j1}\dot{+}\cdots\dot{+}\mathcal{M}_{kj}$, where $\mathcal{M}_{ij} \subset \mathcal{Y}_i$ $(i = 1,\dots,k)$; moreover, $\mathcal{M}_{\ell-1} \supset\cdots\supset \mathcal{M}_1$. Obviously, $\mathcal{M}_j$ is $C_{\tilde{L}}$-invariant and $P_j|_{\mathcal{M}_j}$ is invertible. As

$$P_j = \mathrm{col}[P_1 C_{\tilde{L}}^{i}]_{i=0}^{j-1}$$

and $(P_1, C_{\tilde{L}})$ is a right spectral pair of $\tilde{L}(\lambda)$, by Theorem 2.7.3 there is a factorization $\tilde{L}(\lambda) = (\lambda I+X_1)\cdots(\lambda I+X_\ell)$, where $X_1,\ldots,X_\ell \in L(\mathcal{X})$. In view of (2.8.3), Theorem 2.8.1 is proved. ■

The method of proof of Theorem 2.8.1 can be used to prove more general statements about denseness of factorable operator polynomials. A set $\mathcal{K}$ of operators in $L(\mathcal{X}^\ell)$ will be called *admissible* if $\mathcal{K}$ is invariant under similarity (i.e., if $A \in \mathcal{K}$ and $S \in L(\mathcal{X}^\ell)$ is invertible, then $S^{-1}AS \in \mathcal{K}$) and the simple operators are dense in $\mathcal{K}$ (i.e., for every $A \in \mathcal{K}$ there is a sequence of simple operators $\{A_m\}_{m=1}^{\infty}$, $A_m \in \mathcal{K}$ such that $A_m \to A$ as $m \to \infty$). Many important classes of operators are admissible, at least in case $\mathcal{X}$ is a separable Hilbert space (e.g., the set of all indexless operators, or the set of all operators similar to self-adjoint operators; the former is admissible by Theorem 2.8.2, and the admissibility of the latter follows without difficulty from the spectral theorem for self-adjoint operators). Given an admissible set $\mathcal{K}$, a monic operator polynomial $L(\lambda)$ of degree ℓ on $\mathcal{X}$ is called *associated* with $\mathcal{K}$ if its global linearization (or, equivalently, the companion operator) belongs to $\mathcal{K}$.

THEOREM 2.8.6. *Let $\mathcal{X}$ be a separable Hilbert space, and let $\mathcal{K}$ be an admissible set in $L(\mathcal{X}^\ell)$. Then the set of factorable monic operator polynomials is dense in the set of all operator polynomials associated with $\mathcal{K}$. More precisely, given a monic operator polynomial L associated with $\mathcal{K}$ there is a sequence of factorable monic operator polynomials $\{L_m\}_{m=1}^{\infty}$ associated with $\mathcal{K}$ such that $|||L-L_m||| \to 0$ as $m \to \infty$.*

The proof of Theorem 2.8.5 is the same as that of Theorem 2.8.1.

We conclude this section with a conjecture inspired by Example 2.8.2.

CONJECTURE 2.8.1. *Let* Ind_ℓ *be the set of all monic operator polynomials* $L(\lambda)$ *of degree* ℓ *on a separable Hilbert space such that for every* $\lambda_0 \in \mathbb{C}$ *with semifredholm* $L(\lambda_0)$ *the index of* $L(\lambda_0)$ *is finite and is an integer multiple of* ℓ. *Then the set of factorable monic operator polynomials of degree* ℓ *is dense in* Ind_ℓ.

2.9 Description of the left quotients

In Section 2.7 we have characterized the right divisors $L_1(\lambda)$ of a given monic operator polynomial $L(\lambda)$ in terms of the supporting subspaces. Here we obtain a formula for the quotient $L(\lambda)L_1(\lambda)^{-1}$. At the same time we provide a description of the left monic divisors $L_2(\lambda)$ of $L(\lambda)$ (because each such divisor has the form $L(\lambda)L_1^{-1}(\lambda)$ for some right monic divisor $L_1(\lambda)$).

We present the description in terms of spectral triples.

LEMMA 2.9.1. *Let* $L(\lambda)$ *be a monic operator polynomial of degree* ℓ *acting on a Banach space* $\mathcal{X}$, *with a spectral triple,* (X,T,Y), *where* $X \in L(\mathcal{Y},\mathcal{X})$, $T \in L(\mathcal{Y})$, $Y \in L(\mathcal{X},\mathcal{Y})$. *Let* P *be a projection on* $\mathcal{Y}$. *Then the operator*

$$\operatorname{col}[XT^{i-1}]_{i=1}^{k}\Big|_{\operatorname{Im} P} : \operatorname{Im} P \to \mathcal{X}^k \tag{2.9.1}$$

(where $k < \ell$*) is invertible if and only if the operator*

$$(I-P)[T^{\ell-k-1}Y,\ldots,TY,Y]: \mathcal{X}^{\ell-k} \to \operatorname{Ker} P \tag{2.9.2}$$

is invertible.

PROOF. Put $A = \operatorname{col}[XT^{i-1}]_{i=1}^{\ell}$ and $B = [T^{\ell-i}Y,\ldots,TY,Y]$. With respect to the decompositions $\mathcal{Y} = \operatorname{Im} P \dotplus \operatorname{Ker} P$ and $\mathcal{X}^\ell = \mathcal{X}^k \dotplus \mathcal{X}^{\ell-k}$, write

$$A = \begin{bmatrix} A_1 & A_2 \\ A_3 & A_4 \end{bmatrix}, \quad B = \begin{bmatrix} B_1 & B_2 \\ B_3 & B_4 \end{bmatrix}.$$

Thus, $A_1, B_1 \in L(\operatorname{Im} P, \mathcal{X}^k)$; $A_2, B_2 \in L(\operatorname{Ker} P, \mathcal{X}^k)$; $A_3, B_3 \in L(\operatorname{Im} P, \mathcal{X}^{\ell-k})$; $A_4, B_4 \in L(\operatorname{Ker} P, \mathcal{X}^{\ell-k})$.

Observe that A_1 and B_4 coincide with the operators (2.9.1) and (2.9.2), respectively. In view of formula (2.5.1), the product AB has the form

$$AB = \begin{bmatrix} D_1 & 0 \\ * & D_2 \end{bmatrix}$$

with D_1 and D_2 as invertible operators. Recall that A and B are also invertible (by the properties of a spectral triple). But then A_1 is invertible if and only if B_4 is invertible. This may be seen as follows.

Suppose that B_4 is invertible. Then

$$B\begin{bmatrix} I & 0 \\ -B_4^{-1}B_3 & B_4^{-1} \end{bmatrix} = \begin{bmatrix} B_1 & B_2 \\ B_3 & B_4 \end{bmatrix}\begin{bmatrix} I & 0 \\ -B_4^{-1}B_3 & B_4^{-1} \end{bmatrix}$$

$$= \begin{bmatrix} B_1-B_2B_4^{-1}B_3 & B_2B_4^{-1} \\ 0 & I \end{bmatrix}$$

is invertible in view of the invertibility of B, and then also $B_1-B_2B_4^{-1}B_3$ is invertible. The special form of AB implies $A_1B_2+A_2B_4 = 0$. Hence $D_1 = A_1B_1+A_2B_3 = A_1B_1-A_1B_2B_4^{-1}B_3 = A_1(B_1-B_2B_4^{-1}B_3)$, and it follows that A_1 is invertible. A similar argument shows that invertibility of A_1 implies invertibility of B_4. This proves the lemma. ∎

We say that P is a *supporting projection* for the triple (X, T, Y) if Im P is a nontrivial invariant subspace for T and the operator (2.9.1) is invertible for some positive integer k. One checks without difficulty that k is unique and $k < \ell$. We call k the *degree* of the supporting projection. It follows from Theorem 2.7.1 that P is a supporting projection of degree k if and only if its image is a supporting subspace of some right monic divisor of $L(\lambda)$ of degree k.

Let P be a supporting projection for (X,T,Y) of degree k. Define $T_1 \in L(\text{Im } P)$ and $X_1 \in L(\text{Im } P, \mathcal{X})$ by

$$T_1 y = Ty, \quad X_1 y = Xy.$$

The invertibility of (2.9.1) now implies that $\text{col}[X_1 T_1^{i-1}]_{i=1}^{k}$: $\text{Im } P \to \mathcal{X}^k$ is invertible. Hence there exists a unique $Y_1 \in L(\mathcal{X}, \text{Im } P)$ such that the triple (X_1, T_1, Y_1) is a spectral triple of some monic operator polynomial $L_1(\lambda)$ which is uniquely determined by (X_1, T_1). In fact, (Theorem 2.2.1)

$$L_1(\lambda) = \lambda^k I - X_1 T_1^k (V_1 + V_2\lambda + \cdots + V_k \lambda^{k-1}),$$

where $[V_1 \cdots V_k] = [\text{col}[X_1 T_1^i]_{i=0}^{k-1}]^{-1}$.

The triple (X_1, T_1, Y_1) will be called the *right projection* of (X,T,Y) associated with P. It follows from Theorem 2.7.1 that the polynomial $L_1(\lambda)$ defined by (X_1, T_1, Y_1) is a right divisor of $L(\lambda)$, and every monic right divisor of $L(\lambda)$ is generated by the right projection connected with some supporting projector of (X,T,Y).

By Lemma 2.8.1 the operator (2.9.2) is invertible. Define $T_2 \in L(\text{Ker } P)$ and $Y_2 \in L(\mathcal{X}, \text{Ker } P)$ by

$$T_2 y = (I-P)Ty, \quad Y_2 x = (I-P)Yx.$$

Since Im P is an invariant subspace for T, we have $(I-P)T(I-P) = (I-P)T$. This, together with the invertibility of (2.9.2), implies that

$$[Y_2, \ldots, T_2^{\ell-k-1} Y_2] \in L(\mathcal{X}^{\ell-k}, \text{Ker } P)$$

is invertible. Therefore, there exists a unique X_2: Ker P $\to \mathcal{X}$ such that the triple (X_2, T_2, Y_2) is a spectral triple for some monic operator polynomial $L_2(\lambda)$ (which is necessarily of degree ℓ-k). Actually, by Theorem 2.2.1 we have

$$L_2(\lambda) = \lambda^{\ell-k} I - (W_1 + W_2\lambda + \cdots + W_{\ell-k}\lambda^{\ell-k-1}) T_2^{\ell-k} Y_2,$$

where

$$\mathrm{col}[W_i]_{i=1}^{\ell-k} = [Y_2, T_2Y_2, \ldots, T_2^{\ell-k-1}Y_2]^{-1}.$$

The triple (X_2, T_2, Y_2) will be called the *left projection* of (X,T,Y) associated with P.

The next theorem shows that the monic polynomial $L_2(\lambda)$ defined by the left projection is just the quotient $L(\lambda)L_1^{-1}(\lambda)$ where $L_1(\lambda)$ is the right divisor of $L(\lambda)$ defined by the right projection of (X,T,Y) associated with P.

THEOREM 2.9.2. *Let* $L(\lambda)$ *be a monic operator polynomial with spectral triple* (X,T,Y). *Let* P *be a supporting projection of* (X,T,Y) *of degree* k *and let* $L_1(\lambda)$ *(resp.* $L_2(\lambda)$*) be the monic operator polynomial of degree* k *(resp.* $\ell-k$*) generated by the right (resp. left) projection of* (X,T,Y) *associated with* P. *Then*

$$L(\lambda) = L_2(\lambda)L_1(\lambda). \tag{2.9.3}$$

Conversely, every factorization (2.9.3) *of* $L(\lambda)$ *into a product of two monic factors* $L_2(\lambda)$ *and* $L_1(\lambda)$ *of degrees* $\ell-k$ *and* k, *respectively, is obtained by using some supporting projection of* (X,T,Y) *as above.*

PROOF. Let $P' \in L(\mathcal{X}^\ell)$ be defined by

$$P'y = [\mathrm{col}[X_1T_1^{i-1}]_{i=1}^k]^{-1} \cdot [\mathrm{col}[XT^{i-1}]_{i=1}^k]y,$$

where $X_1 = X|_{\mathrm{Im}\,P}$; $T_1 = T|_{\mathrm{Im}\,P}$. Then P' is a projection and $\mathrm{Im}\,P' = \mathrm{Im}\,P$. We shall verify that

$$\mathrm{Ker}\,P' = \left\{\sum_{i=1}^{\ell-k} T^{\ell-k-i}Yx_{i-1} \mid x_0, \ldots, x_{\ell-k-1} \in \mathcal{X}\right\}. \tag{2.9.4}$$

Indeed,

$$\mathrm{Ker}\,P' = \mathrm{Ker}\,\mathrm{col}[XT^{i-1}]_{i=1}^k. \tag{2.9.5}$$

Recall the formula,

$$XT^jY = 0,\ j = 0, \ldots, \ell-2;\ XT^{\ell-1}Y = I \tag{2.9.6}$$

which implies that

$$\text{(2.9.7)} \qquad \operatorname{Ker} \operatorname{col}[XT^{i-1}]_{i=1}^{k} \supset \operatorname{Im}[Y, TY, \dots, T^{\ell-k-1}Y].$$

To prove that actually the equality holds in (2.9.7), pick $z \in \operatorname{Ker} \operatorname{col}[XT^{i-1}]_{i=1}^{k}$. As $[Y, TY, \dots, T^{\ell-1}Y]$ is invertible,

$$z = \sum_{j=0}^{\ell-1} T^j Y x_j$$

for some $x_0, \dots, x_{\ell-1} \in \mathcal{X}$. Now

$$\begin{bmatrix} X \\ XT \\ \vdots \\ XT^{\ell-1} \end{bmatrix} z = \begin{bmatrix} X \\ XT \\ \vdots \\ XT^{\ell-1} \end{bmatrix} [Y, TY, \dots, T^{\ell-1}Y] \begin{bmatrix} x_0 \\ \vdots \\ XT^{\ell-1}x_{\ell-1} \end{bmatrix} = \begin{bmatrix} K\hat{x} \\ * \end{bmatrix},$$

where $\hat{x} = \operatorname{col}[x_{\ell-k}, x_{\ell-k-1}, \dots, x_{\ell-1}]$ and $K \in L(\mathcal{X}^k)$ is an invertible operator (here (2.9.6) was used again). Since $z \in \operatorname{Ker} \operatorname{col}[XT^{i-1}]_{i=1}^{k}$, we have $K\hat{x} = 0$, and hence $\hat{x} = 0$. So actually $z \in \operatorname{Im}[Y, TY, \dots, T^{\ell-k-1}Y]$, and the equality holds in (2.9.7). Formula (2.9.4) now follows in view of (2.9.5).

Define $S \in L(\mathcal{X}^{\ell}, \operatorname{Im} P \dotplus \operatorname{Ker} P)$ by

$$S = \begin{bmatrix} P' \\ I-P \end{bmatrix}$$

where P' and $I-P$ are considered as operators from $\mathcal{X}$ into $\operatorname{Im} P'$ and $\operatorname{Ker} P$, respectively. One verifies easily that S is invertible. We shall show that

$$\text{(2.9.8)} \qquad [X_1 \quad 0]S = X, \quad ST = \begin{bmatrix} T_1 & Y_1X_2 \\ 0 & T_2 \end{bmatrix} S,$$

which in view of Theorem 2.6.1 means that (X,T) is a right spectral pair for the product $L_2(\lambda)L_1(\lambda)$, and since a monic polynomial is uniquely defined by its standard pair, (2.9.3) follows.

Take $y \in \mathcal{X}^{\ell}$. Then $P'y \in \operatorname{Im} P$ and $\operatorname{col}[X_1T_1^{i-1}P'y]_{i=1}^{k} = \operatorname{col}[XT^{i-1}y]_{i=1}^{k}$. In particular, $X_1P'y = Xy$. This proves that $[X_1 \ \ 0]S = X$. The second equality in (2.9.8) is equivalent to the equalities

$$P'T = T_1P'+Y_1X_2(I-P) \tag{2.9.9}$$

and $(I-P)T = T_2(I-P)$. The last equality is immediate from the definition of T_2 and the fact that Im P is an invariant subspace for T. To prove (2.9.9), take $y \in \mathcal{X}^{\ell}$. The case when $y \in \operatorname{Im} P = \operatorname{Im} P'$ is trivial. Therefore, assume that $y \in \operatorname{Ker} P'$. We then have to demonstrate that $P'Ty = Y_1X_2(I-P)y$. Since $y \in \operatorname{Ker} P'$, there exist $x_0,\ldots,x_{\ell-k-1} \in \mathcal{X}$ such that $y = \Sigma_{i=1}^{\ell-k}T^{\ell-k-i}Yx_{i-1}$. Hence

$$Ty = T^{\ell-k}Yx_0+T^{\ell-k-1}Yx_1+\cdots+TYx_{\ell-k-1} = T^{\ell-k}Yx_0+u$$

with $u \in \operatorname{Ker} P'$ and, as a consequence, $P'Ty = P'T^{\ell-k}Yx_0$. But then it follows from the definition of P' that

$$P'Ty = [T_1^{k-1}Y_1, T_1^{k-2}Y_1,\ldots,Y_1] \operatorname{col}[0,\ldots,0,x_0] = Y_1x_0.$$

On the other hand, putting $x = \operatorname{col}[x_{i-1}]_{i=1}^{\ell-k}$, we have

$$(I-P)y = (I-P)[T^{\ell-k-1}Y,\ldots,TY,Y]x = [T_2^{\ell-k-1}Y_2,\ldots,T_2Y_2,Y_2]x,$$

and so $Y_1X_2(I-P)y$ is also equal to Y_1x_0. This completes the proof. ∎

Using Theorem 2.9.2, it is possible to write down the decomposition $L(\lambda) = L_2(\lambda)L_1(\lambda)$, where $L_2(\lambda)$ and $L_1(\lambda)$ are written in one of the possible forms: right canonical, left canonical, or resolvent. We give in the next corollary one such decomposition in which $L_2(\lambda)$ is in the left canonical form and $L_1(\lambda)$ is in the right canonical form.

COROLLARY 2.9.3. *Let* $L(\lambda)$ *and* (X,T,Y) *be as in Theorem 2.9.2. Let* $L_1(\lambda)$ *be a right divisor of degree* k *of* $L(\lambda)$ *with the supporting subspace* $\mathcal{M}$. *Then*

$$L(\lambda) = [I\lambda^{\ell-k}-(Z_1+\cdots+Z_{\ell-1}\lambda^{\ell-k-1})\tilde{P}T^{\ell-k}\tilde{P}Y]$$

$$\cdot\ [I\lambda^{k}-X|\mathcal{M}(T|\mathcal{M})^{k}(W_1+\cdots+W_k\lambda^{k-1})],$$

where

$$[W_1\cdots W_k] = \left[\operatorname{col}[X|\mathcal{M}(T|\mathcal{M})^{i}]_{i=0}^{k-1}\right]^{-1},$$

$\tilde{P}$ *is some projection with* $\operatorname{Ker}\tilde{P} = \mathcal{M}$, *the operator* $[\tilde{P}Y,\tilde{P}T\tilde{P}Y,\ldots,\tilde{P}T^{\ell-k-1}\tilde{P}Y]\colon \mathcal{X}^{\ell-k} \to \operatorname{Im}\tilde{P}$ *is invertible, and*

$$\operatorname{col}[Z_i]_{i=1}^{\ell-k} = [\tilde{P}Y,\tilde{P}T\tilde{P}Y,\ldots,\tilde{P}T^{\ell-k-1}\tilde{P}Y]^{-1}\colon \operatorname{Im}\tilde{P} \to \mathcal{X}^{\ell-k}.$$

PROOF. Let P be the supporting projection for the right divisor $L_1(\lambda)$. Then, by Theorem 2.9.2, (T_2,Y_2) is a left spectral pair for $L_2(\lambda)$, where $T_2\colon \operatorname{Ker}P \to \operatorname{Ker}P$ and $Y_2\colon \mathcal{X} \to \operatorname{Ker}P$ are defined by $T_2y = (I-P)Ty$, $Y_2x = (I-P)Yx$. Choose $P = I-\tilde{P}$; then $Y_2 = \tilde{P}Y$ and $T_2 = \tilde{P}T|_{\operatorname{Im}\tilde{P}}$. So Corollary 2.9.3 will follow from the left canonical form of $L_2(\lambda)$ if we prove that

$$(2.9.10)\qquad \left(\tilde{P}T|_{\operatorname{Im}\tilde{P}}\right)^{i} = \tilde{P}T^{i}|_{\operatorname{Im}\tilde{P}},\quad i = 0,1,2,\ldots .$$

But this follows from the representation of T relative to the decomposition $\mathcal{X}^{\ell} = \mathcal{M} \dotplus \operatorname{Im}\tilde{P}$:

$$T = \begin{bmatrix} T_{11} & T_{12} \\ 0 & T_{22} \end{bmatrix}$$

where $T_{22} = \tilde{P}T|_{\operatorname{Im}\tilde{P}}$. So

$$T^{i} = \begin{bmatrix} T_{11}^{i} & * \\ 0 & T_{22}^{i} \end{bmatrix},\quad i = 0,1,\ldots ,$$

i.e., $\left(\tilde{P}T|_{\operatorname{Im}\tilde{P}}\right)^i = \tilde{P}T^i|_{\operatorname{Im}\tilde{P}}$, which is exactly (2.9.10). ■

Note that Corollary 2.9.3 does not depend on the choice of the projection $\tilde{P}$.

For further reference we state the following immediate corollary of Theorem 2.9.2.

COROLLARY 2.9.4. *Let* $L(\lambda)$ *be a monic operator polynomial with global linearization* T *and let* $L_1(\lambda)$ *be a monic right divisor with supporting subspace* $\mathcal{M}$. *Then for any complementary subspace* $\mathcal{M}'$ *to* $\mathcal{M}$, *the operator* $PT|_{\mathcal{M}'}$ *is a global linearization of* $L(\lambda)L_1^{-1}(\lambda)$, *where* P *is a projection on* $\mathcal{M}'$ *along* $\mathcal{M}$. *In particular,* $\Sigma(LL_1^{-1}) = \sigma(PT|_{\mathcal{M}'})$.

2.10 Spectral divisors

We consider here an important special case of factorization $L(\lambda) = L_2(\lambda)L_1(\lambda)$ of monic operator polynomials in which $L_2(\lambda)$ and $L_1(\lambda)$ have disjoint spectra. As usual, the coefficients of the operator polynomials involved are in $L(\mathcal{X})$, for some Banach space $\mathcal{X}$.

Let Γ be an oriented closed rectifiable contour in the complex plane without self-intersections (but possibly consisting of a finite number of disjoint pieces) and such that Γ is a boundary of a bounded domain Ω (it is assumed that Ω lies on the left-hand side of Γ). Consider a monic operator polynomial $L(\lambda)$ such that $L(\lambda_0)$ is invertible for every $\lambda \in \Gamma$. A monic operator polynomial $L_1(\lambda)$ is called a *Γ-spectral right divisor* of $L(\lambda)$ if $\Sigma(L_1) \subset \Omega$ and $L = L_2L_1$, where L_2 is a monic operator polynomial with $\Sigma(L_2)$ outside $\Omega \cup \Gamma$. If in the above definition $L = L_2L_1$ is replaced by $L = L_1L_2$, we obtain a *Γ-spectral left divisor* L_1.

THEOREM 2.10.1. *If* T *is a global linearization of* $L(\lambda)$ *and if* $L_1(\lambda)$ *is a* Γ*-spectral right divisor of* $L(\lambda)$, *then the supporting subspace corresponding to* L_1 *with respect to* T *is the image of the Riesz projection* R_Γ *corresponding to* T *and* Γ:

$$R_\Gamma = \frac{1}{2\pi i}\int_\Gamma (\lambda I-T)^{-1}d\lambda .$$

PROOF. Let $\mathcal{M}_\Gamma$ be the T-invariant supporting subspace of L_1, and let $\mathcal{M}'_\Gamma$ be some direct complement to $\mathcal{M}_\Gamma$ in $\mathcal{X}^\ell$ (ℓ is the degree of $L(\lambda)$). By Corollary 2.9.4 and the definition of a Γ-spectral divisor, $\sigma(T|\mathcal{M}_\Gamma)$ is inside Γ and $\sigma(\tilde{P}T|\mathcal{M}'_\Gamma)$ is outside Γ, where $\tilde{P}$ is the projection on $\mathcal{M}'_\Gamma$ along $\mathcal{M}_\Gamma$. Write the operator T with respect to the decomposition $\mathcal{X}^\ell = \mathcal{M}_\Gamma \dot{+} \mathcal{M}'_\Gamma$:

$$T = \begin{bmatrix} T_{11} & T_{12} \\ 0 & T_{22} \end{bmatrix}$$

(so that $T_{11} = T|\mathcal{M}_\Gamma$, $T_{12} = (I-\tilde{P})T|\mathcal{M}'_\Gamma$, $T_{22} = \tilde{P}T|\mathcal{M}'_\Gamma$, where $\tilde{P}$ and $I-\tilde{P}$ are considered as operators on $\mathcal{M}'_\Gamma$ and $\mathcal{M}_\Gamma$, respectively).

Then

$$(I\lambda-T)^{-1} = \begin{bmatrix} (I\lambda-T_{11})^{-1} & (I\lambda-T_{11})^{-1}T_{12}(I\lambda-T_{22})^{-1} \\ 0 & (I\lambda-T_{22})^{-1} \end{bmatrix},$$

and

$$R_\Gamma = \frac{1}{2\pi i}\int_\Gamma (I\lambda-T)^{-1}d\lambda = \begin{bmatrix} I & * \\ 0 & 0 \end{bmatrix},$$

and so $\operatorname{Im} R_\Gamma = \mathcal{M}_\Gamma$. ∎

Note that Theorem 2.10.1 ensures the uniqueness of a Γ-spectral right divisor, if one exists.

2.11 Differential and difference equations

We present briefly some fundamental applications to the differential and difference equations with comstant coefficients in Banach space $\mathcal{X}$.

Consider the inhomogeneous equation

$$u^{(\ell)}(t)+A_{\ell-1}u^{(\ell-1)}(t)+\cdots+A_1u^{(1)}(t)+A_0u(t) = f(t) \tag{2.11.1}$$

where $A_0,\ldots,A_{\ell-1} \in L(\mathcal{X})$ and the $\mathcal{X}$-valued function f(t) of the real variable t is given, while the $\mathcal{X}$-valued function u(t) is unknown. Let

(2.11.2) $$L(\lambda) = \lambda^{\ell} I + \sum_{j=0}^{\ell-1} \lambda^j A_j$$

be the corresponding monic operator polynomial.

THEOREM 2.11.1. *Let* (X,T,Y) *be a spectral triple for* $L(\lambda)$, *and suppose (for example) that* $f(t)$ *is piecewise continuous on* $[t_0,\infty)$. *Then every solution of* (2.11.1) *has the form*

(2.11.3) $$u(t) = X e^{(t-t_0)T} c + X \int_{t_0}^{t} e^{(t-\tau)T} Y f(\tau) d\tau$$

for some $c \in \mathcal{Y}$ *(here* $\mathcal{Y}$ *is the Banach space on which* T *acts).*

PROOF. Since the formula (2.11.3) does not change under similarity of spectral triples $(X,T,Y) \to (XS, X^{-1}TS, S^{-1}Y)$, where S is invertible, and since the spectral triple of $L(\lambda)$ is unique up to similarity, it is enough to prove Theorem 2.11.1 for the special case

(2.11.4) $$X = [I0\cdots0], \quad T = C_L, \quad Y = \mathrm{col}[\delta_{\ell i}]_{i=1}^{\ell}.$$

It will be assumed from now on that (X,T,Y) are given by (2.11.4). Observe that (2.11.1) is equivalent to the equation

(2.11.5) $$\frac{dv}{dt} = Tv+Yf,$$

where the ith component of $v \in \mathcal{X}^{\ell}$ is $u^{(i-1)}$ $(i = 1,\dots,\ell)$. It is well known (see, e.g., Sections X.1, X.2 in Schechter [1]) that the general solution of the homogeneous equation

$$\frac{dv_0}{dt} = Tv_0$$

is

$$v_0(t) = e^{(t-t_0)T} c$$

for some $c \in \mathcal{X}^{\ell}$. One checks by a direct computation that

$$v_1(t) = \int_{t_0}^{t} e^{(t-\tau)T} Yf(\tau)d\tau$$

solves

$$\frac{dv_1(t)}{dt} = Tv_1 + Yf.$$

So the general solution of (2.11.5) is

(2.11.6) $$v(t) = e^{(t-t_0 0T}c + \int_{t_0}^{t} e^{(t-\tau)T} Yf(\tau)d\tau, \quad c \in \mathcal{X}^{\ell}.$$

As the first component of $v(t)$ is $u(t)$, the formula (2.11.3) follows. ■

Using Theorem 2.11.1, an explicit formula for the solution of the initial value problem with

(2.11.7) $$u^{(r)}(t_0) = u_r, \quad r = 0,\dots,\ell-1$$

can be obtained. Thus, putting in formula (2.11.6) $t = t_0$, one obtains

$$\begin{bmatrix} u_0 \\ \vdots \\ u_{\ell-1} \end{bmatrix} = \begin{bmatrix} u(t_0) \\ u'(t_0) \\ \vdots \\ u^{(\ell-1)}(t_0) \end{bmatrix} = v(t_0) = c = \begin{bmatrix} X \\ XT \\ \vdots \\ XT^{\ell-1} \end{bmatrix} c$$

(the last equality follows because

$$X = [I0\cdots 0], \ T = C_L$$

in the formula (2.11.6)), so

(2.11.8) $$c = \{\mathrm{col}[XT^i]_{i=0}^{\ell-1}\}^{-1} \ \mathrm{col}[u_j]_{j=0}^{\ell-1}.$$

At this point we want to use the following equality:

$$(2.11.9)\qquad \begin{bmatrix} X \\ XT \\ \vdots \\ XT^{\ell-1} \end{bmatrix} [Y, TY, \dots, T^{\ell-1}Y] \begin{bmatrix} A_1 & A_2 & \cdots & A_{\ell-1} & I \\ A_2 & & & I & \\ \vdots & & \cdot^{\cdot^{\cdot}} & & \\ A_{\ell-1} & I & & 0 & \\ I & & & & \end{bmatrix} = I.$$

As (2.11.9) is independent of the choice of the spectral triple (X,T,Y), for the verification of (2.11.9) we may assume that

$$X = [I0\cdots0], \quad T = C_L, \quad Y = \mathrm{col}[\delta_{\ell i}]_{i=1}^{\ell}.$$

Then (2.11.9) becomes

$$(2.11.10)\qquad [Y, C_L Y, \dots, C_L^{\ell-1}Y]\, B = I,$$

where by B we denote the third factor in the left-hand side of (2.11.9). Now write down another spectral triple

$$X' = [0\cdots0I], \quad T' = \begin{bmatrix} 0 & \cdots\cdot & 0 & -A_0 \\ I & 0\cdots & 0 & -A_1 \\ \vdots & \ddots & & \vdots \\ & & & -A_{\ell-2} \\ 0 & \cdots\cdot & I & -A_{\ell-1} \end{bmatrix}, \quad Y' = \begin{bmatrix} I \\ 0 \\ \vdots \\ 0 \end{bmatrix}.$$

This is indeed a spectral triple, because

$$X' = XB^{-1}, \quad Y' = BY, \quad T' = BC_2B^{-1}$$

(the last equality follows from (2.2.4)). As by inspection $[Y', T'Y', \dots, T'^{\ell-1}Y'] = I$, the desired formula (2.11.10) is established.

Going back to the expression for c, and using (2.11.9), we obtain

$$c = [Y, TY, \dots, T^{\ell-1}Y] \begin{bmatrix} A_1 & A_2 & \cdots & A_{\ell-1} & I \\ A_2 & & & I & \\ \vdots & & \cdot^{\cdot^{\cdot}} & & \\ A_{\ell-1} & I & & 0 & \\ I & & & & \end{bmatrix} \begin{bmatrix} u_0 \\ \vdots \\ u_{\ell-1} \end{bmatrix}.$$

So the solution of the initial value problem (2.11.1), (2.11.7) is given by the formula

$$u(t) = Xe^{(t-t_0)T}[Y, TY, \dots, T^{\ell-1}Y]\begin{bmatrix} A_1 & A_2 & \cdots & A_{\ell-1} & I \\ A_2 & & \cdot & I & \\ \vdots & \cdot & & & \\ & & 0 & & \\ I & & & & \end{bmatrix}\begin{bmatrix} u_0 \\ u_1 \\ \vdots \\ u_{\ell-1} \end{bmatrix}$$

$$+ X\int_{t_0}^{t} e^{(t-\tau)T}Yf(t)d\tau .$$

This formula is clearly independent of the choice of the spectral triple (X, T, Y) of $L(\lambda)$.

Next we comment on the difference equation

$$(2.11.11) \qquad A_0u_r+A_1u_{r+1}+\dots+A_{\ell-1}u_{r+\ell-1}+u_{r+\ell} = f_r, \quad r = 1,2,\dots$$

where $\{f_r\}_{r=1}^{\infty}$, $f_r \in \mathcal{X}$ is given and solution sequences $\{u_r\}_{r=1}^{\infty}$, $u_r \in \mathcal{X}$ are sought. Again, the operator coefficients $A_0, \dots, A_{\ell-1}$ define the monic operator polynomial $L(\lambda)$ of (2.11.2).

THEOREM 2.11.2. *Let* (X, T, Y) *be a spectral triple for* L. *Then every solution of* (2.11.11) *has the form*

$$(2.11.12) \qquad u_1 = Xc, \ u_r = XT^{r-1}c + \sum_{k=1}^{r-1} XT^{r-k-1}f_k,$$

$$r = 2,3,\dots ,$$

for some $c \in \mathcal{X}$. *The solution of the initial value problem* (2.11.11) *with* $u_1 = v_1, \dots , u_\ell = v_\ell$ *is given by* (2.11.12) *where*

$$c = [Y, TY, \dots, T^{\ell-1}Y]\begin{bmatrix} A_1 & A_2 & \cdots & A_{\ell-1} & I \\ A_2 & & \cdot & I & \\ \vdots & \cdot & & & \\ & & 0 & & \\ I & & & & \end{bmatrix}\begin{bmatrix} v_1 \\ v_2 \\ \vdots \\ v_\ell \end{bmatrix}.$$

The proof is given by considering the special spectral triple (2.11.4), analogously to the proof of Theorem 2.11.1. We omit the details.

2.12 Exercises

Ex. 2.1. Let $L_1(\lambda)$ and $L_2(\lambda)$ be monic operator polynomials of the same degree with coefficients in $L(\mathcal{X})$. Express the left and right spectral pairs of the polynomial

$$L(\lambda) = \begin{bmatrix} L_1(\lambda) & 0 \\ 0 & L_2(\lambda) \end{bmatrix}$$

in terms of the left and right spectral pairs for $L_1(\lambda)$ and $L_2(\lambda)$.

Ex. 2.2. Let $L_1(\lambda)$, $L_2(\lambda)$ and $L(\lambda)$ be as in Ex. 2.1. Express the spectral triple of $L(\lambda)$ in terms of the spectral triples of $L_1(\lambda)$ and $L_2(\lambda)$.

Ex. 2.3. Extend the results of Ex. 2.1 and 2.2 to monic operator polynomials of the form

$$L(\lambda) = L_1(\lambda) \oplus \cdots \oplus L_k(\lambda).$$

Ex. 2.4. Let (X,T) be a right spectral pair for the monic operator polynomial $L(\lambda)$. Show that

$$[X \ 0 \ 0], \quad \begin{bmatrix} 0 & I & 0 \\ 0 & 0 & I \\ T & 0 & 0 \end{bmatrix}$$

is a right spectral pair for $L(\lambda^3)$.

Ex. 2.5. Given a left spectral (T,Y) for $L(\lambda)$, find a left spectral pair of $L(\lambda^3)$.

Ex. 2.6. Let $L(\lambda)$ be a monic operator polynomial with right spectral pair (X,T) and left spectral pair (T,Y). Find a left and right spectral pair of $L(p(\lambda))$, where $p(\lambda)$ is a given monic scalar polynomial.

<u>Ex.2.7</u>. Verify that if (X,T,Y) is a spectral triple for $L(\lambda) = \sum_{j=0}^{\ell} \lambda^j A_j$, $A_\ell = I$, then (Y^*,T^*,X^*) is a spectral triple for the operator polynomial $\sum_{j=0}^{\ell} \lambda^j A_j^*$ (assume that A_j are Hilbert space operators).

<u>Ex. 2.8</u>. Let $\mathcal{X}$ be a separable Hilbert space, and let $L(\lambda)$ be a monic operator polynomial of degree ℓ with coefficients in the $A = \{\alpha I_{\mathcal{X}}+K: \alpha \in \mathbb{C}$ and $K \in L(\mathcal{X})$ is a finite rank operator$\}$. Show that $L(\lambda)$ can be approximated by factorable polynomials with coefficients in A, i.e., for every $\varepsilon > 0$ there is a factorable monic polynomial $L_\varepsilon(\lambda)$ of degree ℓ with coefficients in A such that

$$\|A_j - A_{j\varepsilon}\| < \varepsilon,$$

where A_j (resp. $A_{j\varepsilon}$) is the jth coefficient of $L(\lambda)$ (resp. $L_\varepsilon(\lambda)$). Hint: Use Theorem 2.8.6 with $\mathcal{K}$ being the set of all operators in $\mathcal{X}^\ell$ similar to an operator of the form

$$\begin{bmatrix} 0 & I & 0 & \cdots & 0 \\ 0 & 0 & I & \cdots & 0 \\ \vdots & \vdots & \vdots & & \vdots \\ 0 & 0 & 0 & \cdots & I \\ B_0 & B_1 & B_2 & \cdots & B_{\ell-1} \end{bmatrix}$$

where $B_j \in A$, $j = 0,\dots,\ell-1$.

<u>Ex. 2.9</u>. Extend the result of Ex. 2.8 to the algebra $\{\alpha I+K: \alpha \in \mathbb{C}$ and $K \in L(\mathcal{X})$ is compact$\}$.

<u>Ex. 2.10</u>. Prove Theorem 2.11.2.

<u>Ex. 2.11</u>. Let $L(\lambda) = \lambda^\ell I + \sum_{j=0}^{\ell-1} \lambda^j A_j$ be a monic operator polynomial of degree ℓ with companion operator C_L. Verify the following formula for the resolvent of C_L:

$$(\lambda I - C_L)^{-1} = [Q_{uv}(\lambda)]_{u,v=1}^{\ell},$$

where

$$Q_{uv} = -\lambda^{u-v-1}I+\lambda^{u-v+\ell}L(\lambda)^{-1}\sum_{m=1}^{\ell+1-v}\lambda^{-m}A_{\ell-m+1}$$

if $u > v$, and

$$Q_{uv} = \lambda^{u-v+\ell}L(\lambda)^{-1}\sum_{m=1}^{\ell+1-v}\lambda^{-m}A_{\ell-m+1}$$

if $u \le v$.

Ex. 2.12. Let $\lambda I-Z$ be a right divisor of a monic operator polynomial $L(\lambda)$ such that $\sigma(Z)$ is inside a contour Γ while $\Sigma(L)\backslash\sigma(Z)$ is outside Γ. Prove that $\lambda I-Z$ is a right Γ-spectral divisor of $L(\lambda)$ if and only if

$$(1) \qquad \int_\Gamma \lambda^k L(\lambda)^{-1}d\lambda = Z^k\int_\Gamma L(\lambda)^{-1}d\lambda, \quad k = 1,2,\dots .$$

Hint: Denote $L_1(\lambda) = L(\lambda)(\lambda I-Z)^{-1}$. If $\lambda I-Z$ is Γ-spectral divisor, then

$$(\lambda^k I-Z^k)L(\lambda)^{-1} = (\sum_{j=0}^{k-1}\lambda^j Z^{k-1-j})L_1(\lambda)^{-1}$$

is analytic inside Γ. Conversely, assume (1) holds. Write

$$L_1(\lambda) = \sum_{j=0}^{\ell-2}\lambda^j X_j+\lambda^{\ell-1}I,$$

where

$$X_j = \sum_{m=j+1}^{\ell} A_m Z^{m-j-1}, \quad j = 0,\dots,\ell-2$$

and $A_0,\dots,A_{\ell-1},A_\ell = I$ are the coefficients of $L(\lambda)$. Use the formula for the resolvent of the companion operator C_{L_1} (obtained in Ex. 2.11) to show that

$$\int_\Gamma (\lambda I-C_{L_1})^{-1}d\lambda = 0.$$

Ex. 2.13. Let $\lambda I-Z$ and $L(\lambda)$ be as in Ex. 2.12. Prove that if (1) holds for $k = 1,2,\dots,\ell-1$, where ℓ is the degree of $L(\lambda)$, then $\lambda I-Z$ is a right Γ-spectral divisor of $L(\lambda)$.

Ex. 2.14. Let Γ be a contour that does not intersect $\Sigma(L)$. Assume that

$$\operatorname{Ker}\left(\int_\Gamma L(\lambda)^{-1}d\lambda\right)^* = \{0\}.$$

Show that if for some operator Z holds

$$\int_\Gamma \lambda^k L(\lambda)^{-1}d\lambda = Z^k \int_\Gamma L(\lambda)^{-1}d\lambda, \qquad k = 1,\dots,\ell,$$

where ℓ is the degree of $L(\lambda)$, then $\lambda I - Z$ is a Γ-spectral right divisor of $L(\lambda)$. Hint: Use Ex. 2.13.

2.13 Notes

Most of the results in Sections 2.1-2.7 are taken from Gohberg-Lancaster-Rodman [4]. In the exposition, especially that of Section 2.7, we made use of the exposition in Gohberg-Lancaster-Rodman [2]. The proof of (i) ⇒ (ii) in Theorem 2.8.5 is based on Marcus-Matsaev [1] (the result itself is due to Gurarie [1]). Theorem 2.8.1 was proved in Rodman [5]. The results and proofs in Sections 2.9 and 2.11 are essentially the same as in the finite-dimensional case, and the exposition follows mainly parts from Gohberg-Lancaster-Rodman [2]. In Gohberg-Lerer-Rodman [1], necessary and sufficient conditions are given for existence of Γ-spectral divisors, as well as formulas for these divisors, in terms of block operator matrices with entries of the form $\int_\Gamma \lambda^k L(\lambda)^{-1}d\lambda$. A criterion for existence of both left and right spectral divisors in terms of invertibility of certain block Toeplitz operator matrices is given in Gohberg-Lerer-Rodman [1,2].

The description of right divisors of monic operator polynomials in terms of subspaces was studied also in Langer [1] and Kabak-Markus-Mereutsa [1].

It should be emphasized that although the exposition in this chapter (with the exception of Sections 2.4 and 2.8) is close to the developments in Gohberg-Lancaster-Rodman [2], the underlying Banach space being infinite dimensional does make a difference. One important source of these differences is the nonequivalence of invertibility and one-sided invertibility for operators in $L(\mathcal{X})$, where $\mathcal{X}$ is an infinite-dimensional Banach space (cf. Section 2.4).

An interesting application of the concept of spectral pair to the study of elliptic differential equations with operator coefficients is given in Thijsse [1].

For connections between the resolvent form of monic operator polynomials (Theorem 2.5.2) and the theories of characteristic operator functions, see Bart-Gohberg-Kaashoek [2,3].

Some of the exercises are adapted from Gohberg-Lancaster-Rodman [3]. The results of exercises 2.11-2.14 appear in Markus-Mereutsa [1].

CHAPTER 3: VANDERMONDE OPERATORS AND COMMON MULTIPLES

Let $L_1(\lambda),\dots,L_r(\lambda)$ be monic operator polynomials acting on a Banach space $\mathcal{X}$. An operator polynomial $L(\lambda)$ acting on $\mathcal{X}$ is called *monic left common multiple* of $L_1,\dots,L_r$ if $L(\lambda)$ is monic and $L(\lambda) = M_1(\lambda)L_1(\lambda) = \cdots = M_r(\lambda)L_r(\lambda)$ for some (necessarily monic) operator polynomials $M_1(\lambda),\dots,M_r(\lambda)$. In this chapter we shall study monic left common multiples. The main tool of our investigation will be the Vandermonde operator and its properties. This operator is introduced in Section 3.1.

3.1 Definition and basic properties of the Vandermonde operator

Let $(X_1,T_1),\dots,(X_r,T_r)$ be right spectral pairs of monic operator polynomials $L_1,\dots,L_r$, respectively, and let $U_j = \{\mathrm{col}[X_jT_j^i]_{i=0}^{k_j-1}\}^{-1}$. The following theorem is a starting point for our investigation.

THEOREM 3.1.1. *A monic operator polynomial* $L(\lambda) = \lambda^m I+\lambda^{m-1}A_{m-1}+\cdots+\lambda A_1+A_0$, $A_j \in L(\mathcal{X})$ *is a left common multiple of* $L_1,\dots,L_r$ *if and only if*

$$[A_0A_1\cdots A_{m-1}]\cdot\begin{bmatrix} X_1U_1 & X_2U_2 & & X_rU_r \\ X_1T_1U_1 & X_2T_2U_2 & & X_rT_rU_r \\ \vdots & \vdots & & \vdots \\ X_1T_1^{m-1}U_1 & X_2T_2^{m-1}U_2 & \cdots & X_rT_r^{m-1}U_r \end{bmatrix}$$

$$= -[X_1T_1^mU_1, X_2T_2^mU_2,\dots,X_rT_r^mU_r] .$$

The proof follows immediately from Corollary 2.6.3.

Motivated by this theorem, we introduce the following definition. Let $L_1,\dots,L_r$ be monic operator polynomials of degrees $k_1,\dots,k_r$, and with right spectral pairs $(X_1,T_1),\dots,(X_r,T_r)$, respectively. The operator

$$\begin{bmatrix} X_1U_1 & X_2U_2 & \cdots & X_rU_r \\ X_1T_1U_1 & X_2T_2U_2 & \cdots & X_rT_rU_r \\ \vdots & \vdots & & \vdots \\ X_1T_1^{m-1}U_1 & X_2T_2^{m-1}U_2 & \cdots & X_rT_r^{m-1}U_r \end{bmatrix} \in L(\mathcal{X}^{k_1}\oplus\cdots\oplus\mathcal{X}^{k_r},\mathcal{X}^m)$$

will be called the *Vandermonde operator* of order m of the polynomials $L_1,\dots,L_r$ and will be denoted $V_m(L_1,\dots,L_r)$.

Since the right spectral pair of a monic operator polynomial is determined uniquely up to similarity (Theorem 2.1.1), the operator $V_m(L_1,\dots,L_r)$ actually does not depend on the choices of right spectral pairs $(X_1,T_1),\dots,(X_r,T_r)$. In fact, the operator entries of the Vandermonde operator can be computed directly in terms of the coefficients of $L_1,\dots,L_r$, using the following proposition.

PROPOSITION 3.1.2. *Let* (X,T) *be a right spectral pair of a monic operator polynomial* $L(\lambda) = \lambda^\ell I+\sum_{j=0}^{\ell-1}\lambda^jA_j$, *and let*

$$[V_1V_2\cdots V_\ell] = \{\mathrm{col}[XT^i]_{i=0}^{\ell-1}\}^{-1}.$$

Then the products XT^jV_β *for* $j = 0,\dots,1$, *and* $1 \le \beta \le \ell$ *can be calculated in terms of the coefficients* $A_0,\dots,A_{\ell-1}$, *namely*

$$XT^\alpha V_\beta = \begin{cases} 0 & \text{if } 0 \le \alpha \le \ell-1, \quad \alpha \ne \beta-1 \\ I & \text{if } 0 \le \alpha \le \ell-1, \quad \alpha = \beta-1 \end{cases} \tag{3.1.2}$$

$$XT^\ell V_\beta = -A_{\beta-1} \tag{3.1.3}$$

and

(3.1.4)

$$XT^{\ell+\rho}V_\beta = \sum_{k=1}^{\rho}\left[\sum_{q=1}^{k}\ \sum_{\substack{i_1+\cdots+i_q=k \\ i_j>0}}\ \prod_{j=1}^{q}(-A_{\ell-i_j})\right](-A_{\beta+k-\rho-1})+(-A_{\beta-\rho-1})$$

for $\rho \geq 1$, *where by definition* $A_i = 0$ *for* $i < 0$.

Here, as well as in the proof, the product $(-A_{\ell-i_1})(-A_{\ell-i_2})\cdots(-A_{\ell-i_q})$ is denoted $\prod_{j=1}^{q}(-A_{\ell-i_j})$.

PROOF. Equalities (3.1.2) follow from the definition of V_β, and (3.1.3) is a restatement of (2.2.1). Further, formula (2.6.5) implies that

$$XT^{\alpha+1}V_\beta = (XT^\alpha V_\ell)(-A_{\beta-1})+XT^\alpha V_{\beta-1} \tag{3.1.5}$$

for $1 \leq \beta \leq \ell$ and $\alpha \geq 0$. If in (3.1.5) we take $\alpha = \ell$ and use formula (3.1.3), we obtain

$$XT^{\ell+1}V_\beta = (-A_{\ell-1})(-A_{\beta-1})+(-A_{\beta-2}).$$

It follows that (3.1.4) is proved for $\rho = 1$ and $1 \leq \beta \leq \ell$.

Next we observe that

$$B \overset{\text{def}}{=} \sum_{k=1}^{\rho}\left[\sum_{q=1}^{k}\ \sum_{\substack{i_1+\cdots+i_q=k \\ i_j>0}}\ \prod_{j=1}^{q}(-A_{\ell-i_j})\right]\cdot(-A_{\ell+k-\rho-1})$$

$$= \sum_{k=1}^{\rho}\ \sum_{q=1}^{k}\ \sum_{\substack{i_1+\cdots+i_q+i_{q+1}=\rho+1 \\ i_{q+1}=(\rho+1)-k \\ i_j>0}}\ \prod_{j=1}^{q+1}(-A_{\ell-i_j}).$$

By interchanging the order of the first two summations in the last part of this identity and replacing q+1 by q, we see that

$$B = \left[\sum_{q=1}^{\rho+1}\ \sum_{\substack{i_1+\cdots+i_q=\rho+1 \\ i_j>0}}\ \prod_{j=1}^{q}(-A_{\ell-i_j})\right]-(-A_{\ell-\rho-1}). \tag{3.1.6}$$

The proof is completed by induction on ρ. Suppose (3.1.4) is true for some $\rho \geq 1$ and $1 \leq \beta \leq \ell$. Let B be as above. Using formula (3.1.5), we have that $XT^{\ell+\rho+1}V_\beta$ is equal to

$$B(-A_{\beta-1})+\sum_{k=1}^{\rho}\left[\sum_{q=1}^{k}\sum_{\substack{i_1+\cdots+i_q=k\\ i_j>0}}\prod_{j=1}^{q}(-A_{\ell-i_j})\right]\cdot(-A_{\beta+k-\rho-2})+(A_{\beta-\rho-2}).$$

Inserting the expression for B given by (3.1.6), we obtained the desired formula for $XT^{\ell+\rho+1}V_\beta$. ■

If all the polynomials L_j are of the first degree, $L_j(\lambda) = \lambda I-X_j$ for $j = 1,\ldots,r$, then

$$V_m(L_1,\ldots,L_r) = \begin{bmatrix} I & I & \cdots & I \\ X_1 & X_2 & \cdots & X_r \\ \vdots & \vdots & & \vdots \\ X_1^{m-1} & X_2^{m-1} & \cdots & X_r^{m-1} \end{bmatrix}.$$

Thus, the Vandermonde operator is a generalization of the usual Vandermonde matrix

$$\begin{bmatrix} 1 & 1 & \cdots & 1 \\ x_1 & x_2 & \cdots & x_m \\ \vdots & \vdots & & \vdots \\ x_1^{m-1} & x_2^{m-1} & \cdots & x_m^{m-1} \end{bmatrix}$$

where $x_1,\ldots,x_m$ are complex numbers.

The next property of the Vandermonde operator will be useful later.

PROPOSITION 3.1.3. *Let* $R_1,\ldots,R_s$ *be monic operator polynomials on* $\mathfrak{X}$, *and for each* $j(1 \le j \le s)$ *let* $L_{j1},\ldots,L_{jr_j}$ *be monic operator polynomials which are right divisors of* R_j. *Assume that*

$$m_j = \sum_{i=1}^{r_j} \text{degree } (L_{ji}), \quad (1 \le j \le s)$$

where m_j *is the degree of* R_j. *Then*

$$V_m(L_{11},\ldots,L_{1r_i},L_{21},\ldots,L_{2r_2},\ldots,L_{s1},\ldots,L_{sr_s})$$

$$= V_m(R_1,\ldots,R_s) \begin{bmatrix} V_{m_1}(L_{11},\ldots,L_{1r_1}) & & 0 \\ & \ddots & \\ 0 & & V_{m_s}(L_{s1},\ldots,L_{sr_s}) \end{bmatrix}.$$

PROOF. It is sufficient to verify that

$$(3.1.7) \qquad V_m(L_{11},\ldots,L_{1r_1}) = \text{col}[XT^i]_{i=0}^{m-1} \, UV_{m_1}(L_{11},\ldots,L_{1r_1}),$$

where (X,T) is a right spectral pair for R_1 and

$$U = (\text{col}[XT^i]_{i=0}^{m_1-1})^{-1}.$$

Taking the ith block column of both sides of (3.1.7) (that corresponds to L_{1i}), the equality (3.1.7) amounts to the following: if L is a right divisor of R_1, and if (X_1,T_1) is a right spectral pair for L, then

$$(3.1.8) \qquad X_1T_1^j = XT^jU \, \text{col}[X_1T_1^i]_{i=0}^{m_1-1} \quad \text{for } j = 0,1,\ldots .$$

But these equalities follow from Theorem 2.7.1 and its proof. ∎

For scalar polynomials Proposition 3.1.3 allows us to compute the determinant of a square Vandermonde operator (matrix) as follows.

COROLLARY 3.1.4. *Let* $R_1,\ldots,R_s$ *be monic scalar polynomials, and let*

$$R_j(\lambda) = \prod_{i=1}^{r_j} (\lambda-\lambda_{ji}), \quad 1 \le j \le s.$$

Then

$$(3.1.9) \qquad \det[V_m(R_1,\ldots,R_s)] = \prod_{1\le j_1<j_2\le s} (\lambda_{j_2i_2}-\lambda_{j_1i_1}),$$

where the product is taken over all possible pairs (j_1,i_1) *and* (j_2,i_2) *such that* $j_1 < j_2$, *and* $m = r_1+\cdots+r_s$.

PROOF. For each $1 \le j \le s$, let $L_{ji}((\lambda)) = \lambda-\lambda_{ji}$. Then $V_{r_j}(L_{j_1},\ldots,L_{jr_j}) = [\lambda_{jh}^{i-1}]_{i,h=1}^{r_j}$ and hence

$$\det V_{r_j}(L_{j_1},\ldots,L_{jr_j}) = \prod_{1\le i_1<i_2\le r_j}(\lambda_{ji_2}-\lambda_{ji_1}).$$

Analogously

$$\det V_m(L_{11},\ldots,L_{1r_1},\ldots,L_{s1},\ldots,L_{sr_s}) = \prod(\lambda_{j_2i_2}-\lambda_{j_1i_1}),$$

where the product is taken over all pairs $(j_1,i_1),(j_2,i_2)$ such that either $j_1 < j_2$ or $j_1 = j_2$ and $i_1 < i_2$. Applying Proposition 3.1.3 we see that

$$\det V_m(R_1,\ldots,R_s) = \det V_m(L_{11},\ldots,L_{1r_1},\ldots,L_{s1},\ldots,L_{sr_s}).$$

$$\cdot\ [\prod_{j=1}^{s} \det V_{r_j}(L_{j_1},\ldots,L_{jr_j})]^{-1},$$

and (3.1.9) follows. ∎

In particular, $V_m(R_1,\ldots,R_s)$ is invertible provided the scalar polynomials $R_1,\ldots,R_s$ have disjoint zeros.

3.2 Existence of common multiples

Let $L_1,\ldots,L_r$ be monic operator polynomials of degrees $k_1,\ldots,k_r$, respectively. It follows from Theorem 3.1.1 that a necessary condition for existence of a monic left common multiple of degree m of $L_1,\ldots,L_r$ is that

$$\text{(3.2.1)}\qquad \operatorname{Ker} V_m(L_1,\ldots,L_r) \subset \operatorname{Ker}[X_1T_1^mU_1,\ldots,X_rT_r^mU_r],$$

where (X_j,T_j) is a right spectral pair of L_j and

$$U_j = \{\operatorname{col}[X_jT_j^i]_{i=0}^{k_j-1}\}^{-1}.$$

As

$$[X_1T_1^mU_1,\dots,X_rT_r^mU_r]$$

is the last operator row in $V_{m+1}(L_1,\dots,L_r)$, the inclusion (3.2.1) is clearly equivalent to

$$\text{Ker } V_m(L_1,\dots,L_r) = \text{Ker } V_{m+1}(L_1,\dots,L_r). \tag{3.2.2}$$

Actually, if (3.2.2) holds, then

$$\text{Ker } V_m(L_1,\dots,L_r) = \text{Ker } V_{m+p}(L_1,\dots,L_r) \quad \text{for } p = 1,2,\dots\ . \tag{3.2.3}$$

Indeed, let $(X_1,T_1),\dots,(X_r,T_r)$ be right spectral pairs for $L_1,\dots,L_r$, respectively. The equality (3.2.3) is equivalent to the following:

$$\text{Ker}[X_1T_1^{m+p-1},\dots,X_rT_r^{m+p-1}] \subset \text{Ker}\begin{bmatrix} X_1 & \cdots & X_r \\ X_1T_1 & \cdots & X_rT_r \\ \vdots & & \vdots \\ X_1T_1^{m-1} & \cdots & X_rT_r^{m-1} \end{bmatrix}, \quad p = 1,2,\dots\ . \tag{3.2.4}$$

We prove (3.2.4) by induction on p. For $p = 1$ this is equivalent to (3.2.2), hence true by assumption. Assume (3.2.4) is already proved with p replaced by p-1, and let

$$\text{col}[x_i]_{i=1}^r \in \text{Ker}[X_1T_1^{m+p-1},\dots,X_rT_r^{m+p-1}].$$

Then

$$\text{col}[T_ix_i]_{i=1}^r \in \text{Ker}[X_1T_1^{m+p-2},\dots,X_rT_r^{m+p-2}],$$

and by the induction hypothesis

$$\text{col}[T_ix_i]_{i=1}^r \in \text{Ker}[X_1T_1^m,\dots,X_rT_r^m],$$

and since (3.2.4) holds with $p = 1$, the vector $\text{col}[x_i]_{i=1}^r$ belongs to the right-hand side of (3.2.4).

We introduce the following definition:

$$\operatorname{ind}_1(L_1,\dots,L_r) = \inf\{m > 0 \mid \operatorname{Ker} V_m(L_1,\dots,L_r) = \operatorname{Ker} V_{m+1}(L_1,\dots,L_r)\}.$$

Here, as usual, $\inf \phi = \infty$. We call this number the (first) *index of stabilization* of the family $L_1,\dots,L_r$. From the preceding discussion, we conclude that $\operatorname{ind}_1(L_1,\dots,L_r) < \infty$ is a necessary condition for the existence of a monic left common multiple. It is now clear how to construct two monic operator polynomials which do not have a monic left common multiple.

EXAMPLE 3.2.1. For $k = 1,2,\dots$ define operators A_k and B_k on $\mathbb{C}^k$ by setting

$$A_k = \begin{bmatrix} 0 & 0 & \cdots & 0 & 0 \\ 1 & 0 & \cdots & 0 & 0 \\ 0 & 1 & \cdots & 0 & 0 \\ \cdots & \cdots & \cdots & \cdots & \cdots \\ 0 & 0 & \cdots & 1 & 0 \end{bmatrix}, \quad B_k = \begin{bmatrix} 0 & 0 & \cdots & 0 & 1 \\ 1 & 0 & \cdots & 0 & 0 \\ 0 & 1 & \cdots & 0 & 0 \\ \cdots & \cdots & \cdots & \cdots & \cdots \\ 0 & 0 & \cdots & 1 & 0 \end{bmatrix}.$$

Let $\mathcal{X}$ be the Hilbert space direct sum $\mathbb{C} \oplus \mathbb{C}^2 \oplus \mathbb{C}^3 \oplus \cdots$, and put $A = A_1 \oplus A_2 \oplus A_3 \oplus \cdots$, $B = B_1 \oplus B_2 \oplus B_3 \oplus \cdots$. Then A and B are operators on $\mathcal{X}$ and

$$V_m \overset{\text{def}}{=} V_m(\lambda I-A,\lambda I-B) = \operatorname{col}[A^i \ B^i]_{i=0}^{m-1}.$$

So $\operatorname{Ker} V_m$ consists of all pairs $((x_1,x_2,\dots),(-x_1,-x_2,\dots))$ in $\mathcal{X} \oplus \mathcal{X}$ such that for $k = 1,2,\dots$ the last $m-1$ coordinates of x_k (as element of $\mathbb{C}^k$) are zero. It follows that $\operatorname{Ker} V_m \neq \operatorname{Ker} V_{m+1}$ for all $m \geq 1$. Therefore, $\operatorname{ind}_1(\lambda I-A,\lambda I-B) = \infty$, and the monic operator polynomials $\lambda I-A$ and $\lambda I-B$ do not have a monic left common multiple. In other words, there is no monic operator polynomial for which A and B are right operator roots. ■

THEOREM 3.2.1. *Let $L_1,\dots,L_r$ be monic operator polynomials. A necessary condition for the existence of a monic common left multiple of $L_1,\dots,L_r$ of degree m is that*

$\text{ind}_1(L_1,\ldots,L_r) \leq m$. *This condition is also sufficient if the range of* $V_m(L_1,\ldots,L_r)$ *is closed and complemented.*

Recall that a (closed) subspace $\mathcal{M}$ in a Banach space $\mathcal{Y}$ is called *complemented* if there exists a (closed) subspace $\mathcal{N} \subset \mathcal{Y}$ such that $\mathcal{M} \cap \mathcal{N} = \{0\}$ and $\mathcal{M}+\mathcal{N} = \mathcal{Y}$. Equivalently, $\mathcal{M}$ is complemented if and only if $\mathcal{M}$ is a range of some bounded projector on $\mathcal{Y}$.

PROOF. The necessity of $\text{ind}_1(L_1,\ldots,L_r) \leq m$ was shown above. Assume now that $\text{ind}_1(L_1,\ldots,L_r) \leq m$ and that the range of $V_m(L_1,\ldots,L_r$ is closed and complemented. Define the operators $A_0,\ldots,A_{m-1}$ in the following way. Take $x \in \text{Im}\, V_m(L_1,\ldots,L_r)$, and put

$$[A_0 A_1 \cdots A_{m-1}]x = [X_1 T_1^m U_1,\ldots,X_r T_r^m U_r]y,$$

where y is chosen so that $V_m(L_1,\ldots,L_r)y = x$ (we continue to use the notation introduced in the beginning of this section). The condition $\text{ind}_1(L_1,\ldots,L_r) \leq m$ ensures that $[A_0 \cdots A_{m-1}]$ is well defined on $\text{Im}\, V_m(L_1,\ldots,L_r)$. Next, we define $[A_0 \cdots A_{m-1}]$ to be zero on some direct complement of $\text{Im}\, V_m(L_1,\ldots,L_r)$ in $\mathcal{X}^m$. The operators $A_0,\ldots,A_{m-1}$ defined in this way satisfy (3.1.1), and it remains to appeal to Theorem 3.1.1. ∎

COROLLARY 3.2.2. *If* $V_m(L_1,\ldots,L_r)$ *is left invertible, then there exists a monic left common multiple of* $L_1,\ldots,L_r$ *of degree* m.

This corollary is just a particular case of Theorem 3.2.1.

If we require that $V_m(L_1,\ldots,L_r)$ be only left invertible modulo compacts, then the existence of a monic common left multiple is guaranteed; however, its degree may be bigger than m:

COROLLARY 3.2.3. *If* $V_m(L_1,\ldots,L_r)$ *is left invertible modulo the compacts (i.e., for some operator* Z *and some compact operator* S *we have* $Z\, V_m(L_1,\ldots,L_r) = I+S$*), then* $\text{ind}_1(L_1,\ldots,L_r)$ *is finite and* $L_1,\ldots,L_r$ *have a common monic left multiple of degree* $\max(m, \text{ind}_1(L_1,\ldots,L_r))$.

For the proof of Corollary 3.2.3, we need, besides Theorem 3.2.1, the next well-known proposition, the proof of which is provided for the reader's convenience. An operator $Y \in L(\mathcal{Y},\mathcal{X})$ is called a *generalized inverse* of $X \in L(\mathcal{X},\mathcal{Y})$ if $XYX = X$, $YXY = Y$. It is not difficult to see (the proof of this fact can be found, for instance, in Schechter [1]) that X has a generalized inverse if and only if Ker X is a complemented subspace and Im X is closed and complemented. If $\mathcal{X}$ is a Hilbert space, then, of course, Ker X is always complemented.

PROPOSITION 3.2.4. *Let* $X \in L(\mathcal{X},\mathcal{Y})$, $Z \in L(\mathcal{Y},\mathcal{X})$ *be such that* $Y = ZX-I$ *is compact. Then* X *has a generalized inverse.*

PROOF. Since Y is compact, both Im(I+Y) and Ker(I+Y) are closed and complemented, and moreover, $\dim \operatorname{Ker}(I+Y) < \infty$. Write

$$\mathcal{X} = \operatorname{Ker}(I+Y) \dot{+} \mathcal{X}_1 = \mathcal{X}_2 \dot{+} \operatorname{Im}(I+Y), \tag{3.2.5}$$

and

$$I+Y = \begin{bmatrix} 0 & 0 \\ 0 & V \end{bmatrix}$$

with respect to the direct sum decompositions (3.2.5), so $V \in L(\mathcal{X}_1, \operatorname{Im}(I+Y))$. Analogously decompose

$$X = [X_1 \; X_2], \quad Z = \begin{bmatrix} Z_1 \\ Z_2 \end{bmatrix},$$

where $X_1 \in L(\operatorname{Ker}(I+Y),\mathcal{Y}), X_2 \in L(\mathcal{X}_1,\mathcal{Y})$, $Z_1 \in L(\mathcal{Y},\mathcal{X}_2)$, $Z_2 = L(\mathcal{Y},\operatorname{Im}(I+Y))$. Since V is one-to-one and onto, it is invertible, and $Z_2X_2 = V$. Hence, X_2 is left invertible and consequently, Im X_2 is closed and complemented. As

$$\operatorname{Im} X = \operatorname{Im} X_2 + \operatorname{Im} X_1$$

and $\dim \operatorname{Im} X_1 < \infty$ (this is so because Ker(I+Y) is finite dimensional), the subspace Im X is closed and complemented as well. ∎

PROOF OF COROLLARY 3.2.3. From the conditions of the Corollary, it follows that $\dim \operatorname{Ker} V_m(L_1,\ldots,L_r)$ is finite.

Hence $\mathrm{ind}_1(L_1,\dots,L_r) < \infty$. Further, as $V_m(L_1,\dots,L_r)$ is left invertible modulo the compacts, the same is true for $V_p(L_1,\dots,L_r)$ for each $p \geq m$. By Proposition 3.2.4, the operator $V_p(L_1,\dots,L_r)$ has a generalized inverse for each $p \geq m$. So we may apply Theorem 3.2.1 to get the desired result. ∎

The next example shows that the condition that $V_m(L_1,\dots,L_r)$ has a closed and complemented rank is not necessary to guarantee the existence of a common multiple of degree m.

EXAMPLE 3.2.2. Let B be a compact operator acting on the infinite-dimensional Banach space $\mathcal{X}$, and suppose that $\mathrm{Ker}\, B = \{0\}$. Put $L_1(\lambda) = \lambda I$ and $L_2(\lambda) = \lambda I-B$, where I is the identity operator on $\mathcal{X}$. As

$$\mathrm{Ker} \begin{bmatrix} I & I \\ 0 & B \end{bmatrix} = \{0\},$$

we have $\mathrm{ind}_1(L_1,L_2) = 2$, and $\mathrm{Im}\, \mathrm{col}[B^j]_{j=1}^{m-1}$ is not closed for any m. On the other hand, $L_1(\lambda)$ and $L_2(\lambda)$ commute, and thus $\lambda^2 I-\lambda B$ is a common left multiple of L_1 and L_2. Note that the degree of the multiple is precisely $\mathrm{ind}_1(L_1,L_2)$. ∎

We conclude this section with a result on existence of a common left multiple under certain spectral conditions. The *left spectrum* $\Sigma_\ell(L)$ of a monic operator polynomial $L(\lambda)$ is defined as follows:

$$\Sigma_\ell(L) = \{\lambda_0 \in \mathbb{C} \mid L(\lambda_0) \text{ has no (bounded) left inverse}\}.$$

Observe that $\Sigma_\ell(L) \subset \Sigma(L)$.

THEOREM 3.2.5. *Let* $L_1,\dots,L_r$ *be monic operator polynomials such that* $\Sigma_\ell(L_i) \cap \Sigma_\ell(L_j) = \phi$ *for* $i \neq j$. *Then there exists a monic left common multiple of* $L_1,\dots,L_r$.

The proof of this theorem is quite involved and will not be presented here. We refer the reader to Rodman [1] for the proof.

As a particular case of Theorem 3.2.5, it follows that the existence of a monic common left multiple of $L_1,\dots,L_r$ is ensured provided the spectra of $L_1,\dots,L_r$ are disjoint.

PROBLEM 3.2.1. *Find spectral conditions, as mild as possible, that ensure existence of a monic common left multiple of* $L_1,\dots,L_r$. In particular, in view of Theorem 3.2.5 and Corollary 3.2.3, the following question arises: *is it true that if the essential left spectra of* $L_1,\dots,L_r$ *do not intersect, then* $L_1,\dots,L_r$ *have a monic left common multiple?* Here the essential left spectrum of L_j is the set of all $\lambda_0 \in \mathbb{C}$ such that $L(\lambda_0)$ is not left invertible modulo compacts.

3.3 Common multiples of minimal degree

We study here the minimal degree of a common monic left multiple of $L_1,\dots,L_r$ (here, as in the preceding section, $L_1,\dots,L_r$ are monic operator polynomials on a Banach space $\mathcal{X}$). Define $\operatorname{ind}_2(L_1,\dots,L_r)$ to be the least positive integer k such that $V_m(L_1,\dots,L_r)$ has closed and complemented range for $m \geq k$ (if no such k exists put $\operatorname{ind}_2(L_1,\dots,L_r) = \infty$).

The following theorem is the main result of this section.

THEOREM 3.3.1. *If* $\operatorname{ind}_1(L_1,\dots,L_r)$ *and* $\operatorname{ind}_2(L_1,\dots,L_r)$ *are both finite, then* $L_1,\dots,L_r$ *have a common monic left multiple and the least degree of such a multiple is equal to* $\max(\operatorname{ind}_1(L_1,\dots,L_r), \operatorname{ind}_2(L_1,\dots,L_r))$.

We need some preparation for the proof of this theorem, and we start with general statements on operators with closed and complemented ranges.

PROPOSITION 3.3.2. *Let* $X_1 \in L(\mathcal{X},\mathcal{Y})$, $X_2 \in L(\mathcal{X},\mathcal{Z})$ *be Banach spaces operators such that* $\operatorname{Ker} X_1 \subset \operatorname{Ker} X_2$, *and* X_1 *has closed and complemented range. Then*

$$\begin{bmatrix} X_1 \\ X_2 \end{bmatrix} \in L(\mathcal{X},\mathcal{Y} \oplus \mathcal{Z})$$

has also closed and complemented range.

PROOF. Let $P \in L(\mathcal{Y})$ be a (bounded) projection with $\operatorname{Im} P \cap \operatorname{Im} X_1 = \{0\}$ and $\mathcal{Y} = \operatorname{Im} P + \operatorname{Im} X_1$ (the existence of such a projection is equivalent to X_1 having closed and complemented range). We claim that

$$\text{(3.3.1)} \qquad \operatorname{Im} \begin{bmatrix} P & 0 \\ 0 & I_{\mathcal{Z}} \end{bmatrix} \cap \operatorname{Im} \begin{bmatrix} X_1 \\ X_2 \end{bmatrix} = \{0\},$$

$$\text{(3.3.2)} \qquad \operatorname{Im} \begin{bmatrix} P & 0 \\ 0 & I_{\mathcal{Z}} \end{bmatrix} + \operatorname{Im} \begin{bmatrix} X_1 \\ X_2 \end{bmatrix} = \mathcal{Y} \oplus \mathcal{Z}.$$

Indeed, let

$$\begin{bmatrix} y \\ z \end{bmatrix} = \begin{bmatrix} P & 0 \\ 0 & I_{\mathcal{Z}} \end{bmatrix} \begin{bmatrix} x \\ z \end{bmatrix} = \begin{bmatrix} X_1 \\ X_2 \end{bmatrix} x'$$

for some $x, x' \in \mathcal{X}$, $z \in \mathcal{Z}$. Then $Px = X_1x'$, so by the choice of P, $Px = X_1x' = 0$. It follows that $y = 0$. By the condition $\operatorname{Ker} X_1 \subset \operatorname{Ker} X_2$ we have also $X_2x' = 0$, $z = 0$ as well, and (3.3.1) is verified. To verify (3.3.2), write for any $y \in \mathcal{Y}$ and $z \in \mathcal{Z}$:

$$\begin{bmatrix} y \\ z \end{bmatrix} = \begin{bmatrix} P & 0 \\ 0 & I \end{bmatrix} \begin{bmatrix} x' \\ z - X_2x'' \end{bmatrix} + \begin{bmatrix} X_1 \\ X_2 \end{bmatrix} x'',$$

where $x', x'' \in \mathcal{X}$ are such that $y = Px' + X_1x''$. Now the closedness and complementedness of $\operatorname{Im} \begin{bmatrix} X_1 \\ X_2 \end{bmatrix}$ follow from (3.3.1) and (3.3.2). ■

PROPOSITION 3.3.3. *Let* $X_1 \in L(\mathcal{X}, \mathcal{Y})$ *and* $X_2 \in L(\mathcal{Y}, \mathcal{Z})$. *If the operator*

$$X = \begin{bmatrix} X_1 \\ X_2X_1 \end{bmatrix} \in L(\mathcal{X}, \mathcal{Y} \oplus \mathcal{Z})$$

has the closed complemented range, then X_1 *has the closed complemented range as well.*

PROOF. Write

$$X = \begin{bmatrix} I & 0 \\ X_2 & I \end{bmatrix} \begin{bmatrix} X_1 \\ 0 \end{bmatrix}.$$

As the operator $\begin{bmatrix} I & 0 \\ X_2 & I \end{bmatrix}$ is invertible and hence transforms closed complemented subspaces onto closed complemented subspaces, the proof of Proposition 3.3.3 is reduced to the case when $X = \begin{bmatrix} X_1 \\ 0 \end{bmatrix}$, i.e., $X_2 = 0$. In this case it is easily seen that if $\mathcal{M}$ is a direct complement to Im X in $\mathcal{Y} \oplus \mathcal{Z}$, then $\mathcal{M} \cap \mathcal{Y}$ is a direct complement to Im X_1 in $\mathcal{Y}$. ∎

LEMMA 3.3.4. *Let* $L_1, \ldots, L_r$ *be monic operator polynomials on* $\mathcal{X}$. *If* $\text{ind}_1(L_1, \ldots, L_r) \leq k$ *and* $V_k(L_1, \ldots, L_r)$ *has closed and complemented range,* then $\text{ind}_2(L_1, \ldots, L_r) \leq k$.

PROOF. Take $m > k$ and write

$$V_m(L_1, \ldots, L_r) = \begin{bmatrix} V_k(L_1, \ldots, L_r) \\ B \end{bmatrix}.$$

The condition $\text{ind}_1(L_1, \ldots, L_r) \leq k$ implies that Ker $V_k(L_1, \ldots, L_r) \subset$ Ker B. Hence, we can apply Proposition 3.3.2 to show that $V_m(L_1, \ldots, L_r)$ has closed and complemented range. But then $\text{ind}_2(L_1, \ldots, L_r) \leq k$. ∎

PROOF OF THEOREM 3.3.1. In view of Theorem 3.2.1 we may suppose without loss of generality that $\text{ind}_1(L_1, \ldots, L_r) < m = \text{ind}_2(L_1, \ldots, L_r)$. Suppose that $L_1, \ldots, L_r$ have a common monic left multiple of degree strictly less than m. Then there exists such a multiple of degree m-1, and so one can find (see Theorem 3.1.1) an operator X such that $XV_{m-1}(L_1, \ldots, L_r) = S$, where S is the bottom operator row in $V_m(L_1, \ldots, L_r)$, i.e.,

$$V_m(L_1, \ldots, L_r) = \begin{bmatrix} V_{m-1}(L_1, \ldots, L_r) \\ S \end{bmatrix}.$$

But then we may apply Proposition 3.3.3 to show that $V_{m-1}(L_1, \ldots, L_r)$ has a closed and complemented range. However, in view of Lemma 3.3.4, this contradicts the fact that $m = \text{ind}_2(L_1, \ldots, L_r)$. So $L_1, \ldots, L_r$ have no common monic left multiple of degree m, and the theorem is proved. ∎

3.4 Fredholm Vandermonde operators

Recall that an operator $A \in L(\mathcal{X},\mathcal{Y})$ is called *Fredholm* if dim Ker $A < \infty$ and Im A is closed with finite dimensional direct complement. It is well known that an operator A is Fredholm if and only if A is invertible modulo the compacts, i.e., there is $B \in L(\mathcal{Y},\mathcal{X})$ such that $AB-I_{\mathcal{Y}}$ and $BA-I_{\mathcal{X}}$ are compact operators. So, as a particular case of Corollary 3.2.3, we obtain the following:

THEOREM 3.4.1. *If* $V_m(L_1,\ldots,L_r)$ *is Fredholm, then* $\mathrm{ind}_1(L_1,\ldots,L_r) < \infty$ *and* $L_1,\ldots,L_r$ *have a monic common left multiple of degree* $\max(m,\mathrm{ind}_1(L_1,\ldots,L_r))$.

This result will be applied in a particular situation when the essential spectra of $L_1,\ldots,L_r$ are disjoint. We say that $\lambda_0 \in \mathbb{C}$ is an *essential point* of the spectrum of a monic operator polynomial $L(\lambda)$ if $L(\lambda_0)$ is not Fredholm. The set of all essential points of the spectrum of L will be denoted by $\sigma_e(L)$.

THEOREM 3.4.2. *Let* $L_1,\ldots,L_r$ *be monic operator polynomials of the form*

$$L_i(\lambda) = \lambda^{k_i}I + \sum_{j=0}^{k_i-1} \lambda^j(\alpha_{ij}I+S_{ij}), \quad i = 1,\ldots,r,$$

where α_{ij} *are complex scalars and* S_{ij} *are compact operators. Suppose that* $\sigma_e(L_i) \cap \sigma_e(L_j) = \phi$ *for* $i \neq j$. *Then* $\mathrm{ind}_1(L_1,\ldots,L_r)$ *is finite and there exists a common monic left multiple of* $L_1,\ldots,L_r$ *of degree* $\max(\Sigma_{j=1}^r k_j,\ \mathrm{ind}_1(L_1,\ldots,L_r))$.

PROOF. Consider first the case when $\mathcal{X}$ is finite dimensional. Then the condition $\sigma_e(L_i) \cap \sigma_e(L_j) = \phi$ is trivially satisfied. Also, $\mathrm{ind}_1(L_1,\ldots,L_r)$ is obviously finite. As $\mathrm{ind}_2(L_1,\ldots,L_r)$ is finite as well, our assertion follows from Theorem 3.3.1. In the sequel it will be assumed that $\mathcal{X}$ is infinite dimensional. Put $f_i(\lambda) = \lambda^{k_i}+\Sigma_{j=1}^{k_i-1} \alpha_{ij}\lambda^j$, $i = 1,\ldots,r$. As the space $\mathcal{X}$ is infinite dimensional, $\sigma_e(L_i)$ consists of the roots of $f_i(\lambda)$, and hence the condition $\sigma_e(L_i) \cap \sigma_e(L_j) = \phi$ for $i \neq j$ implies in view of Corollary 3.1.4 that

$\det V_\ell(f_1,\dots,f_r) \neq 0$. Here $\ell = \Sigma_{j=1}^r k_j$. Let $\tilde{L}_i(\lambda) = f_i(\lambda)I$, $i = 1,\dots,r$. Then $V_\ell(\tilde{L}_1,\dots,\tilde{L}_r)$ is invertible too. From Proposition 3.1.2, it follows that

$$V_\ell(L_1,\dots,L_r) = V_\ell(\tilde{L}_1,\dots,\tilde{L}_r) + S$$

for some compact operator S. So $V_\ell(L_1,\dots,L_r)$ is a Fredholm operator and we may apply Theorem 3.4.1 to complete the proof. ∎

The setting of Theorem 3.4.2 can be extended to include matrices with operator entries, as follows. Let $\mathcal{X} = \mathcal{Y}^m$ for some Banach space $\mathcal{Y}$ (it will be assumed that $\mathcal{Y}$ is infinite dimensional). Then every $A \in L(\mathcal{X})$ can be naturally identified with an $m\times m$ matrix $[A_{ij}]_{i,j=1}^m$ whose entries belong to $L(\mathcal{Y})$. Consider the class $\mathcal{C}$ of all operators $A \in L(\mathcal{X})$ such that each A_{ij} has the form $\alpha I+C$, where $\alpha \in \mathbb{C}$ and $C \in L(\mathcal{Y})$ is compact (both α and C depend on A and on the entry (i,j)). The operator $A = [\alpha_{ij}I+C_{ij}]_{i,j=1}^n \in \mathcal{C}$ is Fredholm if and only if $\det[\alpha_{ij}]_{i,j=1}^n \neq 0$. Indeed, if $\det[\alpha_{ij}]_{i,j=1}^n \neq 0$, then A is invertible modulo compacts and hence Fredholm. Conversely, if $\det[\alpha_{ij}]_{i,j=1}^n = 0$, then $\dim \operatorname{Ker}[\alpha_{ij}I]_{i,j=1}^n = \infty$, and hence A cannot be invertible modulo compacts.

THEOREM 3.4.3. *Let $L_1(\lambda),\dots,L_r(\lambda)$ be monic operator polynomials with coefficients in the class $\mathcal{C}$. If*

$$\Sigma_e(L_i) \cap \Sigma_e(L_j) = \phi \text{ for } i \neq j, \tag{3.4.1}$$

then $L_1,\dots,L_r$ have a common monic left multiple.

PROOF. Write

$$L_j(\lambda) = \lambda^{k_j}I + \sum_{i=0}^{k_j-1} \lambda^i[\alpha_{ji}^{(pq)}I+C_{ji}^{(pq)}]_{p,q=1}^n$$

where $\alpha_{ji}^{(pq)}$ are complex numbers and $C_{ji}^{(pq)}$ are compact operators, and put

$$\tilde{L}_j(\lambda) = \lambda^{k_j} I + \sum_{i=0}^{k_j-1} \lambda^i [\alpha_{ji}^{(pq)}]_{p,q=1}^n .$$

The condition (3.4.1) implies that $\sigma(\tilde{L}_i) \cap \sigma(\tilde{L}_j) = \phi$ for $i \neq j$. As the operator polynomials $\tilde{L}_1, \ldots, \tilde{L}_r$ act on a finite dimensional space $\mathbb{C}^n$, by Theorem 3.2.1 there exists a monic common left multiple $\tilde{L}$ of $\tilde{L}_1, \ldots, \tilde{L}_r$.

Consider the Vandermonde operator $V_m(\tilde{L}_1, \ldots, \tilde{L}_r)$, where m is the degree of $\tilde{L}$. Appeal to Proposition 3.5.1 in the next section and deduce that $V_m(\tilde{L}_1, \ldots, \tilde{L}_r)$ is left invertible. But then $V_m(\hat{L}_1, \ldots, \hat{L}_r)$ is left invertible as well, where

$$\hat{L}_j(\lambda) = \lambda^{k_j} I_{\mathcal{X}} + \sum_{i=0}^{k_j-1} \lambda^i [\alpha_{ji}^{(pq)} I_{\mathcal{Y}}]_{p,q=1}^n .$$

As $V_m(L_1, \ldots, L_r) - V_m(\hat{L}_1, \ldots, \hat{L}_r)$ is compact, the operator $V_m(L_1, \ldots, L_r)$ is left invertible modulo compacts, and it remains to apply Corollary 3.2.3. ■

3.5 Vandermonde operators of divisors

Let $L, L_1, \ldots, L_r$ be monic operator polynomials on a Banach space $\mathcal{X}$ with degrees $\ell, k_1, \ldots, k_r$, respectively. In this section we assume that $L_1, \ldots, L_r$ are right divisors of L. In that case one may associate with each L_j a supporting subspace $\mathcal{M}_j$ (see Section 2.7). Our aim is to describe the invertibility properties of the Vandermonde operator $V_m(L_1, \ldots, L_r)$ in terms of the geometric properties of the spaces $\mathcal{M}_1, \ldots, \mathcal{M}_r$.

First, let us recall the definition of the supporting subspaces. Let (X,T), $(X_1,T_1), \ldots, (X_r,T_r)$ be right spectral pairs of $L, L_1, \ldots, L_r$, respectively. For each j put

$$K_j = \{\mathrm{col}[XT^i]_{i=0}^{\ell-1}\}^{-1} \mathrm{col}[X_j T_j^i]_{i=0}^{\ell-1} \in L(\mathcal{X}^{k_j}, \mathcal{X}^{\ell}).$$

Observe that $\mathrm{col}[XT^j]_{j=0}^{\ell-1}$ is invertible, so K_j is correctly defined. As $k_j \le \ell$ and $\mathrm{col}[X_jT_j]_{i=0}^{k_j-1}$ is invertible, the operator K_j is left invertible. The following properties of K_j follow from Theorem 2.7.1 and its proof.

(i) $\mathcal{M}_j = \mathrm{Im}\, K_j$ is invariant under T and $TK_j = K_jT_j$;

(ii) $XT^\alpha K_j = X_jT_j^\alpha$ for $\alpha = 0,1,2,\ldots$;

(iii) the operator $\mathrm{col}[XT^{i-1}]_{i=1}^{k_j}|\mathcal{M}_j\colon \mathcal{M}_j \to \mathcal{X}^{k_j}$ is invertible and its inverse is given by K_jU_j, where

$$U_j = [\mathrm{col}[X_jT_j^{i-1}]_{i=1}^{k_j}]^{-1}.$$

The subspace $\mathcal{M}_j$ is called the *supporting subspace* of the right divisor L_j corresponding to the right spectral pair (X,T) of L. Observe that each $\mathcal{M}_j$ is a complemented subspace of $\mathcal{X}$, i.e., $\mathcal{M}_j+\mathcal{N}_j = \mathcal{X}$, $\mathcal{M}_j \cap \mathcal{N}_j = \{0\}$ for some (closed) subspace $\mathcal{N}_j$ of $\mathcal{X}$. The notations introduced above will be used in the next two sections without further explanation.

By property (ii) mentioned above, the Vandermonde operator $V_m(L_1,\ldots,L_r)$ admits the following representation:

$$V_m(L_1,\ldots,L_r) = [\mathrm{col}[XT^{i-1}]_{i=1}^{m}] \cdot [K_1,K_2,\ldots,K_r]\ \mathrm{diag}\ [U_j]_{j=1}^{r}. \tag{3.5.1}$$

As $\mathrm{col}[XT^{i-1}]_{i=1}^{\ell}$ is invertible, the first factor in the right-hand side of (3.5.1) will be left invertible for $m \ge \ell$. Hence from (3.5.1) one can see that for $m \ge \ell$

$$Z_mV_m(L_1,\ldots,L_r) = V_m(L_1,\ldots,L_r) \cdot \mathrm{diag}[C_j]_{j=1}^{r}, \tag{3.5.2}$$

where $C_1,\ldots,C_r$ are the companion operators of $L_1,\ldots,L_r$ respectively, and $Z_m = \mathrm{col}[XT^i]_{i=1}^m\, Y$, where Y is some left inverse of $\mathrm{col}[XT^{i-1}]_{i=1}^m$. Indeed, for every j $(1 \le j \le r)$ we have $T_jU_j = U_jC_j$ (see formula (2.1.5)), hence $X_jT_j^kU_j = X_jT_j^{k-1}U_jC_j$

for $k = 1,2,\ldots$. Using these equalities, we obtain after premultiplication of (3.5.1) by Z_m (the second equality below follows from (ii)):

$$Z_m V_m(L_1,\ldots,L_r) = \mathrm{col}[XT^i]_{i=1}^m \,[K_1 \cdots K_r]\ \mathrm{diag}\,[U_j]_{j=1}^r$$

$$= [\mathrm{col}[X_1T_1^i]_{i=1}^m,\ldots,\mathrm{col}[X_rT_r^i]_{i=1}^m]\ \mathrm{diag}\,[U_j]_{j=1}^r$$

$$= [\mathrm{col}[X_1T_1^{i-1}]_{i=1}^{m-1},\ldots,\mathrm{col}[X_rT_r^{i-1}]_{i=1}^{m-1}]\ \mathrm{diag}\,[U_jC_j]_{j=1}^r$$

$$= V_m(L_1,\ldots,L_r)\ \mathrm{diag}[C_j]_{j=1}^r.$$

For $m = \ell$ the operator Z_m is merely the companion operator C_L of L, and thus

$$C_L V_\ell(L_1,\ldots,L_r) = V_\ell(L_1,\ldots,L_r) \cdot \mathrm{diag}[C_j]_{j=1}^r.$$

We now clarify the relation between the properties of the Vandermonde operator $V_m(L_1,\ldots,L_r)$ and those of the supporting subspaces.

PROPOSITION 3.5.1. *For* $m \geq \ell$ *we have* $\mathrm{Ker}\ V_m(L_1,\ldots,L_r) = \{0\}$ *if and only if the subspaces* $\mathcal{M}_1,\ldots,\mathcal{M}_r$ *are linearly independent, i.e.,* $x_1+\cdots+x_r = 0$, $x_j \in \mathcal{M}_j$ *for* $j = 1,\ldots,r$ *implies* $x_1 = \cdots = x_r = 0$.

This proposition follows immediately from representation (3.5.1) taking into account the left invertibility of $\mathrm{col}[XT^{i-1}]_{i=1}^m$.

The sum of linearly independent supporting subspaces does not have to be closed, and, when it is closed, it may not have a closed complement. To see this, we consider the following example.

EXAMPLE 3.5.1. Let $A \in L(\mathcal{X})$, where $\mathcal{X}$ is an infinite-dimensional Banach space, and suppose that $\mathrm{Ker}\ A = \{0\}$. Put $L_1(\lambda) = \lambda I$ and $L_2(\lambda) = \lambda I-A$. Then L_1 and L_2 are right divisors of the monic operator polynomial $L(\lambda) = \lambda^2 I-\lambda A$. Let

$$X = [I\ 0],\ C = \begin{bmatrix} 0 & I \\ 0 & A \end{bmatrix}.$$

Then (X,C) is a right spectral pair of L, and the supporting subspaces $\mathcal{M}_1$ and $\mathcal{M}_2$ of L_1 and L_2 corresponding to the pair (X,C) are given by, respectively,

$$\mathcal{M}_1 = \{(x,y) \in \mathcal{X}^2 \mid y = 0\},\ \mathcal{M}_2 = \{(x,Ax) \in \mathcal{X}^2 \mid x \in \mathcal{X}\}.$$

As Ker A = $\{0\}$, it is clear that $\mathcal{M}_1 \cap \mathcal{M}_2 = (0)$, and hence $\mathcal{M}_1$ and $\mathcal{M}_2$ are linearly independent. Note that

$$\mathcal{M}_1 + \mathcal{M}_2 = \mathcal{X} \oplus \text{Im } A.$$

Thus $\mathcal{M}_1 + \mathcal{M}_2$ is closed (has a closed complement) in $\mathcal{X}^2$ if and only if Im A is closed (has a closed complement) in $\mathcal{X}$. Similarly, $\mathcal{M}_1 + \mathcal{M}_2$ is dense in $\mathcal{X}_2$ if and only if Im A is dense in $\mathcal{X}$. ∎

PROPOSITION 3.5.2. *$V_m(L_1, \ldots, L_r)$ is left invertible for $m \geq \ell$ if and only if the supporting subspaces $\mathcal{M}_1, \ldots, \mathcal{M}_r$ are linearly independent and $\mathcal{M} = \mathcal{M}_1 + \cdots + \mathcal{M}_r$ is a (closed) complemented subspace in $\mathcal{X}^\ell$.*

PROOF. In view of formula (3.5.1) we have

$$\text{Im } V_m(L_1, \ldots, L_r) = [\text{col}[XT^{i-1}]_{i=1}^m]\ (\mathcal{M}_1 + \cdots + \mathcal{M}_r).$$

So, taking into account Proposition 3.5.1, we only have to show that

$$[\text{col}[XT^{i-1}]_{i=1}^m](\mathcal{M})$$

is a complemented subspace of $\mathcal{X}^m$ if and only if $\mathcal{M}$ is a complemented subspace of $\mathcal{X}^\ell$. Put $\Omega = \text{col}[XT^{i-1}]_{i=1}^m$. Now, as $m \geq \ell$, the operator Ω is left invertible. In particular, Ω is a topological isomorphism from $\mathcal{X}^\ell$ onto Im Ω. This fact implies that $\mathcal{M}$ is a complemented subspace of $\mathcal{X}^\ell$ if and only if $\Omega(\mathcal{M})$ is a complemented subspace of Im Ω. As Im Ω is complemented in $\mathcal{X}^m$, this completes the proof. ∎

An operator $A \in L(\mathcal{X},\mathcal{Y})$ between Banach spaces $\mathcal{X}$ and $\mathcal{Y}$ is called *regular* if there is $\alpha > 0$ such that $\|Ax\| \geq \alpha\|x\|$ for every $x \in \mathcal{X}$. Clearly, for a regular operator A we have Ker $A = \{0\}$, and it is not difficult to see that Im A is a (closed) subspace. However, Im A need not be a complemented subspace. If $\mathcal{Y}$ is a Hilbert space, then A is regular if and only if A is left invertible. In the same way as Proposition 3.5.2, one proves the following.

PROPOSITION 3.5.3. *For* $m \geq \ell$, $V_m(L_1,\ldots,L_r)$ *is regular if and only if* $\mathcal{M}_1,\ldots,\mathcal{M}_r$ *are linearly independent and* $\mathcal{M}_1+\cdots+\mathcal{M}_r$ *is closed.*

We pass now to the right invertibility of $V_m(L_1,\ldots,L_r)$.

PROPOSITION 3.5.4. *For* $m \geq \ell$, $V_m(L_1,\ldots,L_r)$ *is right invertible if and only if* $m = \ell$, $\mathcal{X}^\ell = \mathcal{M}_1+\cdots+\mathcal{M}_r$, *and* Ker $V_m(L_1,\ldots,L_r)$ *is a complemented subspace.*

PROOF. Assume that $m = \ell$ and $\mathcal{X}^\ell = \mathcal{M}_1+\cdots+\mathcal{M}_r$. Then $V_m(L_1,\ldots,L_r)$ is surjective by formula (3.5.1). But, together with Ker $V_m(L_1,\ldots,L_r)$ is complemented, this implies that $V_m(L_1,\ldots,L_r)$ is right invertible. Conversely, suppose that $V_m(L_1,\ldots,L_r)$ is right invertible. Then, trivially, Ker $V_m(L_1,\ldots,L_r)$ is complemented and, again using formula (3.5.1), one sees that $\text{col}[XT^{i-1}]_{i=1}^m$ is surjective. As $m \geq \ell$, we also know that $\text{col}[XT^{i-1}]_{i=1}^m$ is left invertible. So $\text{col}[XT^{i-1}]_{i=1}^m$ is invertible. But, since $\text{col}[XT^{i-1}]_{i=1}^\ell$ is invertible too, this can only happen if $m = \ell$. Further, using the invertibility of $\text{col}[XT^{i-1}]_{i=1}^m$ and the fact that $V_m(L_1,\ldots,L_r)$ is surjective, it follows from formula (3.5.1) that $\mathcal{M}_1+\cdots+\mathcal{M}_r$ must be $\mathcal{X}^\ell$. ∎

3.6 Divisors with disjoint spectra

We continue to use here the framework and notation introduced in the preceding section. Here we shall assume in addition that the right divisors $L_1,\dots,L_r$ have disjoint spectra.

Our first result in this direction is the following.

THEOREM 3.6.1. *Let* $L_1,\dots,L_r$ *be right divisors of the monic operator polynomial* L, *and let* $\mathcal{M}_1,\dots,\mathcal{M}_r$ *be the corresponding supporting subspaces. Suppose that the spectra of* $L_1,\dots,L_r$ *are mutually disjoint. Then* $\mathcal{M}_1,\dots,\mathcal{M}_r$ *are linearly independent and* $\mathcal{M}_1+\cdots+\mathcal{M}_r$ *is closed. Moreover, the Vandermonde operator* $V_m(L_1,\dots,L_r)$ *is regular for* $m \geq$ *degree* (L).

If, in addition, $\bigcup_{j=1}^{r} \Sigma(L_j)$ *is relatively open subset of* $\Sigma(L)$, *then* $\mathcal{M}_1+\cdots+\mathcal{M}_r$ *is a complemented subspace and* $V_m(L_1,\dots,L_r)$ *is left invertible for* $m \geq$ *degree* (L).

For the proof of Theorem 3.6.1 we need an auxiliary result which is independently interesting.

LEMMA 3.6.2. *Let* $S \in L(\mathcal{X})$, *where* $\mathcal{X}$ *is a Banach space, and let* $\mathcal{N}_1,\dots,\mathcal{N}_r$ *be S-invariant subspaces of* $\mathcal{X}$. *Suppose that*

$$\sigma(S|_{\mathcal{N}_i}) \cap \sigma(S|_{\mathcal{N}_j}) = \phi, \quad (i \neq j).$$

Then $\mathcal{N}_1,\dots,\mathcal{N}_r$ *are linearly independent and* $\mathcal{N}_1,\dots,\mathcal{N}_r$ *is closed.*

PROOF. We shall use the notations and results from the Appendix and prove the lemma by induction on r.

For $r = 1$, there is nothing to prove. Therefore, take $r \geq 2$ and suppose that the lemma has been proved for $r-1$ spaces. Let

$$J: \mathcal{N}_1 \oplus\cdots\oplus \mathcal{N}_r \to \mathcal{X}$$

be defined by $J(x_1,\dots,x_r) = x_1+\cdots+x_r$. We have to show that J is regular. Suppose not. Then there exist $(x_{1n},\dots,x_{rn}) \in \mathcal{N}_1 \oplus\cdots\oplus \mathcal{N}_r$, $n = 1,2,\dots$, such that $\|x_{1n}\| +\cdots+ \|x_{rn}\| = 1$ for $n = 1,2,\dots$ and

$$x_{1n}+\cdots+x_{rn} \to 0, \quad (n \to \infty).$$

By passing to the space $\langle\mathcal{X}\rangle$, we see that it suffices to show that the spaces $\langle\mathcal{N}_1\rangle,\dots,\langle\mathcal{N}_r\rangle$ are linearly independent.

It is easy to see that the subspaces $\langle\mathcal{N}_1\rangle,\ldots,\langle\mathcal{N}_r\rangle$ are $\langle S\rangle$-invariant. From Proposition 3.6.5 and conditions of the lemma, it follows that

$$\sigma(\langle S\rangle|_{\langle\mathcal{N}_i\rangle}) \cap \sigma(\langle S\rangle|_{\langle\mathcal{N}_j\rangle}) = \{0\}$$

for $i \neq j$. So, by our induction hypothesis, the spaces $\langle\mathcal{N}_i\rangle,\ldots,\langle\mathcal{N}_{r-1}\rangle$ are linearly independent and $\mathcal{M} = \langle\mathcal{N}_1\rangle+\cdots+\langle\mathcal{N}_{r-1}\rangle$ is closed in $\langle\mathfrak{X}\rangle$. It remains to prove that $\mathcal{M} \cap \langle\mathcal{N}_r\rangle = \{0\}$. Note that $\mathcal{M}$ is $\langle S\rangle$-invariant and

$$\sigma(\langle S\rangle|_{\mathcal{M}}) = \bigcup_{j=1}^{r-1} \sigma(\langle S\rangle|_{\langle\mathcal{N}_j\rangle}).$$

Hence $\sigma(\langle S\rangle|_{\mathcal{M}}) \cap \sigma(\langle S\rangle|_{\langle\mathcal{N}_r\rangle})$ is empty. Let $\mathcal{M}_0 = \mathcal{M} \cap \langle\mathcal{N}_r\rangle$. At this point we will use the following general fact about an operator $A \in L(\mathfrak{X})$, where $\mathfrak{X}$ is a Banach space: for every A-invariant subspace $\mathcal{N}$, the boundary of $\sigma(A|_{\mathcal{N}})$ is contained in $\sigma(A)$. The proof of this fact (in the Banach algebras framework) is given in Rudin [2] (Theorem 10.18). Denoting by $\partial\sigma(\langle S\rangle|_{\mathcal{M}_0})$ the boundary of the spectrum of $\langle S\rangle|_{\mathcal{M}_0}$, we have (in view of the general fact mentioned above) that

$$\partial\sigma(\langle S\rangle|_{\mathcal{M}_0}) \subset \sigma(\langle S\rangle|_{\mathcal{M}})$$

and

$$\partial\sigma(\langle S\rangle|_{\mathcal{M}_0}) \subset \sigma(\langle S\rangle|_{\langle\mathcal{N}_r\rangle}).$$

Hence $\partial\sigma(\langle S\rangle|_{\mathcal{M}_0}) = \phi$, which means that $\mathcal{M}_0 = \{0\}$, and the proof is complete. ■

PROOF OF THEOREM 3.6.1. The linear independence of $\mathcal{M}_1,\ldots,\mathcal{M}_r$ and the closedness of $\mathcal{M}_1+\cdots+\mathcal{M}_r$ follow from Lemma 3.6.2 taking into account that $\Sigma(L_j) = \sigma(T|_{\mathcal{M}_j})$, where (X,T) is the right spectral pair of L. The regularity of $V_m(L_1,\ldots,L_r)$ (for $m \geq \ell$) follows from Proposition 3.5.3. If we know already that $\mathcal{M}_1+\cdots+\mathcal{M}_r$ is also complemented, then the left invertibility of $V_m(L_1,\ldots,L_r)$ follows from Proposition 3.5.2.

It remains to show that $\mathcal{M}_1+\cdots+\mathcal{M}_r$ is complemented under the additional hypothesis, that $\Sigma(L_1),\ldots,\Sigma(L_r)$ are open and closed subsets of $\Sigma(L) = \sigma(T)$. Let $Q_1,\ldots,Q_r$ be the corresponding Riesz projections, i.e., for each j

$$Q_j = \frac{1}{2\pi i}\int_{\Gamma_j} (\lambda I-T)^{-1}d\lambda,$$

where the contour Γ_j consists of regular points of T and separates $\Sigma(L_j)$ from the rest of the spectrum of $\Sigma(L)$. Note that $\mathcal{M}_j \subset \operatorname{Im} Q_j$ because $\operatorname{Im} Q_j$ is the maximal T-invariant subspace $\mathcal{N}$ of $\mathcal{X}^\ell$, where ℓ = degree(L), such that $\sigma(T|\mathcal{N})$ lies inside Γ_j. Since $\mathcal{M}_j$ is a complemented subspace of $\mathcal{X}^\ell$, it follows that $\mathcal{M}_j$ is complemented in $\operatorname{Im} Q_j$. The fact that $\Sigma(L_1),\ldots,\Sigma(L_r)$ are mutually disjoint implies that $\operatorname{Im} Q_1,\ldots,\operatorname{Im} Q_r$ are linearly independent and $\operatorname{Im} Q_1+\cdots+\operatorname{Im} Q_r$ is a complemented subspace of $\mathcal{X}^\ell$. As $\mathcal{M}_1+\cdots+\mathcal{M}_r$ is complemented in $\operatorname{Im} Q_1+\cdots+\operatorname{Im} Q_r$, it follows that it is complemented in $\mathcal{X}^\ell$. ■

Combining Theorem 3.6.1 with Theorem 3.2.5, we obtain the following fact.

COROLLARY 3.6.3. *Let* $L_1,\ldots,L_r$ *be monic operator polynomials with disjoint spectra. Then for* m *large enough the Vandermonde operator* $V_m(L_1,\ldots,L_r)$ *is regular.*

In connection with Lemma 3.6.2 and Corollary 3.6.3 a conjecture is suggested.

CONJECTURE 3.6.1. *Let* $S \in L(\mathcal{X})$ *and let* $\mathcal{N}_1,\ldots,\mathcal{N}_r$ *be* S*-invariant complemented subspaces. If* $\sigma(S|_{\mathcal{N}_j}) \cap \sigma(S|_{\mathcal{N}_i}) = \phi$ *for* $i \neq j$, *then the sum* $\mathcal{N}_1+\cdots+\mathcal{N}_r$ *is a complemented subspace as well.*

If the conjecture is true, then in Corollary 3.6.3 the regularity of $V_m(L_1,\ldots,L_r)$ can be replaced by left invertibility.

Appendix: Hulls of Operators

In this appendix we present a brief description of a well-known construction for operators in Banach spaces, which is needed for the proof of Lemma 3.6.2.

Let $\mathfrak{X}$ be a Banach space. The Banach space $\ell_\infty(\mathfrak{X})$ consists, by definition, of all bounded sequences $\{x_n\}_{n=0}^\infty$ from $\mathfrak{X}$ with the norm

$$\|\{x_n\}_{n=0}^\infty\| = \sup_{n\geq 0} \|x_n\|.$$

It is not difficult to check that $\ell_\infty(\mathfrak{X})$ is indeed a Banach space. The set

$$c_0(\mathfrak{X}) = \left\{\{x_n\}_{n=0}^\infty \in \ell_\infty(\mathfrak{X}) \mid \lim_{n\to\infty} x_n = 0\right\}$$

is a (closed) subspace in $\ell_\infty(\mathfrak{X})$. So we can define the quotient space

$$\langle\mathfrak{X}\rangle = \ell_\infty(\mathfrak{X})/c_0(\mathfrak{X})$$

which consists of all classes of bounded sequences $\{x_n\}_{n=0}^\infty$, $x_n \in \mathfrak{X}$ (two bounded sequences $\{x_n\}_{n=0}^\infty$ and $\{y_n\}_{n=0}^\infty$ are in the same class if and only if $\|x_n-y_n\| \to 0$ as $n \to \infty$) and is endowed with the quotient norm

$$\|\langle\{x_n\}_{n=0}^\infty\rangle\|_{\langle\mathfrak{X}\rangle} = \inf \{\sup_{n\geq 0} \|x_n-z_n\| \mid \lim_{k\to\infty} z_k = 0\}.$$

Here $\langle\{x_n\}_{n=0}^\infty\rangle$ stands for the class which contains $\{x_n\}_{n=0}^\infty$. Then one easily sees that E: $\mathfrak{X} \to \langle\mathfrak{X}\rangle$ defined by $Ex = \langle(x,x,\ldots)\rangle$ is an isometry, i.e., $\|Ex\| = \|x\|$ for every $x \in \mathfrak{X}$.

If $\mathfrak{X}_1$ and $\mathfrak{X}_2$ are complex Banach spaces, and T: $\mathfrak{X}_1 \to \mathfrak{X}_2$ a bounded linear operator, then the *hull* of the operator T is the operator $\langle T\rangle$: $\langle\mathfrak{X}_1\rangle \to \langle\mathfrak{X}_2\rangle$, given by

$$\langle T\rangle \langle\{x_n\}_{n=0}^\infty\rangle = \langle\{Tx_n\}_{n=0}^\infty\rangle.$$

One easily checks that this operator is well defined and bounded. If $T \in L(\mathfrak{X}_1,\mathfrak{X}_2)$ and $S \in L(\mathfrak{X}_2,\mathfrak{X}_3)$, then

$$\langle ST\rangle = \langle S\rangle \langle T\rangle. \tag{3.6.1}$$

Obviously, $\langle I_\mathfrak{X}\rangle = I_{\langle\mathfrak{X}\rangle}$.

PROPOSITION 3.6.4. *The map* Φ: $T \to \langle T\rangle$ *is a continuous linear transformation from* $L(\mathfrak{X}_1,\mathfrak{X}_2)$ *into* $L(\langle\mathfrak{X}_1\rangle,\langle\mathfrak{X}_2\rangle)$.

PROOF. The linearity of Φ is evident, and to prove continuity it is sufficient to verify that

$$\|\langle T\rangle\| \le \|T\|. \tag{3.6.2}$$

Let $\{x_n\}_{n=0}^{\infty}$ be any bounded sequence from $\mathcal{X}_1$, and let $\{y_n\}_{n=0}^{\infty}$ be any sequence from $\mathcal{X}_1$, with $\lim_{n\to\infty} y_n = 0$. We have

$$\|\langle T\rangle \langle\{x_n\}_{n=0}^{\infty}\rangle\| = \|\langle\{Tx_n\}_{n=0}^{\infty}\rangle\|$$

$$\le \sup_n \|Tx_n+Ty_n\| \le \|T\| \sup_n \|x_n+y_n\|.$$

Taking infimum in the right-hand side over all sequences $\{y_n\}_{n=0}^{\infty}$ with $\lim_{n\to\infty} y_n = 0$, we obtain

$$\|\langle T\rangle \langle\{x_n\}_{n=0}^{\infty}\rangle\| \le \|T\| \langle\{x_n\}_{n=0}^{\infty}\rangle.$$

which implies (3.6.2). ∎

PROPOSITION 3.6.5. For $T \in L(\mathcal{X})$ we have $\sigma(\langle T\rangle) = \sigma(T)$.

PROOF. If $X \in L(\mathcal{X})$ is the inverse of $\lambda I-T$, then by (3.6.1), $\langle X\rangle$ is the inverse of $\lambda I-\langle T\rangle$. So $\sigma(\langle T\rangle) \subset \sigma(T)$.

To prove the opposite inclusion, let $\lambda_0 \in \sigma(T)$. Consider first the case when

$$\inf_{\|x\|=1} \{\|Tx-\lambda_0 x\|\} = 0. \tag{3.6.3}$$

Then there is a sequence $\{x_n\}_{n=0}^{\infty}$ with $\|x_n\| = 1$ and $\|Tx_n-\lambda_0 x_n\| \to 0$ as $n \to \infty$. Therefore

$$(\lambda_0 I-\langle T\rangle) \langle x_n\rangle_{n=0}^{\infty} = \langle(\lambda_0 I-T)X_n\rangle_{n=0}^{\infty} = 0,$$

so λ_0 is an eigenvalue of $\langle T\rangle$, and hence $\lambda_0 \in \sigma(\langle T\rangle)$. If (3.6.3) is false, then $\lambda_0 I-T$ is regular, i.e., $\|(\lambda_0 I-T)x\| \ge \alpha\|x\|$ for every $x \in \mathcal{X}$, where the constant $\alpha > 0$ is independent of x. In particular, $\mathrm{Ker}(\lambda_0 I-T) = \{0\}$ and $\mathrm{Im}(\lambda_0 I-T)$ is closed. As $\lambda_0 \in \sigma(T)$, we have $\mathrm{Im}(\lambda_0 I-T) \ne \mathcal{X}$, hence there is $x \in \mathcal{X}$ such that $\inf\{\|x-y\| \mid y \in \mathrm{Im}(\lambda_0 I-T)\} > 0$. It is easily seen that $\langle x,x,\ldots\rangle$ does not belong to $\mathrm{Im}(\lambda_0 I-T\rangle)$, so $\lambda_0 I-\langle T\rangle$ is not invertible. Again, $\lambda_0 \in \sigma(\langle T\rangle)$. ∎

It follows from the proof of Proposition 3.6.5 that every $\lambda_0 \in \sigma(T)$ for which $\lambda_0 I-T$ is not regular, is actually an eigenvalue of $\langle T\rangle$.

Let $\mathcal{M}$ be a (closed) subspace in a Banach space $\mathcal{X}$. Then $\langle\mathcal{M}\rangle$ can be naturally identified with a linear set in $\langle\mathcal{X}\rangle$.

PROPOSITION 3.6.6. *The linear set* $\langle\mathcal{M}\rangle$ *is a closed subspace in* $\langle\mathcal{X}\rangle$.

PROOF. Suppose not. Then there is a sequence $\{x^{(p)}\}_{p=1}^{\infty} \in \langle\mathcal{M}\rangle$ such that

$$\lim_{p\to\infty} \|x^{(p)}-x\| = 0 \tag{3.6.4}$$

for some $x \in \langle\mathcal{X}\rangle$, but $x \notin \langle\mathcal{M}\rangle$. Write $x = \langle\{x_n\}_{n=0}^{\infty}\rangle$ where $x_n \in \mathcal{X}$; then there is $\varepsilon_0 > 0$ and an infinite subsequence x_{n_k}, $k = 1,2,\dots$ such that

$$\inf_{y\in\mathcal{M}} \|x_{n_k}-y\| \geq \varepsilon_0 \tag{3.6.5}$$

for all k. On the other hand, write

$$x^{(p)} = \langle\{x_n^{(p)}\}_{n=0}^{\infty}\rangle, \quad x_n^{(p)} \in \mathcal{M},$$

and observe that (3.6.4) implies the inequality

$$\sup_n \|x_n^{(p)}-x_n+z_n\| < \frac{\varepsilon_0}{2}$$

for some p and some sequence $\{z_n\}_{n=0}^{\infty}$ such that $\lim_{n\to\infty} z_n = 0$. In particular,

$$\|x_{n_k}^{(p)}-x_{n_k}+z_{n_k}\| < \frac{\varepsilon_0}{2} \tag{3.6.6}$$

for all k. Taking in (3.6.6) k sufficiently large, we obtain a contradiction with (3.6.5). ■

3.7 Application to differential equations

Consider the differential equation

$$u^{(\ell)}(t)+A_{\ell-1}u^{(\ell-1)}(t)+\cdots+A_1u^{(1)}(t)+A_0u(t) = 0, \tag{3.7.1}$$

where $A_0,\ldots,A_{\ell-1} \in L(\mathcal{X})$ and the $\mathcal{X}$-valued function $u(t)$ of the real variable t is unknown (as usual, $\mathcal{X}$ stands for a Banach space). Let

$$L(\lambda) = \lambda^{\ell}I+\sum_{j=0}^{\ell-1}\lambda^{j}A_j$$

be the corresponding monic operator polynomial.

If $L_1(\lambda) = \lambda^kI+\sum_{j=0}^{k-1}\lambda^jB_j$ is a right divisor of L, then clearly every solution of

$$v^{(k)}(t)+\sum_{j=0}^{k-1}B_jv^{(j)}(t) = 0 \tag{3.7.2}$$

is also a solution of (3.7.1). The differential equation (3.7.2) will be called *associated* with the right divisor L_1 of L.

We say that right divisors $L_1,\ldots,L_r$ of L form a *complete set* if every solution of (3.7.1) is a sum of solutions of differential equations associated with $L_1,\ldots,L_r$. The right divisors $L_1,\ldots,L_r$ are said to form a *minimal set* if the equality

$$v_1(t)+\cdots+v_r(t) \equiv 0$$

where $v_j(t)$ is a solution of the differential equation associated with L_j for $j = 1,\ldots,r$, implies that each $v_j(t)$ is identically zero.

Let (X,T) be a right spectral pair of $L(\lambda)$, where $T = C_L$ is the companion operator for L, and let $\mathcal{M}_j$ be the T-invariant supporting subspace associated with the right divisor L_j of L. Then $(X|_{\mathcal{M}_j},T|_{\mathcal{M}_j})$ is a right spectral pair of L_j (see Theorem 2.7.1). Since by Theorem 2.11.1 every solution of the differential equation associated with L (resp. with L_j) is of the form

$$u(t) = Xe^{tT}c,$$

(resp. $u_j(t) = X|_{\mathcal{M}_j} \exp(tT|_{\mathcal{M}_j})c_j$), it follows that $L_1, \ldots, L_r$ form a complete set if and only if

$$\mathcal{M}_1 + \cdots + \mathcal{M}_r = \mathcal{X}^\ell,$$

and $L_1, \ldots, L_r$ form a minimal set if and only if $\mathcal{M}_1, \ldots, \mathcal{M}_r$ are linearly independent. Now appeal to Proposition 3.5.1 and 3.5.4 in order to derive the descriptions of minimal and complete sets in terms of Vandermonde operators:

THEOREM 3.7.1. (a) *The right divisors* $L_1, \ldots, L_r$ *form a minimal set if and only if*

$$\operatorname{Ker} V_\ell(L_1, \ldots, L_r) = \{0\};$$

where ℓ *is the degree of* L.

(b) *If* $V_\ell(L_1, \ldots, L_r)$ *is right invertible, then the right divisors* $L_1, \ldots, L_r$ *form a complete set; in case* $\mathcal{X}$ *is a Hilbert space the right invertibility of* $V_\ell(L_1, \ldots, L_r)$ *is equivalent to the completeness of the set* $L_1, \ldots, L_r$.

(c) *The set of right divisors* $L_1, \ldots, L_r$ *is minimal and complete if and only if* $V_\ell(L_1, \ldots, L_r)$ *is invertible.*

An important particular case appears when all divisors $L_1, \ldots, L_r$ are of the first degree:

COROLLARY 3.7.2. *Let* $Z_1, \ldots, Z_r \in L(\mathcal{X})$ *be such that*

$$Z_j^\ell + \sum_{k=0}^{\ell-1} A_k Z_j^k = 0, \quad j = 1, \ldots, r.$$

Then, (i) *if the operator*

$$(3.7.3) \qquad \begin{bmatrix} I & I & \cdots & I \\ Z_1 & Z_2 & \cdots & Z_r \\ \vdots & & & \vdots \\ Z_1^{\ell-1} & Z_2^{\ell-1} & \cdots & Z_r^{\ell-1} \end{bmatrix}$$

is right invertible, then every solution $u(t)$ *of* (3.7.1) *can be written in the form*

$$u(t) = e^{tZ_1}c_1 + \cdots + e^{tZ_r}c_r \tag{3.7.4}$$

for some $c_1,\ldots,c_r \in \mathcal{X}$. *Conversely, if* $\mathcal{X}$ *is a Hilbert space and every solution of* (3.7.1) *has the form* (3.7.4), *then the operator* (3.7.3) *is right invertible.*

(ii) *The equality*

$$e^{tZ_1}c_1 + \cdots + e^{tZ_r}c_r \equiv 0$$

for some $c_1,\ldots,c_r \in \mathcal{X}$ *implies that* $c_1 = \cdots = c_r = 0$ *if and only if the operator* (3.7.3) *is left invertible.*

(iii) *Every solution* $u(t)$ *of* (3.7.1) *can be written in the form* (3.7.4) *with uniquely determined* $c_1,\ldots,c_r$ *if and only if the operator* (3.7.3) *is invertible.*

3.8 Interpolation problem

Let $L_1,\ldots,L_r$ be monic operator polynomials with coefficients in $L(\mathcal{X})$, of degrees $k_1,\ldots,k_r$, respectively. We study here the following interpolation problem: Given operators $R_{j0},\ldots,R_{j,k_j-1} \in L(\mathcal{X})$ find a monic operator polynomial $L(\lambda)$ such that for some (necessarily monic) operator polynomials $M_1,\ldots,M_r$ we have

$$L(\lambda) = M_j(\lambda)L_j(\lambda) + \sum_{p=0}^{k_j-1} \lambda^p R_{jp} \tag{3.8.1}$$

for $j = 1,\ldots,r$. The problem of finding monic left common multiple is a particular case of this problem when all operators R_{jp} are zeros.

Write

$$L(\lambda) = \lambda^m I + \sum_{p=1}^{m-1} \lambda^p A_p.$$

By Theorem 2.6.2, the equalities (3.8.1) hold if and only if

(3.8.2) $$[A_0 \cdots A_{m-1}]V_m(L_1,\dots,L_r) = [R_1 R_2 \cdots R_r] - W_{m+1}(L_1,\dots,L_r),$$

where

$$R_j = [R_{j0} R_{j1} \cdots R_{jk_j-1}]$$

and $W_{m+1}(L_1,\dots,L_r)$ is the last operator row in $V_{m+1}(L_1,\dots,L_r)$ (so $W_{m+1}(L_1,\dots,L_r) \in L(\mathcal{X}^{k_1+\cdots+k_r},\mathcal{X})$). Because of the formula (3.8.2), one can study the interpolation problem in terms of the Vandermonde operators $V_m(L_1,\dots,L_r)$. We present here several results based on this approach.

THEOREM 3.8.1. *If* $V_m(L_1,\dots,L_r)$ *is left invertible, then for any choice of operators* $R_{j0},\dots,R_{jk_j-1}$, $j = 1,\dots,r$ *there exists a monic operator polynomial* $L(\lambda)$ *of degree* m *such that* (3.8.1) *holds.*

Conversely, if for any choice of R_{jp} *there exists a monic operator polynomial* $L(\lambda)$ *of degree* m *satisfying* (3.8.1), *then* $V_m(L_1,\dots,L_r)$ *is regular. If, in addition, the space* $\mathcal{X}^k$ (*where* $k = k_1+\cdots+k_r$) *is isomorphic to* $\mathcal{X}$, *i.e., there exists an invertible* $C \in L(\mathcal{X}^k,\mathcal{X})$, *then* $V_m(L_1,\dots,L_r)$ *is left invertible.*

In view of (3.8.2), the theorem is just an application of the following lemma.

LEMMA 3.8.2. *Let* $\mathcal{X},\mathcal{Y}$ *be Banach spaces and let* $A \in L(\mathcal{X}^k,\mathcal{Y})$, *where* k *is a fixed positive integer. If the equation*

(3.8.3) $$XA = B$$

has a solution $X \in L(\mathcal{Y},\mathcal{X})$ *for any* $B \in L(\mathcal{X}^k,\mathcal{X})$, *then* A *is regular. Moreover, if, in addition, the Banach space* $\mathcal{X}$ *is such that* $\mathcal{X}^k$ *is isomorphic to* $\mathcal{X}$, *then* A *is left invertible. Conversely, if* A *is left invertible, then* (3.8.3) *has a solution* $X \in (\mathcal{Y},\mathcal{X})$ *for every* $B \in L(\mathcal{X}^k,\mathcal{X})$.

PROOF. The converse statement is evident: if $\hat{A}$ is a left inverse of A, then $X = B\hat{A}$ is a solution of (3.8.3).

Assume now that (3.8.3) has a solution for every B. Arguing by contradiction, assume A is not regular, so there is sequence $\{x_n\}_{n=0}^{\infty}$ in $\mathcal{X}^k$ such that $\|x_n\| = 1$, but $\|Ax_n\| \to 0$. Denote by x_{nj} $(1 \leq j \leq k)$ the jth coordinate of x_n. There exist $\varepsilon > 0$ and j_0 $(1 \leq j_0 \leq k)$ such that

$$\inf_p \|x_{n_p j_0}\| \geq \varepsilon$$

for some infinite subsequence $n_1, n_2, \ldots$. Let $B_0 \in L(\mathcal{X}^k, \mathcal{X})$ be the projection on the j_0^{th} coordinate, and let $X_0 \in L(\mathcal{Y}, \mathcal{X})$ be such that $X_0 A = B_0$. Then

$$\|B_0 x_{n_p}\| \geq \varepsilon_0 \quad \text{for all } p,$$

but $\|X_0 A x_{n_p}\| \to 0$, a contradiction.

Assume now, in addition, that $\mathcal{X}^k$ is isomorphic to $\mathcal{X}$. Then without loss of generality, we can assume $k = 1$. Taking $B = I$, it follows that A is left invertible. ∎

Many important Banach spaces $\mathcal{X}$ (e.g., $L_p[0,1]$, $1 \leq p \leq \infty$) have the property that $\mathcal{X}^k$ is isomorphic to $\mathcal{X}$ for any positive integer k. However, there are infinite dimensional Banach spaces $\mathcal{X}$ which are not isomorphic to $\mathcal{X}^2$ (see Figiel [1]).

Using Theorem 3.8.1 and results exposed in this chapter, one can deduce various results concerning solutions of the interpolation problem. For example:

THEOREM 3.8.2. *Let $\mathcal{X}$ be a Hilbert space, and assume that $\sigma(L_i) \cap \sigma(L_j) = \phi$ for $i \neq j$. Then for any choice of R_{jp} $(0 \leq p \leq k_j - 1,\ 1 \leq j \leq r)$ there is a monic operator polynomial $L(\lambda)$ satisfying (3.8.1). Moreover, the degree m of $L(\lambda)$ can be chosen independent of R_{jp}.*

PROOF. Combine Corollary 3.6.3 and Theorem 3.8.1. ∎

3.9 Exercises

<u>Ex. 3.1</u>. Let $L_1(\lambda) = \lambda I - X$, $L_2(\lambda) = \lambda I - Y$, where the operators X and Y are given by the following infinite matrices:

$$X = \begin{bmatrix} 0 & 1 & & & & & \\ 1 & 0 & & & & & \\ & & 0 & 1 & & & \\ & & 1 & 0 & & & \\ & & & & 0 & 1 & \\ & & & & 1 & 0 & \\ & & & & & & \ddots \end{bmatrix}, \quad Y = \begin{bmatrix} 0 & 0 & 0 & & & & & \\ 1 & 0 & -1 & 0 & 0 & & & \\ & & 0 & 0 & 0 & & & \\ & & 1 & 0 & -1 & 0 & 0 & \\ & & & & 0 & 0 & 0 & \\ & & & & 1 & 0 & -1 & \\ & & & & & & & \ddots \end{bmatrix}.$$

Verify that $V_2(L_1,L_2)$ is right invertible and $\mathrm{Ker}\ V_2(L_1,L_2) \neq \{0\}$ but L_1 and L_2 do not have a common monic left multiple of degree 2. Hint: Show that $\mathrm{ind}_1(L_1,L_2) = 3$ and use Theorem 3.2.1.

Ex. 3.2. Let $L_1(\lambda) = \lambda I-A$, $L_2(\lambda) = \lambda I-B$, where A and B are operators on $\mathcal{X} = L_2(0,1) \oplus L_2(0,1)$ defined as follows:

$$[A(\phi_1,\phi_2)](t) = ((\alpha t+1)\phi_2(t),\beta\phi_1(t)), \quad 0 < t < 1;$$

$$[B(\phi_1,\phi_2)](t) = ((\beta t+1)\phi_2(t),\alpha\phi_1(t)), \quad 0 < t < 1;$$

here $\alpha \neq \beta$ are fixed complex numbers. Prove that $\mathrm{ind}_1(L_1,L_2) = 2$ and $\mathrm{ind}_2(L_1,L_2) = 3$.

Ex. 3.3. Find a common monic left multiple of degree 3 of the polynomials $L_1(\lambda)$ and $L_2(\lambda)$ given in Ex. 3.2. Is there a common monic left multiple of degree 2 of $L_1(\lambda)$ and $L_2(\lambda)$?

Ex. 3.4. Let $A,B \in L(\mathcal{X})$ be such that A-B is invertible. Show that there exists a common monic left multiple of degree 2 of $\lambda I-A$ and $\lambda I-B$.

Ex. 3.5. Let $\mathcal{X}_n = L_2(0,1) \oplus\cdots\oplus L_2(0,1)$ (n times, $n \geq 2$) and let α_n,β_n be two complex numbers such that

$$\alpha_n(\beta_n t+1)^{n-1} \neq \beta_n(\alpha_n t+1)^{n-1}, \quad 0 \leq t \leq 1.$$

Define $A_n,B_n \in L(\mathcal{X}_n)$ by

$$A_n(\phi_1,\ldots,\phi_n)(t) = ((\beta_n t+1)\phi_2(t),\ldots,(\beta_n t+1)\phi_n(t),\alpha_n\phi_1(t)),$$
$$0 < t < 1;$$

$$B_n(\phi_1,\dots,\phi_n)(t) = ((\alpha_n t+1)\phi_2(t),\dots,(\alpha_n t+1)\phi_n(t),\beta_n\phi_1(t)),$$
$$0 < t < 1.$$

Prove that $\mathrm{ind}_1(\lambda I-A_n,\lambda I-B_n) = 2$ and $\mathrm{ind}_2(\lambda I-A_n,\lambda I-B_n) = n+1$.

Ex. 3.6. Let

$$\mathcal{X} = \mathcal{X}_2 \oplus \mathcal{X}_3 \oplus \cdots,$$

where $\mathcal{X}_n$ is the Hilbert space introduced in Ex. 3.5, and let

$$Z_1 = A_2 \oplus A_3 \oplus \cdots ; \qquad Z_2 = B_2 \oplus B_3 \oplus \cdots$$

be operators on $\mathcal{X}$, where A_n and B_n are given as Ex. 3.5 (with the additional constraint that the sequences $\{\alpha_n\}_{n=2}^{\infty}$ and $\{\beta_n\}_{n=2}^{\infty}$ are bounded, to ensure that Z_1 and Z_2 are bounded operators). Prove that $\mathrm{ind}_1(\lambda I-Z_1,\lambda I-Z_2) = 2$.

Ex. 3.7. Let $Z_1,Z_2 \in L(\mathcal{X})$ be defined as in Ex. 3.6. Prove that Z_1 and Z_2 are not right roots of any monic operator polynomial.

Ex. 3.8. Let $L(\lambda)$ be a monic operator polynomial of degree ℓ, and $L_1,\dots,L_r$ be monic right divisors of L. Prove that if $V_\ell(L_1,\dots,L_r)$ is invertible, then $\Sigma(L) = \bigcup_{j=1}^{r} \Sigma(L_j)$.

Ex. 3.9. Let $Z_1,\dots,Z_m$ be commuting operators in $L(\mathcal{X})$. Prove that $V_m(\lambda I-Z_1,\dots,\lambda I-Z_m)$ is invertible if and only if all the differences Z_i-Z_j $(i \neq j)$ are invertible.

Ex. 3.10. Let $Z_1,Z_2,Z_3 \in L(\mathcal{X})$ be such that all the commutators $Z_iZ_j-Z_jZ_i$ $(1 \leq i.j \leq 3)$ are compact. Assume, in addition, that $Z_2-Z_1, Z_3-Z_1, Z_3-Z_2$ are invertible. Prove that $V_3(\lambda I-Z_1,\lambda I-Z_2,\lambda I-Z_3)$ is Fredholm.

Ex. 3.11. Can the result of Ex. 3.10 be generalized to any finite number of operators (instead of 3)?

Ex. 3.12. Let monic operator polynomials $L_1(\lambda),\dots,L_r(\lambda)$ be direct sums:

$$L_j(\lambda) = \begin{bmatrix} L_{j1}(\lambda) & 0 \\ 0 & L_{j2}(\lambda) \end{bmatrix}, \quad j = 1,\dots,r,$$

with respect to the direct sum decomposition $\mathcal{X} = \mathcal{X}_1 \dot{+} \mathcal{X}_2$. Express $V_m(L_1,\dots,L_r)$ in terms of $V_m(L_{11},\dots,L_{r1})$ and $V_m(L_{12},\dots,L_{r2})$.

Ex. 3.13. Let $L_1,\dots,L_r$ be as in Ex. 3.12. Express $\mathrm{ind}_1(L_1,\dots,L_r)$ in terms of $\mathrm{ind}_1(L_{11},\dots,L_{r1})$ and $\mathrm{ind}_1(L_{12},\dots,L_{r2})$. Express $\mathrm{ind}_2(L_1,\dots,L_r)$ in terms of $\mathrm{ind}_2(L_{11},\dots,L_{r1})$ and $\mathrm{ind}_2(L_{12},\dots,L_{r2})$.

Ex. 3.14. For monic operator polynomials $L_1,\dots,L_r$ of degrees $k_1,\dots,k_r$, respectively, with coefficients in $L(\mathcal{X})$, introduce the *left Vandermonde operator* $V_m^{\ell}(L_1,\dots,L_r)$ as follows: Let (T_j, Y_j) be a left spectral pair for $L_j(\lambda)$ $(j = 1,\dots,r)$, and let

$$V_j = \left[Y_j, T_jY, \dots, T_j^{k_j-1}Y_j\right]^{-1}.$$

Then put

$$V_m^{\ell}(L_1,\dots,L_r) = \begin{bmatrix} V_1Y_1 & V_1T_1Y_1 & \cdots & V_1T_1^{m-1}Y_1 \\ V_2Y_2 & V_2T_2Y_2 & \cdots & V_2T_2^{m-1}Y_2 \\ \vdots & \vdots & & \vdots \\ V_rY_r & V_rT_rY_r & \cdots & V_rT_r^{m-1}Y_r \end{bmatrix},$$

where m is a positive integer. Prove that (assuming $\mathcal{X}$ is a Hilbert space)

$$V_m^{\ell}(L_1,\dots,L_r) = (V_m(L_1^*,\dots,L_r^*))^*,$$

where $L_j^*(\lambda)$ is the monic operator polynomial whose coefficients are adjoint to those of $L_j(\lambda)$.

Ex. 3.15. Prove that if $V_m^{\ell}(L_1,\dots,L_r)$ is right invertible, then there is a monic right common multiple of $L_1,\dots,L_r$ of degree m (for the definition of $V_m^{\ell}(L_1,\dots,L_r)$, see Ex. 3.14).

Ex. 3.16. In this and the next two exercises, $L(\lambda) = \lambda^{\ell} I + \sum_{j=0}^{\ell-1} \lambda^j A_j$ is a monic operator polynomial of degree ℓ, and $\lambda I - Z_1,\dots,\lambda I - Z_{\ell}$ are right divisors of $L(\lambda)$. Prove that if

$\sigma(Z_j) \cap \sigma(Z_k) = \phi$ for $j \neq k$ and if $\Sigma(L) = \bigcup_{j=1}^{\ell} \sigma(Z_j)$, then the operator

$$V \overset{\text{def}}{=} V_\ell(\lambda I - Z_1, \ldots, \lambda I - Z_\ell)$$

is left invertible.

Ex. 3.17. Verify that under the hypotheses of Ex. 3.16, one left inverse H of V is given by the formula

$$H = [H_{uv}]_{u,v=1}^{\ell},$$

where

$$H_{uv} = \sum_{m=1}^{\ell+1-v} \frac{1}{2\pi i} \int_{\Gamma_u} \lambda^{\ell+1-v-m} L(\lambda)^{-1} d\lambda A_{\ell-m+1},$$

and Γ_u is a suitable contour such that $\sigma(Z_u)$ is inside Γ_u and $\Sigma(L) \backslash \sigma(Z_u)$ is outside Γ_u.

Ex. 3.18. Prove that if $\sigma(Z_i) \cap \sigma(Z_j) = \phi$ for $i \neq j$, and V is invertible, then each Z_i is a right Γ_i-spectral divisor of $L(\lambda)$, where Γ_i is a contour that separates $\sigma(Z_i)$ from $\Sigma(L) \backslash \sigma(Z_i)$. Hint: Use the formulas for V^{-1} given in Ex. 3.17 to verify the equalities

$$\int_{\Gamma_i} \lambda^s L(\lambda)^{-1} d\lambda = Z_i^s \int_{\Gamma_i} L(\lambda)^{-1} d\lambda$$

for $i = 1, \ldots, \ell$ and $s = 1, \ldots, \ell-1$. Now use the result of Ex. 2.13.

3.10 Notes

This chapter (with the exception of Theorem 3.2.5, the Appendix to Section 3.6 and Section 3.7) is based on Gohberg-Kaashoek-Rodman [1,2]. In the Appendix to Section 3.6, the well known and widely used concept of hull of operators is exposed (see Hirschfeld [1], Berberian [1]). The results of Ex. 3.1-3.7 appear in Gohberg-Kaashoek-Rodman [2]. The results of Ex. 3.17 and 3.18 appear in Markus-Mereutsa [1].

The notion of Vandermonde operators was used in the study of operator polynomials by Markus-Mereutsa [1], Mereutsa [1]. More recently, Vandermonde operators are widely used in the analysis of matrix polynomials and in the linear systems theory (see, e.g., Gohberg-Kaashoek-Lerer-Rodman [1], Gohberg-Lancaster-Rodman [2], Fuhrmann [1], Helton-Rodman [1]).

CHAPTER 4: STABLE FACTORIZATIONS OF MONIC OPERATOR POLYNOMIALS

Let $L(\lambda)$ be a monic operator polynomial with factorization

$$L(\lambda) = L_2(\lambda)L_1(\lambda),$$

where $L_1(\lambda)$ and $L_2(\lambda)$ are monic operator polynomials. In this chapter, we address the problem of stability of such factorizations, that is, when any operator polynomial sufficiently close to $L(\lambda)$ admits a factorization with factors as close to $L_2(\lambda)$ and $L_1(\lambda)$, respectively, as we wish. Here the closeness of monic operator polynomials is understood as closeness of their respective coefficients.

Our approach to this problem is based on the description of factorizations of monic operator polynomials in terms of supporting invariant subspaces (Section 2.7). Thus, we translate the problem of stability of factorizations into the problem of stability of supporting invariant subspaces. To make this approach work, we develop the necessary background on subspaces in a Banach space and on stability of invariant subspaces in the first five sections of this chapter.

4.1 The metric space of subspaces in a Banach space

Let $\mathcal{X}$ be a Banach space. Given (closed) non-zero subspaces $\mathcal{M},\mathcal{N}$ in $\mathcal{X}$ define the *spherical gap* $\theta_s(\mathcal{M},\mathcal{N})$ between $\mathcal{M}$ and $\mathcal{N}$ by the following formula:

$$\theta_s(\mathcal{M},\mathcal{N}) = \max\left\{ \sup_{\substack{x\in\mathcal{M}\\ \|x\|=1}} \inf_{\substack{y\in\mathcal{N}\\ \|y\|=1}} \|y-x\|,\ \sup_{\substack{x\in\mathcal{N}\\ \|x\|=1}} \inf_{\substack{y\in\mathcal{M}\\ \|y\|=1}} \|y-x\| \right\}.$$

If one (or both) of $\mathcal{M},\mathcal{N}$ is the zero subspace we define

$$\theta_s(\mathcal{M},\{0\}) = \theta_s(\{0\},\mathcal{M}) = \begin{cases} 0 & \text{if } \mathcal{M} = \{0\}, \\ 1 & \text{if } \mathcal{M} \neq \{0\}. \end{cases}$$

Clearly, $\Theta_s(\mathcal{M},\mathcal{N})$ is symmetric: $\Theta_s(\mathcal{M},\mathcal{N}) = \Theta_s(\mathcal{N},\mathcal{M})$ for all $\mathcal{M},\mathcal{N}$; and non-negative $\Theta_s(\mathcal{M},\mathcal{N}) \geq 0$ for all $\mathcal{M},\mathcal{N}$. Moreover, one checks easily that $\Theta_s(\mathcal{M},\mathcal{N}) = 0$ if and only if $\mathcal{M} = \mathcal{N}$. Observe also that always

$$\Theta_s(\mathcal{M},\mathcal{N}) \leq 2. \tag{4.1.1}$$

The spherical gap satisfies the triangle inequality:

$$\Theta_s(\mathcal{L},\mathcal{N}) \leq \Theta_s(\mathcal{L},\mathcal{M}) + \Theta_s(\mathcal{M},\mathcal{N}) \tag{4.1.2}$$

for any subspaces $\mathcal{M},\mathcal{L},\mathcal{N}$ in $\mathcal{X}$. Indeed, (4.1.2) is trivial in view of (4.1.1) if at least one of $\mathcal{M},\mathcal{L},\mathcal{N}$ is the zero subspace. So we can assume that all $\mathcal{M},\mathcal{L},\mathcal{N}$ are non-zero subspaces. Given $x \in \mathcal{N}$, $\|x\| = 1$, let $z_x \in \mathcal{M}$, $\|z_x\| = 1$ be such that

$$\|x-z_x\| \leq \inf_{\substack{v\in\mathcal{M}\\ \|v\|=1}} \|x-v\|+\delta,$$

where $\delta > 0$ is a positive number chosen in advance. Then for every $y \in \mathcal{L}$, $\|y\| = 1$, we have

$$\|x-y\| \leq \|x-z_x\|+\|z_x-y\| = \inf_{\substack{v\in\mathcal{M}\\ \|v\|=1}} \|x-v\|+\delta+\|z_x-y\|.$$

Taking the infimum with respect to y it follows that

$$\inf_{\substack{y\in\mathcal{L}\\ \|y\|=1}} \|x-y\| \leq \Theta_s(\mathcal{N},\mathcal{M})+\Theta_s(\mathcal{M},\mathcal{L})+\delta.$$

Taking the supremum with respect to x and repeating the argument with the roles of $\mathcal{L}$ and $\mathcal{N}$, we obtain

$$\Theta_s(\mathcal{L},\mathcal{N}) \leq \Theta_s(\mathcal{N},\mathcal{M})+\Theta_s(\mathcal{M},\mathcal{L})+\delta,$$

and since $\delta > 0$ was chosen arbitrarily, the inequality (4.1.2) is proved.

Thus, the spherical gap is a metric on the set of all subspaces in $\mathcal{X}$, thereby making this set into the metric space.

THEOREM 4.1.1. *The metric space of all subspaces in* $\mathcal{X}$ *is complete. In other words, if* $\{\mathcal{M}_n\}_{n=1}^{\infty}$ *is a sequence of subspaces in* $\mathcal{X}$ *with*

$$\theta_s(\mathcal{M}_m,\mathcal{M}_n) \to 0 \quad \text{as } m,n \to \infty, \tag{4.1.3}$$

then there is a subspace $\mathcal{M}$ *in* $\mathcal{X}$ *such that*

$$\lim_{n\to\infty} \theta_s(\mathcal{M}_n,\mathcal{M}) = 0.$$

PROOF. Let $\{\mathcal{M}_n\}_{n=1}^{\infty}$ be a sequence of subspaces satisfying (4.1.3). A standard argument (using the triangle inequality (4.1.2)) shows that it will suffice to exhibit a converging subsequence of $\{\mathcal{M}_n\}_{n=1}^{\infty}$. Passing to a subsequence we can assume that

$$\theta_s(\mathcal{M}_n,\mathcal{M}_{n+1}) < 2^{-n} \quad (n = 1,2,\ldots). \tag{4.1.4}$$

Given $x_1 \in \mathcal{M}_1$ with $\|x_1\| = 1$, we can construct, using (4.1.4), a sequence of vectors $\{x_n\}_{n=1}^{\infty}$ such that

$$x_n \in \mathcal{M}_n,\ \|x_n\| = 1 \text{ and } \|x_n - x_{n+1}\| < 2^{-n} \tag{4.1.5}$$

for $n = 1,2,\ldots$. Clearly, $\{x_n\}_{n=1}^{\infty}$ is a Cauchy sequence in $\mathcal{X}$, and hence there is a limit $x = \lim_{n\to\infty} x_n \in \mathcal{X}$. The properties (4.1.5) imply that

$$\|x\| = 1 \text{ and } \|x_n - x\| \le 2^{-n+1} \quad (n = 1,2,\ldots). \tag{4.1.6}$$

Let $\mathcal{N} \subset \mathcal{X}$ be the set of limit vectors x of all possible sequences $\{x_n\}$ satisfying the properties (4.1.5), let $\overline{\mathcal{N}}$ be the closure of $\mathcal{N}$ (in the norm topology in $\mathcal{X}$), and by $\mathcal{M}$ we denote the set of all finite linear combinations of elements from $\overline{\mathcal{N}}$.

First, we show that

$$\{x \in \mathcal{M} \mid \|x\| = 1\} = \overline{\mathcal{N}}. \tag{4.1.7}$$

Indeed, the inclusion $\supset$ follows from the definitions above. To prove the opposite inclusion we have to show that if $w = \alpha x+\beta y+\cdots+\gamma z$, where $x,y,\ldots,z \in \overline{\mathcal{N}}$, $\|w\| = 1$ and $\alpha,\beta,\ldots,\gamma$ are complex numbers, then $w \in \overline{\mathcal{N}}$. For simplicity of notation assume $w = \alpha x+\beta y$. Let $\{x_n\}_{n=1}^{\infty}$, $\{y_n\}_{n=1}^{\infty}$ be sequences of vectors from $\mathcal{N}$ such that $\lim_{n\to\infty} x_n = x$, $\lim_{n\to\infty} y_n = y$. For each n, let $\{x_n^{(m)}\}_{m=1}^{\infty}$ be a sequence with the properties that

$$x_n^{(m)} \in \mathcal{M}_m, \ \|x_n^{(m)}\| = 1, \ \|x_n^{(m)}-x_n^{(m+1)}\| < 2^{-m}$$

and $\lim_{m\to\infty} x_n^{(m)} = x_n$. Let $\{y_n^{(m)}\}_{m=1}^{\infty}$ be a sequence with analogous properties for y_n.

Denote $w_n = \alpha x_n^{(n)}+\beta y_n^{(n)}$. Then

$$\lim_{n\to\infty} w_n = w, \tag{4.1.8}$$

because

$$\|w-w_n\| \leq \|w-(\alpha x_n+\beta y_n)\|+|\alpha| \ \|x_n-x_n^{(n)}\|+|\beta| \ \|y_n-y_n^{(n)}\|$$

$$\leq \|w-(\alpha x_n+\beta y_n)\|+(|\alpha|+|\beta|)2^{-n+1},$$

where in the last inequality we have used the property (4.1.6).

Using (4.1.4), choose for each n a sequence $\{t_k^{(n)}\}_{k=1}^{\infty}$ such that $\|t_k^{(n)}\| = 1$, $t_k^{(n)} \in \mathcal{M}_k$, $\|t_k^{(n)}-t_{k+1}^{(n)}\| < 2^{-k}$ for $k = 1,2,\ldots$, and $t_n^{(n)} = \|w_n\|^{-1}w_n$. Let $t^{(n)} = \lim_{k\to\infty} t_k^{(n)}$. As $\|w\| = 1$, it follows from (4.1.8) that $\lim_{n\to\infty} t_n^{(n)} = w$. Consequently, also $\lim_{n\to\infty} t^{(n)} = w$ (because $\|t_n^{(n)}-t^{(n)}\| \leq 2^{-n+1}$ for $n = 1,2,\ldots$). This proves that $w \in \overline{\mathcal{N}}$, and (4.1.7) follows.

In particular, the set $\{x \in \mathcal{M} \mid \|x\| = 1\}$ is closed, and therefore $\mathcal{M}$ is a (closed) subspace in $\mathcal{X}$.

Finally, let us prove that $\lim_{n\to\infty} \Theta_s(\mathcal{M}_n, \mathcal{M}) = 0$. Let $y \in \mathcal{M}_n$, $\|y\| = 1$. Choose a sequence $\{x_k\}_{k=1}^{\infty}$ such that $x_k \in \mathcal{M}_k$, $\|x_k\| = 1$, $\|x_k - x_{k+1}\| < 2^{-k}$ for $k = 1, 2, \ldots,$ and $x_n = y$. Let $x = \lim_{k\to\infty} x_k$. Then $x \in \mathcal{N}$; so $x \in \mathcal{M}$ and $\|x\| = 1$. We have $\|y-x\| \leq 2^{-n+1}$, hence

$$\sup_{\substack{y\in\mathcal{M} \\ \|y\|=1}} \inf_{\substack{x\in\mathcal{M} \\ \|x\|=1}} \|y-x\| \leq 2^{-n+1}. \tag{4.1.9}$$

On the other hand, let $x \in \mathcal{N}$ with a sequence $\{x_n\}_{n=1}^{\infty}$ such that $x_n \in \mathcal{M}_n$, $\|x_n\| = 1$, $\|x_n - x_{n+1}\| < 2^{-n}$ and $\lim_{n\to\infty} x_n = x$. Then $\|x - x_n\| \leq 2^{-n+1}$, so

$$\sup_{x\in\mathcal{N}} \inf_{\substack{y\in\mathcal{M}_n \\ \|y\|=1}} \|x-y\| \leq 2^{-n+1}.$$

Because of (4.1.7) we have also

$$\sup_{\substack{x\in\mathcal{M} \\ \|x\|=1}} \inf_{\substack{y\in\mathcal{M}_n \\ \|y\|=1}} \|x-y\| \leq 2^{-n+1}. \tag{4.1.10}$$

The inequalities (4.1.9) and (4.1.10) imply $\Theta_s(\mathcal{M}_n, \mathcal{M}) \leq 2^{-n+1}$, and we are done. ∎

Often another measure of distance between subspaces in a Banach space, called the *gap*, is used. Given subspaces $\mathcal{L}, \mathcal{M}$ in a Banach space $\mathcal{X}$, the gap $\Theta(\mathcal{L}, \mathcal{M})$ is defined by $\Theta(\mathcal{L}, \mathcal{M}) =$

$$\max\left\{ \sup_{\substack{x\in\mathcal{M} \\ \|x\|=1}} \inf_{y\in\mathcal{L}} \|y-x\|, \sup_{\substack{x\in\mathcal{L} \\ \|x\|=1}} \inf_{y\in\mathcal{M}} \|y-x\| \right\}$$

if both $\mathcal{L}$ and $\mathcal{M}$ are non-zero, and

$$\Theta(\{0\}, \mathcal{M}) = \Theta(\mathcal{M}, \{0\}) = \begin{cases} 1 & \text{if } \mathcal{M} \neq \{0\}, \\ 0 & \text{if } \mathcal{M} = \{0\}. \end{cases}$$

One checks immediately that $\Theta(\mathcal{L}, \mathcal{M}) = \Theta(\mathcal{M}, \mathcal{L})$ and that $\Theta(\mathcal{M}, \mathcal{L}) = 0$ if and only if $\mathcal{M} = \mathcal{L}$. However, the gap does not satisfy in general the triangle inequality (a simple explicit example of this situation with two-dimensional Banach space $\mathcal{X}$ is found in Gohberg-Lancaster-Rodman [3] (Example 13.8.3)), so one cannot always introduce a metric on the set of subspaces in $\mathcal{X}$ based on the function $\Theta(\mathcal{L},\mathcal{M})$.

It turns out that the gap and the spherical gap are not far away from each other. Namely,

$$\Theta(\mathcal{L},\mathcal{M}) \leq \Theta_s(\mathcal{L},\mathcal{M}) \leq 2\Theta(\mathcal{L},\mathcal{M}) \tag{4.1.11}$$

for any two subspaces $\mathcal{L},\mathcal{M} \subset \mathcal{X}$. Indeed, the left-hand inequality of (4.1.11) is evident from the definitions of $\Theta(\mathcal{L},\mathcal{M})$ and $\Theta_s(\mathcal{L},\mathcal{M})$. To prove the right-hand inequality in (4.1.11), it is sufficient to verify that for every $v \in \mathcal{X}$ with $\|v\| = 1$ and every subspace $\mathcal{M} \subset \mathcal{X}$, we have

$$\inf_{\substack{x\in\mathcal{M}\\ \|x\|=1}} \|v-x\| \leq 2 \inf_{x\in\mathcal{M}}\|v-x\|. \tag{4.1.12}$$

For a given $\varepsilon > 0$ there exists a $x \in \mathcal{M}$ such that

$$\|v-x\| < \inf_{y\in m}\|v-y\|+\varepsilon, \tag{4.1.13}$$

and we can assume that $x \neq 0$. (Otherwise, replace x by a nonzero vector sufficiently close to zero so that (4.1.13) still holds.) Then $x_0 \overset{\text{def}}{=} x/\|x\|$ has norm one and hence

$$\inf_{\substack{y\in\mathcal{M}\\ \|y\|=1}} \|v-y\| \leq \|v-x_0\| \leq \|v-x\|+\|x-x_0\|.$$

But

$$\|x-x_0\| = \big|\,\|x\|-1\,\big| = \big|\,\|x\|-\|v\|\,\big| \leq \|x-v\|,$$

and we have

$$\inf_{\substack{y\in\mathcal{M}\\ \|y\|=1}} \|v-y\| \leq 2\|x-v\| < 2 \inf_{y\in\mathcal{M}}\|v-y\|+2\varepsilon.$$

As $\varepsilon > 0$ is arbitrary, the desired inequality (4.1.11) follows.

If $\mathcal{X}$ is a Hilbert space we have

$$(4.1.14) \qquad \Theta(\mathcal{M},\mathcal{N}) = \|P_{\mathcal{M}}-P_{\mathcal{N}}\|,$$

where $P_{\mathcal{M}}$ (resp. $P_{\mathcal{N}}$) is the orthogonal projector on $\mathcal{M}$ (resp. $\mathcal{N}$), so in this case the gap is clearly a metric. Since (4.1.14) will not be used in the sequel, we do not prove (4.1.14) here; it is proved, for example, in Gohberg-Lancaster-Rodman [3] (Theorem 13.1.1), and in Kato [1].

The inequalities (4.1.11) show that the metric $\Theta_s(\mathcal{L},\mathcal{M})$ and the gap $\Theta(\mathcal{L},\mathcal{M})$ define the same topology on the set of all subspaces of a Banach space.

We conclude this section with the following important fact.

THEOREM 4.1.2. *Let $\mathcal{X}$ be a Banach space and $T \in L(\mathcal{X})$. The set* Inv(T) *of all* T*-invariant subspaces is closed in the spherical gap metric, i.e., if $\lim_{m\to\infty} \Theta_s(\mathcal{M}_m,\mathcal{M}) = 0$, where $\mathcal{M}_m \in$ Inv(T), then also $\mathcal{M} \in$ Inv(T).*

PROOF. We use the description of $\mathcal{M}$ given in the proof of Theorem 4.1.1. Without loss of generality assume

$$\Theta_s(\mathcal{M}_n,\mathcal{M}_{n+1}) < 2^{-n}, \quad n = 1,2,\dots .$$

In the notation of the proof of Theorem 4.1.1, we have only to verify that $Tx \in \mathcal{M}$ for every vector $x \in \mathcal{M}$. We can assume that $Tx \neq 0$.

Let $x = \mathcal{N}$, $\|x\| = 1$, and let $\{x_n\}_{n=1}^{\infty}$ be a sequence with the properties that $x_n \in \mathcal{M}_n$, $\|x_n\| = 1$, $\|x_n-x_{n+1}\| < 2^{-n}$ and $x = \lim_{n\to\infty} x_n$. We obviously have $Tx = \lim_{n\to\infty} Tx_n$, $Tx_n \in \mathcal{M}_n$. Let $y_n = \frac{Tx_n}{\|Tx_n\|}$ (observe that $Tx \neq 0$ implies $Tx_n \neq 0$, at least for sufficiently large n). Pick an integer p satisfying $\frac{2\|T\|}{\|Tx\|} \leq 2^p$, and let $\mathcal{N}_n = \mathcal{M}_{n+p}$. Then $y_{n+p} \in \mathcal{N}_n$, $\|y_{n+p}\| = 1$ and, for sufficiently large n, we have

$$\|y_{n+p}-y_{n+p+1}\| = \left\| \frac{Tx_{n+p}}{\|Tx_{n+p}\|} - \frac{Tx_{n+p+1}}{\|Tx_{n+p+1}\|} \right\|$$

$$\le \frac{2\|T\|}{\|Tx\|} \|x_{n+p}-x_{n+p+1}\| < 2^{-n}.$$

So, again by the proof of Theorem 4.1.1, the vector $y = \lim_{n\to\infty} y_{n+p}$ belongs to $\lim_{n\to\infty} \mathcal{N}_n = \mathcal{M}$, where the limit is understood in the spherical gap metric. As $y = \frac{Tx}{\|Tx\|}$, the inclusion $Tx \in \mathcal{M}$ follows. ■

4.2 Spherical gap and direct sums

An important property of the spherical gap metric is that it behaves well with respect to direct sums:

THEOREM 4.2.1. *Let* $\mathcal{M} \dotplus \mathcal{N} = \mathcal{X}$ *be a direct sum of two subspaces* $\mathcal{M}$ *and* $\mathcal{N}$. *Then there exists* $\varepsilon > 0$ *such that*

$$\mathcal{M}' \dotplus \mathcal{N}' = \mathcal{X}$$

for any subspaces $\mathcal{M}'$ *and* $\mathcal{N}'$ *satisfying*

$$\theta_s(\mathcal{M},\mathcal{M}')+\theta_s(\mathcal{N},\mathcal{N}') < \varepsilon.$$

We shall actually prove a quantitative version of this result which not only establishes the existence of ε with the properties required in Theorem 4.2.1, but also shows a way to compute one such ε. To this end we need the notion of a minimal angle between subspaces in $\mathcal{X}$.

Given two subspaces $\mathcal{L},\mathcal{M} \subset \mathcal{X}$, the *minimal angle* $\phi_{min}(\mathcal{L},\mathcal{M})$ $(0 \le \phi_{min}(\mathcal{L},\mathcal{M}) \le \pi/2)$ between $\mathcal{L}$ and $\mathcal{M}$ is determined by

$$\sin \phi_{min}(\mathcal{L},\mathcal{M}) = \inf\{\|x+y\| \mid x \in \mathcal{L},\ y \in \mathcal{M},\ \max\{\|x\|,\|y\|\} = 1\}.$$

PROPOSITION 4.2.2. *For subspaces* $\mathcal{M},\mathcal{N} \subset \mathcal{X}$ *the inequality* $\phi_{min}(\mathcal{M},\mathcal{N}) > 0$ *holds if and only if* $\mathcal{M} \cap \mathcal{N} = \{0\}$ *and the sum* $\mathcal{M}+\mathcal{N}$ *is closed.*

PROOF. If $\phi_{min}(\mathcal{M},\mathcal{N}) > 0$, then $\mathcal{M} \cap \mathcal{N} = \{0\}$ (otherwise there is $x \in \mathcal{M} \cap \mathcal{N}$, $\|x\| = 1$; hence

$$\sin \phi_{min}(\mathcal{M},\mathcal{N}) \leq \|x+(-x)\| = 0,$$

a contradiction). So without loss of generality we can assume that $\mathcal{M} \cap \mathcal{N} = \{0\}$. Introduce a norm $|\cdot|$ in the sum $\mathcal{M}+\mathcal{N}$ by the formula

$$|z| = \sup\{\|x\|, \|y\|\},$$

where the supremum is taken over all pairs $x \in \mathcal{M}$, $y \in \mathcal{N}$ such that $z = x+y$. It is not difficult to check that $\mathcal{M}+\mathcal{N}$ is complete with respect to the norm $|\cdot|$. On the other hand, $\|z\| \leq 2|z|$ for all $z \in \mathcal{M}+\mathcal{N}$. Therefore, the set $\mathcal{M}+\mathcal{N}$ is complete with respect to the original norm $\|\cdot\|$ if and only if the norms $|\cdot|$ and $\|\cdot\|$ are equivalent, i.e., there is constant $C > 0$ such that

$$(4.2.1) \qquad |z| \leq C\|z\| \quad \text{for all } z \in \mathcal{M}+\mathcal{N}.$$

It remains to observe that the completeness of $\mathcal{M}+\mathcal{N}$ in the norm $\|\cdot\|$ is equivalent to the closedness of $\mathcal{M}+\mathcal{N}$ in $\mathcal{X}$, while the condition (4.2.1) is equivalent to

$$(4.2.2) \qquad \sin \phi_{min}(\mathcal{M},\mathcal{N}) \geq \frac{1}{C}.$$

Let us verify the latter statement. Assume (4.2.1) holds. Choose $\varepsilon > 0$ and let $x \in \mathcal{M}$, $y \in \mathcal{N}$ be such that $\max\{\|x\|, \|y\|\} = 1$ and

$$\sin \phi_{min}(\mathcal{M},\mathcal{N}) > \|x+y\|-\varepsilon.$$

Denote $z = x+y$. Then obviously $|z| \geq 1$, so

$$\sin \phi_{min}(\mathcal{M},\mathcal{N}) > \|z\|-\varepsilon \geq \frac{\|z\|}{|z|} -\varepsilon \geq \frac{1}{C} - \varepsilon.$$

Letting $\varepsilon \to 0$, the inequality (4.2.2) follows. Conversely, if (4.2.2) holds, then $\|x+y\| \geq \frac{1}{C}$ for every $x \in \mathcal{M}$, $y \in \mathcal{N}$ with $\max\{\|x\|, \|y\|\} = 1$. For any $z \in \mathcal{M}+\mathcal{N}$ and any $\varepsilon > 0$ choose $x \in \mathcal{M}$, $y \in \mathcal{N}$ such that $z = x+y$ and $|z| > \max\{\|x\|, \|y\|\}+\varepsilon$. Then

$$|z|-\varepsilon \leq C\|z\|,$$

and letting $\varepsilon \to 0$ we obtain (4.2.1). ∎

PROPOSITION 4.2.3. *For any three subspaces $\mathcal{L},\mathcal{M},\mathcal{N} \subset \mathcal{X}$, the following inequality holds:*

$$\sin \phi_{min}(\mathcal{L},\mathcal{N}) \geq \sin \phi_{min}(\mathcal{L},\mathcal{M}) - \Theta_s(\mathcal{M},\mathcal{N}). \tag{4.2.3}$$

PROOF. We can assume that all three subspaces $\mathcal{L},\mathcal{M},\mathcal{N}$ are non-zero (otherwise (4.2.3) is trivial). Let $y_1 \in \mathcal{L}$ and $y_3 \in \mathcal{N}$ be arbitrary vectors satisfying $\max\{\|y_1\|,\|y_3\|\} = 1$. Letting ε be any fixed positive number, choose $y_2 \in \mathcal{M}$ such that $\|y_2\| = \|y_3\|$ and

$$\|y_3-y_2\| \leq (\Theta_s(\mathcal{M},\mathcal{N})+\varepsilon)\|y_3\| \leq \Theta_s(\mathcal{M},\mathcal{N})+\varepsilon.$$

Indeed, if $y_3 = 0$, choose $y_2 = 0$; if $y_3 \neq 0$, then the definition of $\Theta_s(\mathcal{M},\mathcal{N})$ allows us to choose a suitable y_2. Now

$$\|y_1+y_3\| \geq \|y_1+y_2\|-\|y_3-y_2\| \geq \sin \phi_{min}(\mathcal{L},\mathcal{M})-(\Theta_s(\mathcal{M},\mathcal{N})+\varepsilon).$$

As $\varepsilon > 0$ was arbitrary, the inequality (4.2.3) follows. ∎

Now we are ready to state and prove the following result of which Theorem 4.2.1 is an immediate corollary.

THEOREM 4.2.4. *Let $\mathcal{M} \dotplus \mathcal{N} = \mathcal{X}$ be a direct sum, where $\mathcal{M},\mathcal{N}$ are non-zero subspaces. Then* $\sin \phi_{min}(\mathcal{M},\mathcal{N}) > 0$, *and for every pair of subspaces $\mathcal{M}_1,\mathcal{N}_1 \subset \mathcal{X}$ such that*

$$\Theta_s(\mathcal{M},\mathcal{M}_1)+\Theta_s(\mathcal{N},\mathcal{N}_1) < \sin \phi_{min}(\mathcal{M},\mathcal{N}) \tag{4.2.4}$$

we have $\mathcal{M}_1 \dotplus \mathcal{N}_1 = \mathcal{X}$.

PROOF. The inequality $\sin \phi_{min}(\mathcal{M},\mathcal{N}) > 0$ follows from Proposition 4.2.2.

Assume now $\mathcal{M}_1,\mathcal{N}_1$ are such that (4.2.4) holds. In view of Proposition 4.2.3 we have

$$\sin \phi_{min}(\mathcal{M}_1,\mathcal{N}_1) \geq \sin \phi_{min}(\mathcal{M}_1,\mathcal{N})-\Theta_s(\mathcal{N},\mathcal{N}_1)$$

and

$$\sin \phi_{min}(\mathcal{M}_1,\mathcal{N}) \geq \sin \phi_{min}(\mathcal{N},\mathcal{M})-\Theta_s(\mathcal{M},\mathcal{M}_1).$$

Adding these inequalities, and using (4.2.4) we find that

$$\sin \phi_{min}(\mathcal{M}_1, \mathcal{N}_1) > 0$$

which implies by Proposition 4.2.2 that $\mathcal{M}_1 \cap \mathcal{N}_1 = \{0\}$ and $\mathcal{M}_1+\mathcal{N}_1$ is closed.

To prove that $\mathcal{M}_1+\mathcal{N}_1 = \mathcal{X}$, suppose first that $\mathcal{M} = \mathcal{M}_1$. Let $\varepsilon > 0$ be so small that

$$\frac{\Theta_s(\mathcal{N},\mathcal{N}_1)+\varepsilon}{\sin \phi_{min}(\mathcal{M},\mathcal{N})} = \delta < 1.$$

If $\mathcal{M}+\mathcal{N}_1 \neq \mathcal{X}$, then there exists a vector $x \in \mathcal{X}$ with $\|x\| = 1$ and $\|x-y\| > \delta$ for all $y \in \mathcal{M}+\mathcal{N}_1$. (Indeed, by Hahn Banach theorem choose a bounded linear functional ϕ on $\mathcal{X}$ such that $\|\phi\| = 1$ and $\phi(y) = 0$ for all $y \in \mathcal{M}+\mathcal{N}_1$; then choose $x \in \mathcal{X}$ so that $\|x\| = 1$ and $|\phi(x)| > \delta$.) We can represent the vector x as $x = y+z$, $y \in \mathcal{M}$, $z \in \mathcal{N}$. It follows from definition of $\sin \phi_{min}(\mathcal{M},\mathcal{N})$ that

$$\|z\| \leq (\sin \phi_{min}(\mathcal{M},\mathcal{N}))^{-1}.$$

Indeed, denoting $u = \max\{\|y\|,\|z\|\}$, we have

$$\begin{aligned}\sin \phi_{min}(\mathcal{L},\mathcal{M}) &\\ &= \inf\{\|x_1+x_2\| \mid x_1 \in \mathcal{L},\ x_2 \in \mathcal{M},\ \max\{\|x_1\|,\|x_2\|\} = 1\} \\ &\leq \|\frac{y}{u} + \frac{z}{u}\| = \frac{1}{u} \leq \frac{1}{\|z\|}.\end{aligned}$$

By the definition of $\Theta_s(\mathcal{L},\mathcal{M})$ we can find a vector z_1 from $\mathcal{N}_1$ with

$$\|z_1\| = \|z\|,\ \|z-z_1\| < [\Theta_s(\mathcal{N},\mathcal{N}_1)+\varepsilon]\|z\| \leq \delta.$$

The last inequality contradicts the choice of x, because

$z-z_1 = x-t$, where $t = y+z_1 \in \mathcal{M}\dot{+}\mathcal{N}_1$, and $\|x-t\| < \delta$.

Now consider the general case. Inequality (4.2.4) implies $\Theta_s(\mathcal{N},\mathcal{N}_1) < \sin \phi_{min}(\mathcal{M},\mathcal{N})$, and, in view of Proposition 4.2.3, $\Theta_s(\mathcal{M},\mathcal{M}_1) < \sin \phi_{min}(\mathcal{M},\mathcal{N}_1)$. Applying the part of Theorem 4.2.4 already proved, we obtain $\mathcal{M}+\mathcal{N}_1 = \mathcal{X}$, and then $\mathcal{M}_1+\mathcal{N}_1 = \mathcal{X}$. ∎

We also need the following result.

THEOREM 4.2.5. *Let* $\mathcal{M}\dot{+}\mathcal{M}' = \mathcal{X}$ *for some subspaces* $\mathcal{M}$ *and* $\mathcal{M}'$. *There exists constant* $K > 0$ *such that for all subspaces* $\mathcal{N}$ *with* $\Theta_s(\mathcal{M},\mathcal{N})$ *sufficiently small we have*

$$\frac{1}{2}\Theta_s(\mathcal{M},\mathcal{N}) \le \|P_{\mathcal{M}}-P_{\mathcal{N}}\| \le K\Theta_s(\mathcal{M},\mathcal{N}), \tag{4.2.5}$$

where $P_{\mathcal{M}}$ *(resp.* $P_{\mathcal{N}}$*) is the projection on* $\mathcal{M}$ *(resp.* $\mathcal{N}$*) along* $\mathcal{M}'$.

Observe that by Theorem 4.2.1 $\mathcal{N}$ is indeed a direct complement to $\mathcal{M}'$ (if $\Theta_s(\mathcal{M},\mathcal{N})$ is small enough), so $P_{\mathcal{N}}$ is correctly defined.

PROOF. In view of (4.1.11) we use $\Theta(\mathcal{M},\mathcal{N})$ in place of $\Theta_s(\mathcal{M},\mathcal{N})$ in the proof of the right-hand inequality in (4.2.5).

We prove first the following inequality (for $\Theta(\mathcal{M},\mathcal{N})$ small enough):

$$\|z+y\| \ge \frac{1}{2}\|P_{\mathcal{M}}\|^{-1}\|z\| \tag{4.2.6}$$

for all $z \in \mathcal{N}$ and $y \in \mathcal{M}'$. Without loss of generality assume $\|z\| = 1$. Suppose $\Theta(\mathcal{M},\mathcal{N}) < \delta$ (where δ is a positive number) and let $x \in \mathcal{M}$. Then

$$\|z+y\| \ge \|x+y\|-\|z-x\| \ge \|P_{\mathcal{M}}\|^{-1}\|x\|-\delta,$$

where in the last inequality we have used the definition of $\Theta(\mathcal{M},\mathcal{N})$ and the fact that $x = P_{\mathcal{M}}(x+y)$ (which implies

$$\|x+y\| \ge \|P_{\mathcal{M}}\|^{-1}\|x\|).$$

Further, $x = (x-z)+z$ implies $\|x\| \ge 1-\delta$, and so

$$\|z+y\| \ge \|P_{\mathcal{M}}\|^{-1}(1-\delta)-\delta,$$

and for δ small enough (4.2.6) follows.

Our second observation is that for any $x \in \mathcal{X}$

$$\|x-P_{\mathcal{M}}x\| \le K_3 d(x,\mathcal{M}), \tag{4.2.7}$$

where K_3 is a positive constant depending on $\mathcal{M}$ and $\mathcal{M}'$ only, and $d(x,\mathcal{M})$ is the distance between x and $\mathcal{M}$:

$$d(x,\mathcal{M}) = \inf_{y\in\mathcal{M}} \|x-y\|.$$

To establish (4.2.7), it is sufficient to consider the case that $x \in \operatorname{Ker} P_{\mathcal{M}} = \mathcal{M}'$ and $\|x\| = 1$. Indeed, assuming (4.2.7) is already proved for all $x \in \mathcal{M}'$, we have for $x \in \mathcal{X}$:

$$\|x-P_{\mathcal{M}}x\| = \|(x-P_{\mathcal{M}}x)-P_{\mathcal{M}}(x-P_{\mathcal{M}}x)\| \leq K_3 d(x-P_{\mathcal{M}}x,\mathcal{M}) = K_3 d(x,\mathcal{M}).$$

To verify (4.2.7) for $x \in \mathcal{M}'$, just put

$$K_3 = \sup_{\substack{x\in\mathcal{M}' \\ \|x\|=1}} \left[\frac{1}{d(x,\mathcal{M})}\right].$$

(As $\mathcal{M}$ and $\mathcal{M}'$ form a direct sum, one verifies easily that $K_3 < \infty$.)

Return now to the proof of (4.2.5). Assume that $\Theta_s(\mathcal{M},\mathcal{N})$ is small enough and let $x \in \mathcal{N}$, $\|x\| = 1$. By (4.2.7) we have (using (4.1.11))

$$\|(P_{\mathcal{M}}-P_{\mathcal{N}})x\| = \|x-P_{\mathcal{M}}x\| \leq K_3 d(x,\mathcal{M}) \leq K_3\Theta(\mathcal{M},\mathcal{N}) \leq K_3\Theta_s(\mathcal{M},\mathcal{N}).$$

Then, if $w \in \mathcal{X}$, $\|w\| = 1$ and $w = y+z$, $y \in \mathcal{N}$, $z \in \mathcal{M}'$,

$$\|(P_{\mathcal{M}}-P_{\mathcal{N}})w\| = \|(P_{\mathcal{M}}-P_{\mathcal{N}})y\| \leq \|y\|K_3\Theta_s(\mathcal{M},\mathcal{N}) \leq 2K_3\|P_{\mathcal{M}}\|\Theta_s(\mathcal{M},\mathcal{N}),$$

where the last inequality follows from (4.1.11). This proves the right-hand side inequality in (4.2.5).

It remains to prove the left-hand inequality in (4.2.5). Assume, for instance, that

$$\Theta(\mathcal{M},\mathcal{N}) = \sup_{\substack{x\in\mathcal{M} \\ \|x\|=1}} \inf_{y\in\mathcal{N}} \|x-y\|.$$

For given $\varepsilon > 0$ let $x_\varepsilon \in \mathcal{M}$, $\|x_\varepsilon\| = 1$ be such that

$$\inf_{y\in\mathcal{N}} \|x_\varepsilon-y\| > \Theta(\mathcal{M},\mathcal{N})-\varepsilon.$$

Then obviously

$$\|P_{\mathcal{M}}x_{\varepsilon}-P_{\mathcal{N}}x_{\varepsilon}\| > \Theta(\mathcal{M},\mathcal{N})-\varepsilon .$$

As

$$\|P_{\mathcal{M}}x_{\varepsilon}-P_{\mathcal{N}}x_{\varepsilon}\| \leq \|P_{\mathcal{M}}-P_{\mathcal{N}}\|,$$

by letting $\varepsilon \to 0$ and taking into account (4.1.11) the desired inequality follows. ■

4.3 Stable invariant subspaces

Let T be a (linear bounded) operator on a Banch space $\mathcal{X}$. A (closed) subspace $\mathcal{M} \subset \mathcal{X}$ is called *stable* T-*invariant* if $\mathcal{M}$ is T-invariant, and for every $\varepsilon > 0$ there is $\delta > 0$ such that every $S \in L(\mathcal{X})$ with $\|T-S\| < \delta$ has an invariant subspace $\mathcal{N}$ with the property that

$$\Theta_{s}(\mathcal{M},\mathcal{N}) < \varepsilon .$$

Clearly, $\{0\}$ and $\mathcal{X}$ both are stable T-invariant for any $T \in L(\mathcal{X})$. Less trivial but still easily understood examples of stable invariant subspaces are given by the following proposition.

PROPOSITION 4.3.1. *Let $\sigma(T)$ be the union of two disjoint sets σ_1 and σ_2, and let Γ be a suitable closed contour in the resolvent set of* T *such that σ_1 is inside Γ and σ_2 is outside Γ. Then the image of the Riesz projection*

$$\mathcal{M} = \operatorname{Im}\left[\frac{1}{2\pi i}\int_{\Gamma}(\lambda I-T)^{-1}d\lambda\right]$$

is a stable T-*invariant subspace.*

PROOF. Clearly, if $S \in L(\mathcal{X})$ and $\|S-T\|$ is small enough, say $\|S-T\| \leq \varepsilon_0$, then Γ lies also in the resolvent set of S. For such S write

$$(\lambda I-T)^{-1}-(\lambda I-S)^{-1} = (\lambda I-T)^{-1}[T-S](\lambda I-S)^{-1}, \quad \lambda \in \Gamma. \tag{4.3.1}$$

Since the function $F(\lambda) = (\lambda I-T)^{-1}$ is continuous on Γ and Γ is compact, the set of operators $\{(\lambda I-T)^{-1} \mid \lambda \in \Gamma\}$ is compact as well and, in particular, bounded. So

$$M \overset{\text{def}}{=} \sup_{\lambda\in\Gamma} \|(\lambda I-T)^{-1}\| < \infty.$$

Now let $S \in L(\mathcal{X})$ be such that

$$\|S-T\| \leq \min\{\varepsilon_0, \tfrac{1}{2M}\}. \tag{4.3.2}$$

Then for $\lambda \in \Gamma$ we have

$$(\lambda I-S)^{-1} = ((\lambda I-T)+(T-S))^{-1} = [I-(\lambda I-T)^{-1}(T-S)]^{-1}(\lambda I-T)^{-1}$$

$$= (I+\sum_{m=1}^{\infty}[(\lambda I-T)^{-1}(T-S)]^m)(\lambda I-T)^{-1},$$

and therefore

$$\|(\lambda I-S)^{-1} \leq \|(\lambda I-T)^{-1}\| \cdot (I+\sum_{m=1}^{\infty}\|(\lambda I-T)^{-1}(T-S)\|^m) \leq 2M.$$

Now the equality (4.3.1) shows that

$$\left\|\frac{1}{2\pi i}\int_\Gamma (\lambda I-T)^{-1}d\lambda - \frac{1}{2\pi i}\int_\Gamma (\lambda I-S)^{-1}d\lambda\right\| \leq \frac{M^2\ell}{\pi}\|T-S\| \tag{4.3.3}$$

for S satisfying (4.3.2), where ℓ is the length of Γ. Denoting by $\mathcal{N}$ the S-invariant subspace

$$\operatorname{Im}\left[\frac{1}{2\pi i}\int_\Gamma (\lambda I-S)^{-1}d\lambda\right],$$

the inequality (4.3.3), together with the definition of $\Theta(\mathcal{M},\mathcal{N})$, implies easily that

$$\Theta(\mathcal{M},\mathcal{N}) \leq \frac{M^2\ell}{\pi}\|T-S\|$$

for any S satisfying (4.3.2). In view of (4.1.11) this proves stability of $\mathcal{M}$. ∎

In the finite-dimensional case a full description of stable invariant subspaces is available.

THEOREM 4.3.2. *Assume* $\dim \mathcal{X} < \infty$, *and let* $\mathcal{M} \subset \mathcal{X}$ *be a* T-*invariant subspace where* $T \in L(\mathcal{X})$. *Then* $\mathcal{M}$ *is stable* T-*invariant if and only if for every eigenvalue* λ_0 *of* T

with $\dim \operatorname{Ker}(\lambda_0 I-T) > 1$ *either* $\mathcal{M} \cap \operatorname{Ker}(\lambda_0 I-T)^n = \{0\}$ *or* $\mathcal{M} \supset \operatorname{Ker}(\lambda_0 I-T)^n$ *holds, where* $n = \dim \mathcal{X}$.

For the proof of Theorem 4.3.2 we refer the reader to Bart-Gohberg-Kaashoek [1,2]; Campbell-Daughtry [1]; Gohberg-Lancaster-Rodman [2,3].

In the infinite-dimensional case not much is known about stably invariant subspaces (some results and examples are found in Apostol-Foias-Salinas [1]). Here we present the following basic properties of stable invariant subspaces.

THEOREM 4.3.3. *Given* $T \in L(\mathcal{X})$, *the set of all stable T-invariant subspaces is closed. In other words, if* $\{\mathcal{M}_m\}_{m=1}^{\infty}$ *are stable T-invariant subspaces and* $\lim_{m\to\infty} \Theta_s(\mathcal{M}_m, \mathcal{M}) = 0$ *for some subspace* $\mathcal{M} \subset \mathcal{X}$, *then* $\mathcal{M}$ *is again stable T-invariant.*

THEOREM 4.3.4. *Let* $\mathcal{M}$ *be a complemented stable T-invariant subspace, and let* $\mathcal{N} \subset \mathcal{M}$ *be a stable* $T|_{\mathcal{M}}$*-invariant subspace (in other words,* $\mathcal{N}$ *is stable S-invariant, where* $S \in L(\mathcal{M})$ *is the operator defined by the property that* $Sx = Tx$, $x \in \mathcal{M}$). *Then* $\mathcal{N}$ *is stable T-invariant.*

We relegate the lengthy proof of Theorems 4.3.3 and 4.3.4 to the next section.

As the following example shows, the statement converse to Theorem 4.3.4 is not true in general, i.e., a stable T-invariant subspace $\mathcal{N} \subset \mathcal{M}$ need not be stable $T|_{\mathcal{M}}$-invariant.

EXAMPLE 4.3.1. Let

$$B = \begin{bmatrix} 0 & 1 & 0 \\ 0 & 0 & 1 \\ 0 & 0 & 0 \end{bmatrix} \in L(\mathbb{C}^3),$$

and let $\{U_n\}_{n=1}^{\infty}$ be a sequence of unitary operators in $\mathbb{C}^3$ which is dense in the unitary group on $\mathbb{C}^3$. Put $\mathcal{X} = \bigoplus_{n=0}^{\infty} \mathbb{C}^3$ (the infinite orthogonal sum of the copies of $\mathbb{C}^3$), and let

$$T = \begin{bmatrix} 0 & I & \frac{1}{2} I & \cdots & \frac{1}{n} I & \cdots \\ 0 & U_1^* B U_1 + I & 0 & \cdots & 0 & \\ 0 & 0 & \frac{1}{4}(U_2^* B U_2 + I) & \cdots & 0 & \\ \vdots & \vdots & \vdots & & \vdots & \\ 0 & 0 & 0 & \cdots & \frac{1}{n^2}(U_n^* B U_n + I) & \cdots \\ \vdots & \vdots & \vdots & & \vdots & \end{bmatrix}$$

be an operator on $\mathcal{X}$ written in the standard matrix representation (here I is the identity operator on $\mathbb{C}^3$). Actually, T is compact. It is easy to see that

$$\sigma(T) = \{0, 1, \tfrac{1}{4}, \ldots, \tfrac{1}{n^2}, \ldots\}$$

with

$$\text{Ker } T = \{(x_0, 0, \ldots, 0, \ldots) \mid x_0 \in \mathbb{C}^3\}$$

(we write the vectors $x \in \mathcal{X}$ in the coordinate form $x = (x_0, x_1, \ldots)$ where $x_j \in \mathbb{C}^3$). Introduce the spectral subspace $\mathcal{X}_n$ corresponding to $\frac{1}{n^2}$:

$$\mathcal{X}_n = \text{Im}\left[\frac{1}{2\pi i} \int_{\Gamma_n} (\lambda I - T)^{-1} d\lambda\right],$$

where Γ_n is small circle around $\frac{1}{n^2}$ such that $\frac{1}{n^2}$ is the only point of spectrum of T inside and on Γ_n. An easy calculation shows that

$$\mathcal{X}_n = \text{Ker}(T - \tfrac{1}{n^2} I)^3 = \{(x_0, x_1, \ldots, x_n, \ldots) \mid x_j = 0$$
$$\text{for } j \neq 0, n;\ x_n = \tfrac{1}{n}(I + U_n^* B U_n) x_0\}.$$

As B is the Jordan form of the restriction $T|_{\mathcal{X}_n}$, it follows from Theorems 4.3.2 and 4.3.4 that the T-invariant subspaces

$$\mathcal{M}_{n1} = \mathrm{Ker}(T - \frac{1}{n^2} I)$$

$$= \{(x_0, x_1, \ldots) \in \mathcal{X}_n \mid x_0 \in U_n^* (\mathrm{Span}\begin{bmatrix}1\\0\\0\end{bmatrix})\},$$

$$\mathcal{M}_{n2} = \mathrm{Ker}(T - \frac{1}{n^2} I)^2$$

$$= \{(x_0, x_1, \ldots) \in \mathcal{X}_n \mid x_0 \in U_n^* (\mathrm{Span}\{\begin{bmatrix}1\\0\\0\end{bmatrix}, \begin{bmatrix}0\\1\\0\end{bmatrix}\})\},$$

and $\mathcal{X}_n$ itself are stable. Next, observe that every non-zero subspace $\mathcal{Y}$ in Ker T is the limit (in the spherical gap metric) of a sequence of subspaces from the set $\{\mathcal{X}_n; \mathcal{M}_{n1}; \mathcal{M}_{n2};\ n = 1,2,\ldots\}$. Indeed, assume for definiteness that $\mathcal{Y} = \mathrm{Span}\{(y_0, 0, \ldots, 0, \ldots), (z_0, 0, \ldots)\}$, where y_0, z_0 are orthonormal vectors in $\mathbb{C}^3$. Let $\{U_{n_k}\}_{k=1}^{\infty}$ be a subsequence such that

$$y_0 = \lim_{k\to\infty} U_{n_k}^* \begin{bmatrix}1\\0\\0\end{bmatrix}, \quad z_0 = \lim_{k\to\infty} U_{n_k}^* \begin{bmatrix}0\\1\\0\end{bmatrix}.$$

Then

$$\lim_{k\to\infty} \theta_s(\mathcal{M}_{n_k 2}, \mathcal{Y}) = 0.$$

By Theorem 4.3.3 every T-invariant subspace contained in Ker T is stable. However, Theorem 4.3.2 shows that the only stable subspaces for the restriction $T|_{\mathrm{Ker}\,T}$ are the trivial ones: $\{0\}$ and Ker T itself. ■

PROBLEM 4.3.1. *Is Theorem 4.3.4 true for T-invariant subspaces $\mathcal{M}$ which are not complemented?*

4.4 Proof of Theorems 4.3.3 and 4.3.4

We start with introducing some notation. Given $T \in L(\mathcal{X})$ and given a subspace $\mathcal{M} \subset \mathcal{X}$, let $\ell(\mathcal{M};T)$ be the infimum of $\theta_s(\mathcal{M},\mathcal{N})$, where $\mathcal{N}$ runs over the set of all T-invariant subspaces.

It is easily seen that $\ell(\mathcal{M};T) = 0$ if and only if $\mathcal{M}$ is T-invariant (indeed, assume $\lim_{m\to\infty} \Theta_s(\mathcal{M},\mathcal{N}_m) = 0$, where $\mathcal{N}_m$ are T-invariant subspaces, then $\mathcal{M}$ is T-invariant by Theorem 4.1.2). Further, for every $\varepsilon > 0$ define

$$s_\varepsilon(T;\mathcal{M}) = \sup\{\ell(\mathcal{M};S) \mid \|S-T\| \leq \varepsilon\},$$

and finally put

$$s(T;\mathcal{M}) = \inf\{s_\varepsilon(T;\mathcal{M}) \mid \varepsilon > 0\}.$$

Because $s_\varepsilon(T;\mathcal{M})$ decreases with ε we actually have

$$s(T;\mathcal{M}) = \lim_{\varepsilon\to 0} s_\varepsilon(T;\mathcal{M}).$$

LEMMA 4.4.1. *A subspace $\mathcal{M} \subset \mathcal{X}$ is stable T-invariant if and only if* $s(T;\mathcal{M}) = 0$.

PROOF. Assume $s(T;\mathcal{M}) = 0$, and given $\varepsilon > 0$, let $\delta > 0$ be such that $s_\delta(T;\mathcal{M}) < \varepsilon$. Then for any $S \in L(\mathcal{X})$ satisfying $\|S-T\| < \delta$ we have

$$\ell(\mathcal{M};S) \leq s_\delta(T;\mathcal{M}) < \varepsilon. \tag{4.4.1}$$

In particular, as $\varepsilon > 0$ was chosen arbitrarily, (4.4.1) implies $\ell(\mathcal{M};T) = 0$, i.e., $\mathcal{M}$ is T-invariant. Further, by the definition of $\ell(\mathcal{M};S)$ there is S-invariant subspace $\mathcal{N}$ such that $\Theta_s(\mathcal{M},\mathcal{N}) < \varepsilon$. This proves the stability of $\mathcal{M}$.

Conversely, assume that for every $\varepsilon > 0$ there is $\delta > 0$ such that any $S \in L(\mathcal{X})$ with $\|S-T\| < \delta$ has an invariant subspace $\mathcal{N}$ satisfying $\Theta_s(\mathcal{M},\mathcal{N}) < \varepsilon$. For such S we have $\ell(\mathcal{M};S) < \varepsilon$, and so

$$s(T;\mathcal{M}) \leq s_{\delta/2}(T;\mathcal{M}) \leq \varepsilon,$$

which shows that $s(T;\mathcal{M}) = 0$. ∎

LEMMA 4.4.2. *For* $T \in L(\mathcal{X})$ *and subspaces* $\mathcal{M},\mathcal{M}' \subset \mathcal{X}$, *the inequality*

$$|s(T;\mathcal{M})-s(T;\mathcal{M}')| \leq \Theta_s(\mathcal{M},\mathcal{M}')$$

holds.

PROOF. It will suffice to prove that

$$|s_\varepsilon(T;\mathcal{M})-s_\varepsilon(T;\mathcal{M}')| \le \Theta_s(\mathcal{M},\mathcal{M}') \tag{4.4.2}$$

for every positive ε. Let $S \in L(\mathcal{X})$ be such that $\|T-S\| \le \varepsilon$. By definition of $\ell(\mathcal{M};S)$ for every $\varepsilon' > 0$ there is an S-invariant subspace $\mathcal{N}$ such that

$$\ell(\mathcal{M};S) \ge \Theta_s(\mathcal{M},\mathcal{N})-\varepsilon'.$$

Using the triangle inequality and the definition of $\ell(\mathcal{M}';S)$ we obtain

$$\ell(\mathcal{M};S) \ge \Theta_s(\mathcal{M},\mathcal{N})-\varepsilon' \ge \Theta_s(\mathcal{M}',\mathcal{N})-\Theta_s(\mathcal{M},\mathcal{M}')-\varepsilon'$$

$$\ge \ell(\mathcal{M}';S)-\Theta_s(\mathcal{M},\mathcal{M}')-\varepsilon',$$

or

$$\ell(\mathcal{M}';S)-\ell(\mathcal{M};S) \le \Theta_s(\mathcal{M},\mathcal{M}')+\varepsilon'. \tag{4.4.3}$$

Interchanging the roles of $\mathcal{M}$ and $\mathcal{M}'$, we have

$$\ell(\mathcal{M};S)-\ell(\mathcal{M}';S) \le \Theta_s(\mathcal{M},\mathcal{M}')+\varepsilon'. \tag{4.4.4}$$

Putting together (4.4.3) and (4.4.4), and letting $\varepsilon' \to 0$, it follows that

$$|\ell(\mathcal{M}';S)-\ell(\mathcal{M};S)| \le \Theta_s(\mathcal{M},\mathcal{M}'). \tag{4.4.5}$$

Finally, given $\varepsilon'' > 0$ find $S \in L(\mathcal{X})$ such that $\|S-T\| \le \varepsilon$ and $s_\varepsilon(T;\mathcal{M}) \le \ell(\mathcal{M};S)+\varepsilon''$. Now by using (4.4.5) we have

$$s_\varepsilon(T;\mathcal{M})-s_\varepsilon(T;\mathcal{M}') \le \ell(\mathcal{M};S)+\varepsilon''-s_\varepsilon(T;\mathcal{M}')$$

$$\le \ell(\mathcal{M};S)+\varepsilon''-\ell(\mathcal{M}';S) \le \Theta_s(\mathcal{M},\mathcal{M}')+\varepsilon''.$$

Interchanging the roles of $\mathcal{M}$ and $\mathcal{M}'$, and letting $\varepsilon'' \to 0$, (4.4.2) follows. ■

We are in the position now to prove Theorem 4.3.3.

PROOF OF THEOREM 4.3.3. Let $\mathcal{M}_m, \mathcal{M}$ be subspaces in $\mathcal{X}$ with the properties described in Theorem 4.3.3. By Lemma 4.4.1, $s(T;\mathcal{M}_m) = 0$ for $m = 1,2,\ldots$. Now Lemma 4.4.2 gives $s(T;\mathcal{M}) = 0$

which in view of the same Lemma 4.4.1 means that $\mathcal{M}$ is stable T-invariant. ∎

For the proof of Theorem 4.3.4, it is convenient to prove first two lemmas.

LEMMA 4.4.3. *Let* $T \in L(\mathcal{X})$ *and let* $\mathcal{M}$ *be a subspace in* $\mathcal{X}$. *If* $\{T_m\}_{m=1}^{\infty}$ *is a sequence of operators on* $\mathcal{X}$, *and* $\lim_{m\to\infty} \|T_m - T\| = 0$, *then*

$$s(T;\mathcal{M}) \geq \limsup_{m\to\infty} \ell(\mathcal{M};T_m).$$

Moreover, there is a sequence $\{T_m\}_{m=1}^{\infty}$ *as above such that for some choice of a* T_m*-invariant subspace* $\mathcal{M}_m$, *for* $m = 1,2,\ldots$ *we have*

$$s(T;\mathcal{M}) = \lim_{m\to\infty} \Theta_s(\mathcal{M},\mathcal{M}_m) = \lim_{m\to\infty} \ell(\mathcal{M};T_m). \tag{4.4.6}$$

PROOF. Choose a sequence $\{\varepsilon_m\}_{m=1}^{\infty}$ tending to 0 such that $\|T-T_m\| < \varepsilon_m$ $(m = 1,2,\ldots)$. Now

$$\ell(\mathcal{M};T_m) \leq s_{\varepsilon_m}(T;\mathcal{M}),$$

so

$$\limsup_{m\to\infty} \ell(\mathcal{M};T_m) \leq \lim_{m\to\infty} s_{\varepsilon_m}(T;\mathcal{M}) = s(T;\mathcal{M}).$$

On the other hand, choosing $T_m \in L(\mathcal{X})$ and $\mathcal{M}_m$ a T_m-invariant subspace such that

$$\|T_m - T\| \leq \frac{1}{m};\quad \ell(\mathcal{M};T_m) > s_{1/m}(T;\mathcal{M}) - \frac{1}{m};$$

$$\Theta_s(\mathcal{M},\mathcal{M}_m) < \ell(\mathcal{M};T_m) + \frac{1}{m}$$

$(m = 1,2,\ldots)$, we have

$$\lim_{m\to\infty} (\ell(\mathcal{M};T_m) - s_{1/m}(T;\mathcal{M}))$$
$$= \lim_{m\to\infty}(\Theta_s(\mathcal{M};\mathcal{M}_m) - \ell(\mathcal{M};T_m)) = 0, \tag{4.4.7}$$

so

(4.4.8) $$\liminf_{m\to\infty} \Theta_s(\mathcal{M},\mathcal{M}_m) = \lim_{m\to\infty} \ell(\mathcal{M};T_m)$$

$$= \lim_{m\to\infty} s_{1/m}(T;\mathcal{M}) = s(T;\mathcal{M}).$$

Further, arguing by contradiction and using (4.4.7), one proves

(4.4.9) $$s(T;\mathcal{M}) \geq \limsup_{m\to\infty} \Theta_s(\mathcal{M},\mathcal{M}_m).$$

The equality (4.4.6) follows now from (4.4.8) and (4.4.9). ∎

LEMMA 4.4.4. *Let $\mathcal{M} \subset \mathcal{X}$ be a complemented subspace. Then there exist positive constants* K *and* ε *with the following property. For every subspace $\mathcal{N} \subset \mathcal{X}$ with $\Theta_s(\mathcal{M},\mathcal{N}) < \varepsilon$ there is an invertible operator $S \in L(\mathcal{X})$ such that $S\mathcal{N} = \mathcal{M}$ and*

$$\|I-S\| \leq K\, \Theta_s(\mathcal{M},\mathcal{N}).$$

PROOF. Let $\mathcal{M}'$ be a direct complement to $\mathcal{M}$ in $\mathcal{X}$. By Theorem 4.2.1 there is $\varepsilon > 0$ such that $\mathcal{N} \dot{+} \mathcal{M}' = \mathcal{X}$ for every subspace $\mathcal{N}$ satisfying $\Theta_s(\mathcal{M},\mathcal{N}) \leq \varepsilon$. For such $\mathcal{N}$, let $P_{\mathcal{N}}$ be the projection on $\mathcal{N}$ along $\mathcal{M}'$. By Theorem 4.2.5 there is a constant $K_1 > 0$ depending on $\mathcal{M}$ and $\mathcal{M}'$ only such that

(4.4.10) $$\|P_{\mathcal{M}}-P_{\mathcal{N}}\| \leq K_1\Theta(\mathcal{M},\mathcal{N}),$$

for every subspace $\mathcal{N}$ with $\Theta(\mathcal{M},\mathcal{N})$ small enough (observe that here we use $\Theta(\mathcal{M},\mathcal{N})$ in place of $\Theta_s(\mathcal{M},\mathcal{N})$).

Put

$$S = P_{\mathcal{M}}P_{\mathcal{N}}+(I-P_{\mathcal{M}})(I-P_{\mathcal{N}}) \in L(\mathcal{X}).$$

Clearly, $S\mathcal{N} \subset \mathcal{M}$ and $S\mathcal{M}' \subset \mathcal{M}'$. Further, rewrite S in the form

$$S = I-(I-2P_{\mathcal{M}})(P_{\mathcal{N}}-P_{\mathcal{M}}),$$

and observe in view of (4.4.10) that for $\Theta(\mathcal{M},\mathcal{N})$ small enough the norm $\|I-S\|$ is less than 1, so S is invertible for such $\mathcal{N}$. Then, clearly $S\mathcal{N} = \mathcal{M}$, and Lemma 4.4.4 holds with

$$K = 2\|P_{\mathcal{M}}\| \left[\sup_{\substack{x\in\mathcal{M}' \\ \|x\|=1}} (d(x,\mathcal{M}))^{-1}\right](1+2\|P_{\mathcal{M}}\|). \qquad \blacksquare$$

We have completed the preparations for the proof of Theorem 4.3.4.

PROOF OF THEOREM 4.3.4. Let $\mathcal{N}$ be a stable $T|_{\mathcal{M}}$-invariant subspace. By Lemma 4.4.3 we can find a sequence $\{T_m\}_{m=1}^{\infty}$, $T_m \in L(\mathcal{X})$ that converges to T and for which

$$s(T;\mathcal{N}) = \lim_{m\to\infty} \ell(\mathcal{N};T_m).$$

Since by Lemma 4.4.1, $s(T;\mathcal{M}) = 0$, the same Lemma 4.4.3 implies $\lim_{m\to\infty} \ell(\mathcal{M};T_m) = 0$. By definition of $\ell(\mathcal{M};T_m)$ for every $m = 1,2,\ldots,$ one can find a T_m-invariant subspace $\mathcal{M}_m$ such that $\lim_{m\to\infty} \Theta_s(\mathcal{M},\mathcal{M}_m) = 0$. Next, we appeal to Lemma 4.4.4 and find a sequence of invertible operators $\{S_m\}_{m=1}^{\infty}$, $S_m \in L(\mathcal{X})$, such that $\lim_{m\to\infty} \|I-S_m\| = 0$ and (at least for m sufficiently large) $S_m\mathcal{M}_m = \mathcal{M}$. Put $T_m' = S_mT_mS_m^{-1}$, $m = 1,2,\ldots$. Then the subspace $\mathcal{M}$ is T_m'-invariant (for sufficiently large m). Because $\mathcal{N}$ is stably $T|_{\mathcal{M}}$-invariant, there is a sequence $\{\mathcal{N}_m'\}_{m=1}^{\infty}$ of subspaces in $\mathcal{M}$ such that $\mathcal{N}_m'$ is T_m'-invariant and $\lim_{m\to\infty} \Theta_s(\mathcal{N},\mathcal{N}_m') = 0$. Put $\mathcal{N}_m = S_m^{-1}\mathcal{N}_m'$. Then $\mathcal{N}_m$ is T_m-invariant and consequently

$$s(T;\mathcal{N}) = \lim_{m\to\infty} \ell(\mathcal{N};T_m) \le \lim_{m\to\infty} \Theta_s(\mathcal{N},\mathcal{N}_m)$$

$$\le \lim_{m\to\infty} \Theta_s(\mathcal{N},\mathcal{N}_m') + \lim_{m\to\infty} \Theta(\mathcal{N}_m',\mathcal{N}_m).$$

The first summand in the right-hand side is zero because of the choice of $\mathcal{N}_m'$ while the second summand is easily verified to be zero by the definition of $\Theta_s(\mathcal{N}_m',\mathcal{N}_m)$ and using the fact that $\|I-S_m\| \to 0$ as $m \to \infty$.

Thus, $s(T;\mathcal{N}) = 0$, and by Lemma 4.4.1 and subspace $\mathcal{N}$ is stable T-invariant. ∎

4.5 Lipschitz stable invariant subspaces and one-sided resolvents

We introduce here a class of invariant subspaces of a Banach space operator that enjoy a stability property (with respect to small perturbations of the operator) which is stronger than the stability property described in the preceding section.

Let $\mathcal{X}$ be a Banach space, and let $T \in L(\mathcal{X})$. A T-invariant subspace $\mathcal{M}$ is called *Lipschitz stable* if there is $K > 0$ (depending on T and $\mathcal{M}$ only) such that for any $S \in L(\mathcal{X})$ there exists an S-invariant subspace $\mathcal{N}$ satisfying

$$\theta_s(\mathcal{M},\mathcal{N}) \leq K\|S-T\|. \tag{4.5.1}$$

Equivalently, $\mathcal{M}$ is Lipschitz stable if (4.5.1) holds for some S-invariant subspace $\mathcal{N}$, where $S \in L(\mathcal{X})$ is any operator with $\|S-T\| \leq \varepsilon$, and $\varepsilon > 0$ is a positive constant depending on T and $\mathcal{M}$ only. Indeed, assume (4.5.1) holds for all $S \in L(\mathcal{X})$ with $\|S-T\| \leq \varepsilon$. As $\theta_s(\mathcal{Y},\mathcal{Z}) \leq 2$ for all subspaces $\mathcal{Y},\mathcal{Z} \subset \mathcal{X}$, we obviously have for $S \in L(\mathcal{X})$ such that $\|S-T\| \geq \varepsilon$:

$$\theta_s(\mathcal{M},\mathcal{X}) \leq 2 \leq \frac{2}{\varepsilon} \cdot \|S-T\|.$$

So (4.5.1) follows for all $S \in L(\mathcal{X})$ with the constant K replaced by the possibly bigger constant $\max(K,\frac{2}{\varepsilon})$.

Proposition 4.3.1 and its proof provide basic examples of Lipschitz stable invariant subspaces. Namely, if $\sigma(T)$ is the union of two disjoint sets σ_1 and σ_2, then

$$\mathcal{M} = \operatorname{Im}\left[\frac{1}{2\pi i}\int_\Gamma (\lambda I-T)^{-1}d\lambda\right]$$

is a Lipschitz stable T-invariant subspace, where Γ is a suitable closed contour such that σ_1 (resp. σ_2) is inside (resp. outside) Γ. In the finite-dimensional case ($\dim \mathcal{X} < \infty$) all Lipschitz stable T-invariant subspaces are of this type (we do not exclude the cases when $\sigma_1 = \phi$ (then $\mathcal{M} = \{0\}$) or $\sigma_2 = \phi$ (then $\mathcal{M} = \mathcal{X}$)). For the proof of this statement, see Kaashoek-van der Mee-Rodman [2], Gohberg-Lancaster-Rodman [3]. This fact shows, in particular, that not every stable invariant subspace is Lipschitz stable, already in the finite-dimensional case.

In the infinite-dimensional case additional examples of Lipschitz stably invariant subspaces are provided by the notion of one-sided resolvent. We describe this class of subspaces in detail.

Let $T \in L(\mathcal{X})$, where $\mathcal{X}$ is a Banach space, and let Ω be an open subset of $\mathbb{C}$. A continuous function $F: \Omega \to L(\mathcal{X})$ is called a *right resolvent* for T on Ω if

$$(\lambda I-T)F(\lambda) = I \tag{4.5.2}$$

and

$$F(\lambda)-F(\mu) = (\mu-\lambda)F(\lambda)F(\mu) \tag{4.5.3}$$

for all $\lambda,\mu \in \Omega$. Clearly, a necessary condition for existence of a right resolvent for T on Ω is that $\lambda I-T$ be right invertible for all $\lambda \in \Omega$. Replacing (4.5.2) by

$$F(\lambda)(\lambda I-T) = I,$$

we obtain the definition of a *left resolvent* for T on Ω; a necessary condition for existence of a left resolvent for T on Ω is that $\lambda I-T$ be left invertible for all $\lambda \in \Omega$.

It is apparent from (4.5.3) that one-sided resolvent $f(\lambda)$ (left or right) for T on Ω is actually analytic in Ω, and

$$F'(\lambda) = -(F(\lambda))^2, \quad \lambda \in \Omega.$$

The next theorem shows, in particular, that one-sided resolvents produce naturally invariant subspaces.

Let Σ be a compact set in $\mathbb{C}$ contained in an open set $\Omega \subset \mathbb{C}$ such that $T \in L(\mathcal{X})$ has a one-sided resolvent F defined on $\Omega\backslash\Sigma$. Denote by $\mathcal{F}(\Sigma)$ the algebra of scalar analytic functions defined on some neighborhood of Σ (this neighborhood may depend on the function). For $f \in \mathcal{F}(\Sigma)$ define the operator

$$f_{\Sigma}(T) = \frac{1}{2\pi i}\int_{\partial\Delta} f(\lambda)F(\lambda)d\lambda, \tag{4.5.4}$$

where $\partial\Delta$ is the boundary of a bounded Cauchy domain Δ such that $\Sigma \subset \Delta \subset \overline{\Delta} \subset \Omega$, and $f(\lambda)$ is defined in a neighborhood of $\overline{\Delta}$. It is easily seen that (4.5.4) does not depend on the choice of Δ (subject to the above conditions).

THEOREM 4.5.1. (i) *The mapping* $f \to f_{\Sigma}(T)$ *is a homomorphism from the algebra* $\mathcal{F}(\Sigma)$ *into* $L(\mathfrak{X})$.

(ii) *If* χ *is the characteristic function of* Ω *and* $F(\lambda)$ *is a right resolvent, then* $P = \chi_{\Sigma}(T)$ *is a projection such that* $T(\text{Im } P) \subset \text{Im } P$. *Moreover,*

$$\sigma(T|_{\text{Im } P}) \subset \Sigma$$

and for every $\lambda \in \Sigma$ *the operator*

$$\lambda I-(I-P)T(I-P) \in L(\text{Ker } P)$$

is right invertible.

(iii) *If* χ *is the characteristic function of* Ω *and* $F(\lambda)$ *is a left resolvent, then* $P = \chi_{E}(T)$ *is a projection such that* $T(\text{Ker } P) \subset \text{Ker } P$. *Moreover,*

$$\sigma(PTP|_{\text{Im } P}) \subset \Sigma \tag{4.5.5}$$

and for every $\lambda \in \Sigma$ *the operator*

$$\lambda I-T|_{\text{Ker } P}$$

is left invertible.

PROOF. We prove the part (iii) only (the proof of (ii) is analogous, while the proof of (i) follows standard arguments (see, e.g., Section VII.3 in Dunford-Schwartz [1]).

Use the relation

$$\frac{1}{2\pi i}\int_{\partial\Delta} F(\lambda)(\lambda I-T)d\lambda = \frac{1}{2\pi i}\int_{\partial\Delta} I\, d\lambda = 0,$$

to obtain

$$PT = \frac{1}{2\pi i}\int_{\partial\Delta} \lambda F(\lambda)d\lambda = \frac{1}{2\pi i}\int_{\partial\Delta} \chi(\lambda)\lambda\chi(\lambda)F(\lambda)d\lambda = PTP,$$

whence we deduce that Ker P is invariant under T.

If $\lambda_0 \notin \Sigma$, we can choose Δ so that $\lambda_0 \notin \overline{\Delta}$; then

$$B(\lambda_0) = \frac{1}{2\pi i} \int_{\partial\Delta} (\lambda_0-\lambda)^{-1} F(\lambda)d\lambda$$

is well defined, and we have

$$P = \frac{1}{2\pi i} \int_{\partial\Delta} F(\lambda)d\lambda = \frac{1}{2\pi i} \int_{\partial\Delta} (\lambda_0-\lambda)^{-1}(\lambda_0-\lambda)F(\lambda)d\lambda$$

$$= B(\lambda_0)P(\lambda_0 I-T) = P(\lambda_0 I-T)PB(\lambda_0).$$

This equality shows that $B(\lambda_0)|_{\mathrm{Im}\, P}$ is the inverse of $\lambda_0 I-PTP$, and (4.5.5) is proved.

On the other hand, define $G: \Sigma \to L(\mathrm{Ker}\, P)$ by the equation

$$G(\varsigma) = \frac{1}{2\pi i} \int_{\partial\Delta} (I-P)(\varsigma-\lambda)^{-1} F(\lambda)d\lambda .$$

Then we have for every $\varsigma \in \Sigma$:

$$G(\varsigma)(\varsigma I-T)(I-P) = \frac{1}{2\pi i} \int_{\partial\Delta} (I-P)(\varsigma-\lambda)^{-1} F(\lambda)(\varsigma I-T)(I-P)d\lambda$$

$$= \frac{1}{2\pi i} \int_{\partial\Delta} (I-P)(\varsigma-\lambda)^{-1} F(\lambda)[(\varsigma-\lambda)+(\lambda I-T)](I-P)d\lambda$$

$$= \frac{1}{2\pi i} \int_{\partial\Delta} (I-P)(\varsigma-\lambda)^{-1}[I+(\varsigma-\lambda)F(\lambda)](I-P)d\lambda$$

$$= (I-P)\frac{1}{2\pi i} \int_{\partial\Delta} (\varsigma-\lambda)^{-1}d\lambda(I-P) + \frac{1}{2\pi i} \int_{\partial\Delta} F(\lambda)d\lambda(I-P)$$

$$= I-P + (I-P)P(I-P) = I-P,$$

so that $G(\varsigma)$ is a left inverse for $\varsigma I-T|_{\mathrm{Ker}\, P}$. ■

Next we show that existence of a one-sided resolvent is preserved under small perturbations, in the following sense.

THEOREM 4.5.2. *Let* $A \in L(\mathcal{X})$, *and assume there exists a right resolvent* $G(\lambda)$ *of* A *on an open set* $\Omega \subset \mathbb{C}$. *Then for every open set* Ω' *such that* $\overline{\Omega'} \subset \Omega$ *there is* $\varepsilon = \varepsilon(\Omega', G)$ *with the following property: For every* $T \in L(\mathcal{X})$ *with* $\|T-A\| < \varepsilon$ *there is a right resolvent* $G_T(\lambda)$ *of* T *on* Ω'. *Moreover,* $G_T(\lambda)$ *can be chosen so that*

$$\sup \frac{\max_{\lambda \in K} \|G_T(\lambda)-G_A(\lambda)\|}{\|T-A\|} < \infty \tag{4.5.6}$$

for every compact set $K \subset \Omega'$, *where the supremum is taken over all* $T \in L(\mathcal{X})$ *with* $\|T-A\| < \varepsilon$.

An analogous result holds, of course, for the left resolvent.

PROOF. Using the fact that $G_A(\lambda) = (\lambda I-A)^{-1}$ on any unbounded connected component of Ω, it is easy to see that without loss of generality Ω (and hence also Ω') can be assumed bounded. Put

$$\varepsilon = \frac{1}{2} \min_{\lambda \in \overline{\Omega'}} \{\|G_A(\lambda)\|^{-1}\}.$$

If $T \in L(\mathcal{X})$ is such that $\|C\| < \varepsilon$, where $C = T-A$, then

$$\|G_A(\lambda)\| \cdot \|C\| < \frac{1}{2}, \quad \lambda \in \Omega',$$

and therefore $I-G_A(\lambda)C$ is invertible for all λ in Ω'. Define $G_T(\lambda) = [I-G_A(\lambda)C]^{-1}G_A(\lambda)$; we have

$$\begin{aligned}(\lambda-T)G_T(\lambda) &= (\lambda I-A-C)G_T(\lambda) \\ &= (\lambda I-A)[I-G_A(\lambda)C]\cdot[I-G_A(\lambda)C]^{-1}G_A(\lambda) = I(\lambda \in \Omega').\end{aligned}$$

Furthermore, using the series expansion

$$(I-G_A(\lambda)C)^{-1} = \sum_{j=0}^{\infty} (G_A(\lambda)C)^j,$$

it is easy to check that

$$(I-G_A(\lambda)C)^{-1}G_A(\lambda) = G_A(\lambda)(I-CG_A(\lambda))^{-1}.$$

So for $\lambda,\mu \in \Omega'$, we have

$$\begin{aligned} G_T(\lambda)-G_T(\mu) &= [I-G_A(\lambda)C]^{-1}G_A(\lambda)-G_A(\mu)[I-CG_A(\mu)]^{-1} \\ &= [I-G_A(\lambda)C]^{-1}[G_A(\lambda)-G_A(\mu)][I-CG_A(\mu)]^{-1} \\ &= [I-G_A(\lambda)C]^{-1}(\mu-\lambda)G_A(\lambda)G_A(\mu)[I-CG_A(\mu)]^{-1} \\ &= (\mu-\lambda)G_T(\lambda)G_T(\mu). \end{aligned}$$

This shows that $G_T(\lambda)$ is a right resolvent of T on Ω'. The inequality (4.5.6) follows easily from the definition of $G_T(\lambda)$, because

$$G_T(\lambda)-G_A(\lambda) = \sum_{j=1}^{\infty} (G_A(\lambda)(T-A))^j G_A(\lambda), \qquad \lambda \in \Omega',$$

and consequently

$$\|G_T(\lambda)-G_A(\lambda)\| \le [\sum_{j=0}^{\infty} (\|G_A(\lambda)\|\cdot\|C\|)^j]\|G_A(\lambda)\|^2\|T-A\|. \qquad \blacksquare$$

Using Theorems 4.5.1 and 4.5.2 we assert now the Lipschitz stability of invariant subspaces associated with one-sided resolvents. This is the main result of this section.

THEOREM 4.5.3. *Let* $A \in L(\mathcal{X})$, *and suppose that there is a one-sided resolvent* $F(\lambda)$ *of* A *on the set* $\Omega\backslash\Sigma$, *where* Ω *is open and* $\Sigma \subset \Omega$ *is compact. Let*

$$P_A = \frac{1}{2\pi i}\int_{\partial\Delta} F(\lambda)d\lambda,$$

where Δ *is a bounded Cauchy domain such that* $\Sigma \subset \Delta \subset \overline{\Delta} \subset \Omega$. *Then the* A*-invariant subspace* Im P *(if* $F(\lambda)$ *is a right resolvent) or the* A*-invariant subspace* Ker P *(if* $F(\lambda)$ *is a left resolvent) are Lipschitz stable.*

PROOF. The A-invariance of the subspaces Im P or Ker P follows from Theorem 4.5.1. Now apply Theorem 4.5.2 where $\Omega' \subset \Omega\backslash\Sigma$ is an open set such that $\overline{\Omega'} \subset \Omega\backslash\Sigma$ and $\partial\Delta \subset \Omega'$. Let ε and $G_T(\lambda)$ be as in Theorem 4.5.2, and put

$$P_T = \frac{1}{2\pi i} \int_{\partial\Delta} G_T(\lambda)d\lambda$$

for $T \in L(\mathcal{X})$ such that $\|T-A\| < \varepsilon$. The inequality (4.5.6) (with $K = \partial\Delta$) implies that

$$\|P_T-P_A\| \le C\|T-A\|, \tag{4.5.7}$$

where the constant C does not depend on T.

Now let

$$S = P_AP_T+(I-P_A)(I-P_T).$$

It is easily seen that $S(\text{Im } P_T) \subset \text{Im } P_A$, $S(\text{Ker } P_T) \subset \text{Ker } P_A$. Further,

$$S = I-(I-2P_A)(P_T-P_A),$$

and for T sufficiently close to A it follows from (4.5.7) that $\|I-S\| < \frac{1}{2}$, and hence, in particular, S is invertible and

$$\|I-S^{-1}\| \le \|S^{-1}\|\cdot\|S-I\| \le 2\|I-2P_A\| \; \|P_T-P_A\|.$$

The definition of the spherical gap now implies

$$\begin{aligned}\theta_s(\text{Im } P_T, \text{ Im } P_A) &\le \max(\|I-S^{-1}\|, \|I-S\|)\\ &\le 2 \; \|I-2P_A\| \; \|P_T-P_A\|\end{aligned}$$

for all T sufficiently close to A. Analogous inequality holds for θ_s $(\text{Ker } P_T, \text{ Ker } P_A)$. This, together with (4.5.7), proves the theorem. ∎

We conclude this section with one result on existence of one-sided resolvents.

THEOREM 4.5.4. *Let $T \in L(\mathcal{X})$, and let $\Omega = \{\lambda \in \mathbb{C} \mid \lambda I-T$ is right invertible and Fredholm$\}$. Then for every $\varepsilon > 0$ there exists a right resolvent of T on the set $\Omega\backslash S$, where S is some denumerable (at most) set without accumulation points in Ω and such that for every $\lambda_0 \in S$ the distance from λ_0 to the boundary of Ω is at most ε.*

Replacing here "right" by "left" we obtain the dual result.

The full proof of Theorem 4.5.4 or of its dual would take us too far afield; it can be found in Herrero [1], Section 3.1.3; see also Zemanek [1].

In connection with Theorem 4.5.4 the following open problems are of interest.

PROBLEM 4.5.1. *Relax (if possible) the hypotheses in Theorem* 4.5.4. *In particular, prove or disprove the following statement. Given* $T \in L(\mathcal{X})$, *there is a right (left) resolvent of* T *on the set*

$$\{\lambda \in \mathbb{C} \mid \textit{is right (left) invertible and Fredholm}\}.$$

PROBLEM 4.5.2. *Let* $\mathcal{X}$ *be an infinite-dimensional Banach space (or an infinite-dimensional separable Hilbert space to start with), and let* $T \in L(\mathcal{X})$. *Describe all Lipschitz stable* T*-invariant subspaces in terms of the spectral structure of* T.

In particular, is it true that all Lipschitz stable invariant subspaces are the spectral subspaces and the subspaces arising from one-sided resolvents (as shown in this section)?

4.6 Lipschitz continuous dependence of supporting subspaces and factorizations

Let $\mathcal{X}$ be a Banach space. Consider a monic operator polynomial $L(\lambda)$ of degree ℓ with coefficients in $L(\mathcal{X})$, and let (X,T) be a right spectral pair of $L(\lambda)$, where $X \in L(\mathcal{Y},\mathcal{X})$, $T \in L(\mathcal{Y})$, and $\mathcal{Y}$ is a Banach space. By Theorem 2.7.1 there is one-to-one correspondence between T-invariant subspaces $\mathcal{M}$ such that

$$\{\mathrm{col}[XT^i]_{i=0}^{k-1}\}\big|_{\mathcal{M}} \in L(\mathcal{M},\mathcal{X}^k)$$

is invertible, and right divisors $L_1(\lambda)$ of $L(\lambda)$ of degree k. This correspondence is given by the following formulas

$$L_1(\lambda) = \lambda^k I - X(T|_{\mathcal{M}})^k(V_1+V_2\lambda+\cdots+V_k\lambda^{k-1}), \tag{4.6.1}$$

where

$$[V_1 V_2 \cdots V_k] = \{\{\operatorname{col}[XT^i]_{i=0}^{k-1}\}|_{\mathcal{M}}\}^{-1} \in L(\mathcal{X}^k, \mathcal{M});$$

$$(4.6.2) \qquad \mathcal{M} = \operatorname{Im}\{\{\operatorname{col}[XT^i]_{i=0}^{\ell-1}\}^{-1}\operatorname{col}[X_1 T_1^i]_{i=0}^{\ell-1}\},$$

where (X_1, T_1) is some right spectral pair of $L_1(\lambda)$.

In this section we study the continuous character of this correspondence. To this end we need a measure of distance between subspaces (for this purpose the spherical gap $\Theta_s(\mathcal{M},\mathcal{N})$ introduced and studied in Section 4.1 will be used), and also a measure of distance between operator polynomials, which will be introduced now.

Let $\mathcal{P}_k$ be the class of all monic operator polynomials of degree k with coefficients in $L(\mathcal{X})$. Define a function σ_k on $\mathcal{P}_k \times \mathcal{P}_k$ by

$$\sigma_k(\lambda^k I + \sum_{j=0}^{k-1} B_j \lambda^j, \quad \lambda^k I + \sum_{j=0}^{k-1} B_j' \lambda^j) = \sum_{j=0}^{k-1} \|B_j - B_j'\|.$$

It is easily verified that σ_k is a metric on $\mathcal{P}_k$. Sometimes it is convenient also to use other metrics on $\mathcal{P}_k$. For a suitable contour Γ (e.g., it suffices to assume that Γ is closed, rectifiable and simple) define

$$\sigma_{k,\Gamma}(L(\lambda), \tilde{L}(\lambda)) = \max_{\lambda\in\Gamma} \|L(\lambda) - \tilde{L}(\lambda)\|$$

for $L, \tilde{L} \in \mathcal{P}_k$. Again, $\sigma_{k,\Gamma}$ is a metric on $\mathcal{P}_k$. All these metrics are equivalent:

PROPOSITION 4.6.1. *For any suitable contour Γ there exist positive constants $C_{k,\Gamma}$ and $K_{k,\Gamma}$ such that*

$$(4.6.3) \qquad C_{k,\Gamma}\sigma_k(L,\tilde{L}) \le \sigma_{k,\Gamma}(L,\tilde{L}) \le K_{k,\Gamma}\sigma_k(L,\tilde{L})$$

for all $L, \tilde{L} \in \mathcal{P}_k$.

PROOF. One checks easily that the choice $K_{k,\Gamma} = \max_{\lambda\in\Gamma} \|\lambda\|^{k-1}$ will ensure the right-hand inequality in (4.6.3).

To prove the left-hand inequality in (4.6.3) assume (without loss of generality) that the point $z_0 = 0$ is inside Γ. Let

$$L(\lambda) = \lambda^k I + \sum_{j=0}^{k-1} \lambda^j B_j, \quad \tilde{L}(\lambda) = \lambda^k I + \sum_{j=0}^{k-1} \lambda^j \tilde{B}_j.$$

By the maximum modulus principle

$$\|B_0 - \tilde{B}_0\| = \|L(0) - \tilde{L}(0)\| \leq \sigma_{k,\Gamma}(L, \tilde{L}). \tag{4.6.4}$$

Put

$$M(\lambda) = \lambda^{k-1} I + \sum_{j=1}^{k-1} \lambda^{j-1} B_j, \quad \tilde{M}(\lambda) = \lambda^{k-1} I + \sum_{j=1}^{k-1} \lambda^{j-1} \tilde{B}_j.$$

Then $M(\lambda) = \lambda^{-1}(L(\lambda) - L(0))$, $\tilde{M}(\lambda) = \lambda^{-1}(\tilde{L}(\lambda) - \tilde{L}(0))$, and hence

$$\sigma_{k-1,\Gamma}(M, \tilde{M}) \leq \frac{1}{\min_{\lambda \in \Gamma} |\lambda|} \left[\max_{\lambda \in \Gamma} \|L(\lambda) - \tilde{L}(\lambda)\| + \|L(0) - \tilde{L}(0)\| \right]$$

$$\leq \frac{2}{\min_{\lambda \in \Gamma} |\lambda|} \sigma_{k,\Gamma}(L, \tilde{L}).$$

Assuming (by induction on k) that (4.6.3) is already proved with k replaced by k-1, we obtain

$$C_{k-1,\Gamma} \sigma_{k-1}(M, \tilde{M}) \leq \sigma_{k-1,\Gamma}(M, \tilde{M}) \leq \frac{2}{\min_{\lambda \in \Gamma} |\lambda|} \sigma_{k,\Gamma}(L, \tilde{L}).$$

Using (4.6.4) and the obvious equation

$$\sigma_k(L, \tilde{L}) = \sigma_{k-1}(M, \tilde{M}) + \|B_0 - \tilde{B}_0\|,$$

the left-hand side inequality in (4.6.3) is proved. ■

Consider the set $\mathcal{W}_\ell$ consisting of all pairs $\{\mathcal{M}, L(\lambda)\}$, where $L(\lambda)$ is a monic operator polynomial of degree ℓ, and $\mathcal{M}$ is an invariant subspace in $\mathcal{X}^\ell$ for the companion operator C_L of L. This set $\mathcal{W}_\ell$ will be provided with the metric induced from the spherical gap metric and the metric σ_ℓ (so an ε-neighborhood of

$\{\mathcal{M}, L\} \in \mathcal{W}_\ell$ consists of all pairs $\{\tilde{\mathcal{M}}, \tilde{L}\} \in \mathcal{W}_\ell$ for which $\Theta_s(\tilde{\mathcal{M}}, \mathcal{M}) + \sigma_\ell(\tilde{L}, L) < \varepsilon$).

Define now the subset $\mathcal{V}_k \subset \mathcal{W}_\ell$ consisting of all pairs $\{\mathcal{M}, L(\lambda)\}$, where $L(\lambda) \in \mathcal{P}_\ell$ and $\mathcal{M}$ is a supporting subspace (with respect to the right spectral pair $([I\ 0 \cdots 0], C_L)$ of $L(\lambda)$), associated with a monic right divisor of L of degree k. The set $\mathcal{V}_k$ will be called the *supporting set* of order k. It follows from Theorem 2.7.1 that $\{\mathcal{M}, L(\lambda)\} \in \mathcal{V}_k$ if and only if $\mathcal{M}$ is C_L-invariant and

$$\mathcal{M} \dot{+} \mathcal{Y}_{\ell-k} = \mathcal{X}^\ell, \tag{4.6.5}$$

where $\mathcal{Y}_{\ell-k}$ is the closed subspace in $\mathcal{X}^\ell$ consisting of all ℓ-tuples of vectors $\mathrm{col}[x_i]_{i=1}^{\ell}$, $x_j \in \mathcal{X}$ with the property that $x_1 = \cdots = x_k = 0$. Theorem 4.2.1 shows that $\mathcal{V}_k$ is open in $\mathcal{W}_\ell$. Define a map $F_k\colon \mathcal{V}_k \to \mathcal{P}_{\ell-k} \times \mathcal{P}_k$ in the following way: the image of $(\mathcal{M}, L) \in \mathcal{V}_k$ is to be the pair of monic operator polynomials (L_2, L_1) where L_1 is the right divisor of L associated with $\mathcal{M}$ and L_2 is the quotient obtained by division of L on the right by L_1. It is evident that F_k is one-to-one and surjective so that the map F_k^{-1} exists.

For (L_2, L_1), $(\tilde{L}_2, \tilde{L}_1) \in \mathcal{P}_{\ell-k} \times \mathcal{P}_k$ put

$$\rho((L_2, L_1), (\tilde{L}_2, \tilde{L}_1)) = \sigma_{\ell-k}(L_2, \tilde{L}_2) + \sigma_k(\tilde{L}_1, \tilde{L}_1);$$

so $\mathcal{P}_{\ell-k} \times \mathcal{P}_k$ is a metric space with the metric ρ.

If X_1, X_2 are topological spaces with metrics ρ_1, ρ_2, defined on each connected component of X_1 and X_2, respectively, the map $G\colon X_1 \to X_2$ is called *locally Lipschitz continuous* if for every $x \in X_1$ there is a deleted neighborhood U_x of x for which

$$\sup_{y \in U_x} (\rho_2(Gx, Gy)/\rho_1(x, y)) < \infty.$$

It is an easily verified fact (and will be used in the proof of Theorem 4.6.2 below) that the composition of locally Lipschitz continuous maps is again locally Lipschitz continuous.

THEOREM 4.6.2. *The maps* F_k *and* F_k^{-1} *are locally Lipschitz continuous.*

PROOF. Let $(\mathcal{M},L) \in \mathcal{V}_k$ and

$$F_k(\mathcal{M},L) = (L_2,L_1).$$

Recall that by (4.6.1), $L_1(\lambda)$ has the following representation:

$$L_1(\lambda) = I\lambda^k - XC_L^k(W_1+W_2\lambda+\cdots+W_k\lambda^{k-1}),$$

where $X = [I\ 0\ \cdots\ 0]$,

$$[W_1\cdots W_k] = (Q_k|_{\mathcal{M}})^{-1},$$

and $Q_k = [I_{\mathcal{X}^k}\ \ 0] \in L(\mathcal{X}^\ell,\mathcal{X}^k)$.

In view of this formula, in order to verify that L_1 is a Lipschitz continuous function of $\mathcal{M}$ and L, it is sufficient to check that for a fixed $\mathcal{M} \subset \mathcal{X}^\ell$ which satisfies (4.6.5) there exist positive constants δ and C such that for any subspace $\mathcal{N} \subset \mathcal{X}^\ell$ with $\Theta_s(\mathcal{M},\mathcal{N}) < \delta$ we have $\mathcal{N}\dot{+}\mathcal{Y}_{\ell-k} = \mathcal{X}^\ell$ and

$$\|P_{\mathcal{M}}-P_{\mathcal{N}}\| \le C\Theta_s(\mathcal{M},\mathcal{N}),$$

where $P_{\mathcal{M}}$ (resp. $P_{\mathcal{N}}$) is the projection on $\mathcal{M}$ (resp. $\mathcal{N}$) along $\mathcal{Y}_{\ell-k}$. But this follows from the inequality (4.2.5).

To prove the local Lipschitz continuity of L_2 (as a function of $\mathcal{M}$ and L) we shall appeal to Proposition 4.6.1. Let Γ be a sufficiently large contour (so that the spectrum of L_1 is inside Γ). Then, for every $(\mathcal{M}',L')$ in a sufficiently small neighborhood of $(\mathcal{M},L)$, and for operator polynomials L_1',L_2' defined by

$$F_k(\mathcal{M}',L') = (L_2',L_1'),$$

the operator polynomial L_1' will be invertible on Γ. Moreover, we have for $\lambda \in \Gamma$:

$$\|L_2'(\lambda)-L_2(\lambda)\| \le \|L(\lambda)-L'(\lambda)\| \, \|L_1(\lambda)^{-1}\|+\|L'(\lambda)\| \, \|L_1(\lambda)^{-1}-L_1'(\lambda)^{-1}\|$$

$$\le \|L(\lambda)-L'(\lambda)\| \, \|L_1(\lambda)^{-1}\|+\|L'(\lambda)\| \, \|L_1(\lambda)^{-1}\| \, \|L_1'(\lambda)^{-1}\|$$

$$\cdot \, \|L_1(\lambda)-L_1'(\lambda)\|.$$

This inequality shows (taking into account the already proved local Lipschitz continuity of L_1) that

$$\sigma_{\ell-k,\Gamma}(L_2,L_2') \le C\{\Theta_s(\mathcal{M},\mathcal{M}')+\sigma_\ell(L,L')\},$$

where the constant C depends on $(\mathcal{M},L)$ only. In view of Proposition 4.6.1 the local Lipschitz continuity of L_2 follows.

To establish the local Lipschitz continuity of F_k^{-1} we consider a fixed $(L_2,L_1) \in \mathcal{P}_{\ell-k}\times\mathcal{P}_k$. It is apparent that the polynomial $L = L_2L_1$ will be a Lipschitz continuous function of L_2 and L_1 in a neighborhood of the fixed pair. To examine the behavior of the spherical gap between supporting subspaces associated with neighboring pairs we observe an explicit construction for $P_\mathcal{M}$, the projection on $\mathcal{M}$ along $\mathcal{Y}_{\ell-k}$ (associated with the pair L_2,L_1). In fact, $P_\mathcal{M}$ has the representation

$$P_\mathcal{M} = \begin{bmatrix} I & 0 \\ F & 0 \end{bmatrix}, \quad F = \begin{bmatrix} P_1C_{L_1}^k \\ \vdots \\ P_1C_{L_1}^{\ell-1} \end{bmatrix}, \tag{4.6.6}$$

with respect to the decomposition $\mathcal{X}^\ell = \mathcal{X}^k \oplus \mathcal{X}^{\ell-k}$, where $P_1 = [I \quad 0\cdots0]$. Indeed, $P_\mathcal{M}$ given by (4.6.6) is obviously a projection along $\mathcal{Y}_{\ell-k}$. Let us check that $\operatorname{Im} P_\mathcal{M} = \mathcal{M}$. The subspace $\mathcal{M}$ is the supporting subspace corresponding to the right divisor $L_1(\lambda)$ of $L(\lambda)$; by formula (4.6.2), $\mathcal{M} = \operatorname{Im} \operatorname{col}[P_1C_{L_1}^i]_{i=0}^{\ell-1} = \operatorname{Im} P_\mathcal{M}$.

The local Lipschitz continuity of $P_\mathcal{M}$ as a function of L_1 is apparent from formula (4.6.6), and the local Lipschitz continuity of F_k^{-1} now follows from the left-hand inequality of (4.2.5). ∎

4.7 Stability of factorizations of monic operator polynomials

We say that a factorization

$$L(\lambda) = L_2(\lambda)L_1(\lambda)$$

of a monic operator polynomial $L(\lambda)$, where $L_1(\lambda)$ and $L_2(\lambda)$ are monic polynomials as well, is *stable* if for any $\varepsilon > 0$ there exists a $\delta > 0$ such that any monic operator polynomial $\tilde{L}(\lambda)$ with $\sigma_\ell(\tilde{L},L) < \delta$ admits a factorization $\tilde{L}(\lambda) = \tilde{L}_2(\lambda)\tilde{L}_1(\lambda)$, where $\tilde{L}_1(\lambda)$ are monic operator polynomials satisfying

$$\sigma_k(\tilde{L}_1,L_1)+\sigma_{\ell-k}(\tilde{L}_2,L_2) < \varepsilon.$$

Here ℓ is the degree of L and $\tilde{L}$, whereas k is the degree of L_1 and $\tilde{L}_1$.

It turns out that stable factorizations are precisely the ones with stable supporting subspaces.

THEOREM 4.7.1. *Let* L, L_1, L_2 *be monic operator polynomials such that*

$$L(\lambda) = L_2(\lambda)L_1(\lambda). \tag{4.7.1}$$

Let $\mathcal{M}$ *be the supporting subspace for the factorization* (4.7.1), *with respect to the right spectral pair* $([I\ 0\ \cdots\ 0],C_L)$ *of* $L(\lambda)$, *where* C_L *is the companion operator of* $L(\lambda)$ *(in particular,* $\mathcal{M}$ *is* C_L*-invariant). Then* (4.7.1) *is a stable factorization if and only if* $\mathcal{M}$ *is a stable* C_L*-invariant subspace.*

PROOF. If $\mathcal{M}$ is stable C_L-invariant subspace, then use Theorem 4.6.2 to show that (4.7.1) is a stable factorization.

Now conversely, suppose the factorization is stable, but $\mathcal{M}$ is not. Then there exists $\varepsilon > 0$ and a sequence of operators $\{C_m\}$ converging to C_L such that for all $\mathcal{V} \in \mathrm{Inv}(C_m)$

$$\Theta_s(\mathcal{V},\mathcal{M}) \geq \varepsilon,\ m = 1,2,\ldots . \tag{4.7.2}$$

Here $\mathrm{Inv}(C_m)$ denotes the collection of all invariant subspaces for C_m. Put $Q = [I\ \ 0\ \cdots\ 0]$ and

$$S_m = \mathrm{col}[QC_m^{i-1}]_{i=1}^{\ell}, \quad m = 1,2,\ldots .$$

Then $\{S_m\}$ converges to $\mathrm{col}[QC_L^{i-1}]_{i=1}^{\ell}$, which is equal to $I_{\mathcal{X}^\ell}$. So without loss of generality we may assume that S_m is invertible for all m, say with inverse $S_m^{-1} = [U_{m1}, U_{m2}, \ldots, U_{m\ell}]$. Note that

$$U_{mi} \to \mathrm{col}[\delta_{ji} I]_{j=1}^{\ell}, \quad i = 1,\ldots,\ell. \tag{4.7.3}$$

A straightforward calculation shows that $S_m C_m S_m^{-1}$ is the companion operator associated with the operator polynomial

$$L_m(\lambda) = \lambda^\ell I - \sum_{i=0}^{\ell-1} \lambda^i QC_m^\ell U_{m,i+1} .$$

From (4.7.3) and the fact that $C_m \to C_L$, it follows that $\sigma_\ell(L_m, L) \to 0$. But then we may assume that for all m the polynomial L_m admits a factorization $L_m = L_{m2}L_{m1}$ with $L_{m1} \in \mathcal{P}_r$, $L_{m2} \in \mathcal{P}_{\ell-r}$, and

$$\sigma_k(L_{m1}, L_1) \to 0, \quad \sigma_{\ell-k}(L_{m2}, L_2) \to 0.$$

Let $\mathcal{M}_m$ be the supporting subspace corresponding to the factorization $L_m = L_{m2}L_{m1}$. By Theorem 4.6.2 we have $\Theta_s(\mathcal{M}_m, \mathcal{M}) \to 0$. Put $\mathcal{V}_m = S_m^{-1}\mathcal{M}_m$. Then $\mathcal{V}_m$ is an invariant subspace for C_m. Moreover, it follows from $S_m \to I$ that

$$\lim_{m\to\infty} \Theta_s(\mathcal{V}_m, \mathcal{M}_m) = 0. \tag{4.7.4}$$

Indeed, by Theorem 4.2.5 we have

$$\Theta_s(\mathcal{V}_m, \mathcal{M}_m) \leq 2\|S_m^{-1}P_m S_m - P_m\|,$$

where P_m is the projection on $\mathcal{M}_m$ along the subspace

$$\mathcal{Y} = \{\mathrm{col}[x_i]_{i=1}^{\ell} \in \mathcal{X}^\ell \mid x_1 = \cdots = x_k = 0\}.$$

Now by Theorem 4.6.2

$$\|P_m - P\| \to 0 \text{ as } m \to \infty,$$

where P is the projector on $\mathcal{M}$ along $\mathcal{Y}$. Hence

$$\|S_m^{-1}P_mS_m - P_m\| \le \|(S_m^{-1}-I)P_mS_m\| + \|P_m(I-S_m)\|$$

$$\le (\max_m \|S_m\| \cdot \|S_m^{-1}-I\| + \|I-S_m\|)\max_m \|P_m\|,$$

which tends to zero as m tends to infinity, and (4.7.4) follows. But (4.7.4) contradicts our assumption (4.7.2), and the proof is complete. ∎

Combining Theorem 4.7.1 with the results of Section 4.3 we obtain the following facts.

THEOREM 4.7.2. *The set of stable factorizations of a monic operator polynomial* $L(\lambda)$ *is closed in the following sense: If*

$$L(\lambda) = L_{2m}(\lambda)L_{1m}(\lambda), \quad m = 1,2,\ldots$$

are stable factorizations of $L(\lambda)$, *and if* $\sigma_k(L_{1m}, L_1) \to 0$ *as* $m \to \infty$ *for some monic operator polynomial* L_1, *then the factorization*

$$L(\lambda) = L_2(\lambda)L_1(\lambda) \tag{4.7.5}$$

is stable as well.

Note that the condition $\lim_{m\to\infty} \sigma_k(L_{1m}, L_1) = 0$ easily implies that L_1 is a right divisor of L, so the factorization (4.7.5) indeed exists.

THEOREM 4.7.3. *Let* $L = L_2L_1$ *be a stable factorization of a monic operator polynomial* L, *and let* $L_1 = M_2M_1$ *be in turn a stable factorization of the monic operator polynomial* L_1. *Then* $L = (L_2M_2)\cdot M_1$ *is a stable factorization of* L.

We pass now to the notion of Lipschitz stable factorizations.

A factorization

$$L(\lambda) = L_2(\lambda)L_1(\lambda) \tag{4.7.6}$$

of the monic operator polynomial $L(\lambda)$, where $L_1(\lambda)$ and $L_2(\lambda)$ are monic operator polynomials as well, is called *Lipschitz stable* if there exist positive constants ε and K such that any monic operator polynomial $\tilde{L}(\lambda)$ with $\sigma_\ell(\tilde{L},L) < \varepsilon$ admits a factorization $\tilde{L}(\lambda) = \tilde{L}_2(\lambda)\tilde{L}_1(\lambda)$ with monic operator polynomials $\tilde{L}_1(\lambda)$ satisfying

$$\sigma_k(\tilde{L}_1,L_1)+\sigma_{\ell-k}(\tilde{L}_2,L_2) \leq K\sigma_\ell(\tilde{L},L).$$

Here ℓ is the degree of L and $\tilde{L}$, and k is the degree of L_1 and $\tilde{L}_1$. Obviously, every Lipschitz stable factorization is stable. The converse is not true in general, already in the finite-dimensional case, as will follow from Theorem 4.7.4 below.

Analogously to the stable factorizations, the Lipschitz stable factorizations are described in terms of supporting invariant subspaces with this property:

THEOREM 4.7.4. *The factorization* (4.7.6) *is Lipschitz stable if and only if the corresponding supporting* C_L*-invariant subspace* $\mathcal{M}$ *is Lipschitz stable. (As usual,* C_L *stands for the companion operator of* $L(\lambda)$*).*

PROOF. The proof follows the pattern of the proof of Theorem 4.7.1. If the subspace $\mathcal{M}$ is Lipschitz stable, then by Theorem 4.6.2 the factorization (4.7.6) is Lipschitz stable. Conversely, assume that the factorization (4.7.6) is Lipschitz stable but the subspace $\mathcal{M}$ is not. Then there exists a sequence $\{C_m\}_{m=1}$ of operators such that $\|C_m-C_L\| < (1/m)$ and for every C_m-invariant subspace $\mathcal{L}$ the inequality

$$\theta_s(\mathcal{M},\mathcal{L}) \geq m\|C_m-C_L\| \tag{4.7.7}$$

holds. Putting $S_m = \mathrm{col}[QC_m^{i-1}]_{i=1}^{\ell}$, where $Q = [I\ 0\ \cdots\ 0]$, we verify that S_m is invertible (at least for large m) and that $S_mC_mS_m^{-1}$ is the companion operator associated with the operator polynomial

$$M_m(\lambda) = \lambda^\ell I - \sum_{i=0}^{\ell-1}\lambda^i QC_m^\ell U_{m,i+1}$$

where $[U_{m1}, U_{m2}, \ldots, U_{m\ell}] = S_m^{-1}$. We assume that S_m is invertible for $m = 1,2,\ldots$. Observe that $\mathrm{col}[QC_L^{i-1}]_{i=1}^{\ell} = I_{\mathcal{X}^\ell}$; so it is not difficult to check that for $m = 1,2,\ldots$

(4.7.8) $$\sigma_\ell(M_m, L) \le K_1\|C_m - C_L\|.$$

Here and in the sequel we denote certain positive constants independent of m by $K_1, K_2, \ldots$. As the factorization (4.7.6) is Lipschitz stable, for m sufficiently large the polynomial $M_m(\lambda)$ admits a factorization

(4.7.9) $$M_m(\lambda) = M_{2m}(\lambda)M_{1m}(\lambda)$$

with monic operator polynomials $M_{1m}(\lambda)$ and $M_{2m}(\lambda)$ such that

(4.7.10) $$\sigma_k(M_{1m}, L_1) + \sigma_{\ell-k}(M_{2m}, L_2) \le K_2\sigma_\ell(M_m, L).$$

Let $\mathcal{M}_m$ be the C_{M_m}-invariant subspace corresponding to the factorization (4.7.10). By Theorem 4.6.2 we have

(4.7.11) $$\Theta_s(\mathcal{M}_m, \mathcal{M}) + \sigma_\ell(M_m, L) \le K_3(\sigma_k(M_{1m}, L_1) + \sigma_{\ell-k}(M_{2m}, L_2)).$$

From (4.7.8), (4.7.10), and (4.7.11) one obtains

(4.7.12) $$\Theta_s(\mathcal{M}_m, \mathcal{M}) \le K_1K_2K_3\|C_m - C_L\|.$$

Put $\mathcal{V}_m = S_m^{-1}\mathcal{M}_m$ for $m = 1,2,\ldots$. Then $\mathcal{V}_m$ is C_m-invariant for each m. Further, the formula for S_m shows that

(4.7.13) $$\|I - S_m\| \le K_4\|C_m - C_L\|.$$

Indeed,

$$\begin{aligned} I - S_m &= \mathrm{col}[Q(C_L^{i-1} - C_m^{i-1})]_{i=1}^{\ell} \\ &= \mathrm{col}[Q(C_L^{i-2}(C_L - C_m) + C_L^{i-3}(C_L - C_m)C_m + \cdots + (C_L - C_m)C_m^{i-2}]_{i=1}^{\ell} \end{aligned}$$

and (4.7.13) follows. Now (cf. the proof of Theorem 4.7.1)

$$\Theta_s(\mathcal{V}_m, \mathcal{M}_m) \le K_5\|C_m - C_L\|.$$

Using this inequality and (4.7.12) we obtain

$$\Theta_s(\mathcal{V}_m,\mathcal{M}) \le \Theta_s(\mathcal{V}_m,\mathcal{M}_m)+\Theta_s(\mathcal{M}_m,\mathcal{M}) \le K_6\|C_m-C_L\|$$

a contradiction with (4.7.7). ∎

Combining Theorem 4.7.4 with known classes of Lipschitz stable invariant subspaces (see Section 4.5) we obtain the following result.

THEOREM 4.7.5. *Let* $L(\lambda) = L_2(\lambda)L_1(\lambda)$ *be a factorization of monic operator polynomials. If* $\Sigma(L_1) \cap \Sigma(L_2) = \phi$, *then this factorization is Lipschitz stable.*

Indeed, the condition $\Sigma(L_1) \cap \Sigma(L_2) = \phi$ means that the spectrum of the companion operator C_L is a disjoint union of $\Sigma(L_1)$ and $\Sigma(L_2)$, and the supporting C_L-invariant subspace for the factorization $L = L_2L_1$ is precisely

$$\operatorname{Im}\left[\frac{1}{2\pi i}\int_{\Gamma}(\lambda I-C_L)^{-1}d\lambda\right],$$

where Γ is a suitable contour that separates $\Sigma(L_1)$ from $\Sigma(L_2)$. It remains to apply Proposition 4.3.1 and Theorem 4.7.4.

An analogous result concerning Lipschitz stability of factorizations can be stated using the Lipschitz stable invariant subspaces related to one-sided resolvents.

THEOREM 4.7.6. *Let* $L = L_2L_1$ *be a factorization with monic operator polynomials* L, L_1, *and* L_2, *and assume that the supporting subspace* $\mathcal{M}$ *of this factorization with respect to a right spectral pair* (X,T) *of* L *is of the following form:*

$$\mathcal{M} = \operatorname{Im}\left[\frac{1}{2\pi i}\int_{\Gamma}F(\lambda)d\lambda\right].$$

Here $F(\lambda)$ *is a one-sided resolvent for* T *with respect to an open set* Ω, *and* Γ *is a simple rectifiable Jordan curve in* Ω. *Then the factorization* $L = L_2L_1$ *is Lipschitz stable.*

For the proof of Theorem 4.7.5 combine Theorems 4.5.3 and 4.7.4.

4.8 Stable sets of invariant subspaces

The notion of stable invariant subspaces can be extended, in a natural way, to sets of subspaces. Let $\mathcal{X}$ be a Banach space, with the set $S(\mathcal{X})$ of all subspaces in $\mathcal{X}$, considered as a complete metric space in the spherical gap metric $\Theta_s(\mathcal{M},\mathcal{N})$. For two closed subsets $S_1, S_2 \subset S(\mathcal{X})$ define the Hausdorff distance

$$d(S_1,S_2) = \max\left\{\sup_{\mathcal{M}\in S_1} \inf_{\mathcal{N}\in S_2} \Theta_s(\mathcal{M},\mathcal{N}),\ \sup_{\mathcal{M}\in S_2} \inf_{\mathcal{N}\in S_1} \Theta_s(\mathcal{M},\mathcal{N})\right\}.$$

It is easy to see that

$$d(S_1,S_2) = d(S_2,S_1);$$

$d(S_1,S_2) \geq 0$; and $d(S_1,S_2) = 0$ if and only if $S_1 = S_2$. Moreover, the triangle inequality is valid;

$$d(S_1,S_2) \leq d(S_1,S_3)+d(S_3,S_2).$$

This is proved in the same way as the triangle inequality (4.1.2) for the spherical gap. So the set of closed subsets in $S(\mathcal{X})$ is a metric space with the metric $d(S_1,S_2)$. As $S(\mathcal{X})$ is complete in the spherical gap metric, the set of closed subsets in $S(\mathcal{X})$ is complete in the metric $d(S_1,S_2)$ (see, e.g., Munkres [1]).

Let $T\colon \mathcal{X} \to \mathcal{X}$ be a linear bounded operator. A closed set $\mathcal{C}$ of T-invariant subspaces will be called *stable* if for every $\varepsilon > 0$ there is $\sigma > 0$ such that any operator $S \in L(\mathcal{X})$ with $\|S-T\| < \delta$ has a closed set $\mathcal{C}_S$ of S-invariant subspaces with the property that

$$d(\mathcal{C}_S,\mathcal{C}) < \varepsilon.$$

If $\mathcal{C}$ consists of one element, a T-invariant subspace $\mathcal{M}$, then the stability of $\mathcal{C}$ as a closed set and the stability of $\mathcal{M}$ as a T-invariant subspace (as defined in Section 4.3) are the same. Thus, the notion of stability of closed sets of invariant subspaces indeed extends the definition of a stable invariant subspace given in Section 4.3.

Clearly, a necessary condition for a closed set $\mathcal{C}$ of T-invariant subspaces to be stable is that every member of $\mathcal{C}$ is a stable T-invariant subspace. The converse is also true provided $\mathcal{C}$ is finite:

THEOREM 4.8.1. *A finite set $\mathcal{C}$ of T-invariant subspaces is stable if and only if every member in $\mathcal{C}$ is a stable T-invariant subspace.*

PROOF. We have to prove the part "if." Let $\mathcal{C} = \{\mathcal{M}_1, \ldots, \mathcal{M}_p\}$ where every $\mathcal{M}_i$ is a stable T-invariant subspace. Given $\varepsilon > 0$ there is $\delta > 0$ such that every $S \in L(\mathcal{X})$ with $\|S-T\| < \delta$ has invariant subspaces $\mathcal{N}_i$ $(i = 1, \ldots, p)$ with the property that

$$\Theta_s(\mathcal{N}_i, \mathcal{M}_i) < \varepsilon, \quad i = 1, \ldots, p.$$

Letting $\mathcal{C}_S = \{\mathcal{N}_1, \ldots, \mathcal{N}_p\}$ we obtain $d(\mathcal{C}, \mathcal{C}_S) < \varepsilon$, and the theorem follows. ■

In the finite-dimensional case this theorem implies a description of all stable sets of invariant subspaces.

COROLLARY 4.8.2. *If $\dim \mathcal{X} < \infty$, then a set $\mathcal{C}$ of T-invariant subspaces is stable if and only if every member of $\mathcal{C}$ is a stable T-invariant subspace.*

PROOF. Assume that every member of $\mathcal{C}$ is a stable T-invariant subspace. It follows from Theorem 4.3.2 that the number of stable T-invariant subspaces is finite (indeed, this number is precisely $p_1 p_2 \cdots p_r$, where $\lambda_1, \ldots, \lambda_r$ are all the distinct eigenvalues of T, and $p_j = 2$ if $\dim \operatorname{Ker}(\lambda_j T - I) > 1$ and $p_j = 1 + \dim \operatorname{Ker}(\lambda_j T - I)^n$ if $\dim \operatorname{Ker}(\lambda_j T - I) = 1$; here $n = \dim \mathcal{X}$). It remains to apply Theorem 4.8.1. ■

The notion of stable sets of invariant subspaces can be naturally applied to the study of stable sets of factorizations of a given monic operator polynomial.

Given two closed (in the metric σ_ℓ) sets S_1 and S_2 of monic operator polynomials of degree ℓ, define the distance $d_\ell(S_1, S_2)$ as follows:

$$d_\ell(S_1, S_2) = \max\left\{\sup_{T \in S_1} \inf_{M \in S_2} \sigma_\ell(L, M), \ \sup_{L \in S_2} \inf_{M \in S_1} \sigma_\ell(L, M)\right\}.$$

Again, d_ℓ is a metric on the set of all closed sets of monic operator polynomials of degree ℓ, and this set is complete in the metric d_ℓ.

Let $L(\lambda)$ be a monic operator polynomial of degree ℓ with coefficients in $L(\mathcal{X})$. A *closed* set of factorizations

$$L(\lambda) = L_{2\alpha}(\lambda)L_{1\alpha}(\lambda), \quad \alpha \in A \tag{4.8.1}$$

where $L_{1\alpha}(\lambda)$ and $L_{2\alpha}(\lambda)$ are monic operator polynomials of degrees ℓ-k and k, respectively, and A is an index set, is called *stable* if, given $\varepsilon > 0$ there is $\delta > 0$ such that for any monic operator polynomial $\tilde{L}(\lambda)$ with $\sigma_\ell(\tilde{L},L) < \delta$ there is a *closed* set of factorizations

$$\tilde{L}(\lambda) = \tilde{L}_{2\beta}(\lambda)\tilde{L}_{1\beta}(\lambda), \quad \beta \in \mathcal{B} \tag{4.8.2}$$

where $\tilde{L}_{1\beta}$ and $\tilde{L}_{2\beta}$ are of degrees k and ℓ-k, respectively, such that

$$d_k(\tilde{S},S)+d_{\ell-k}(\tilde{T},T) < \varepsilon .$$

Here

$$S = \{L_{1\alpha}(\lambda) \mid \alpha \in A\}; \quad \tilde{S} = \{L_{1\beta}(\lambda) \mid \beta \in \mathcal{B}\};$$

$$T = \{L_{2\alpha}(\lambda) \mid \alpha \in A\}; \quad \tilde{T} = \{L_{2\beta}(\lambda) \mid \beta \in \mathcal{B}\};$$

and the closedness of the sets of factorizations (4.8.1) and (4.8.2) is understood in the natural sense: if

$$L(\lambda) = L_{2\alpha_j}(\lambda)L_{1\alpha_j}(\lambda), \quad j = 1,2,\ldots$$

are factorizations from the set (4.8.1), and

$$\lim_{j\to\infty}[\sigma_k(L_{1\alpha_j},L_1)+\sigma_{\ell-k}(L_{2\alpha_j},L_2)] = 0$$

for some operator polynomials L_1 and L_2, then the factorization $L = L_2L_1$ also belongs to the set (4.8.1).

As in the proof of Theorem 4.7.1 one shows that a closed set of factorizations (4.8.1) is stable if and only if the corresponding closed set of C_L-invariant subspaces is stable. In particular, applying Theorem 4.8.1, we see that a finite set of factorizations is stable if and only if each member of this set is a stable factorization.

Analogously, one can define and study Lipschitz stable closed sets of invariant subspaces and Lipschitz stable closed set of factorizations of monic operator polynomials. However, we will not do this here.

4.9 Exercises

Ex. 4.1. Prove the part (ii) of Theorem 4.5.1.

Ex. 4.2. Prove the result for left resolvents analogous to Theorem 4.5.2.

Ex. 4.3. Show that the notions of spherical gap, gap and minimal angle are unitary invariant (assuming $\mathcal{X}$ is a Hilbert space)

$$\Theta_s(\mathcal{M},\mathcal{N}) = \Theta_s(U\mathcal{M},U\mathcal{N});$$

$$\phi_{min}(\mathcal{M},\mathcal{N}) = \phi_{min}(U\mathcal{M},U\mathcal{N});$$

$$\Theta(\mathcal{M},\mathcal{N}) = \Theta(U\mathcal{M},U\mathcal{N})$$

for any unitary $U \in L(\mathcal{X})$.

Ex. 4.4. Let $A \in L(\mathcal{X})$ where $\mathcal{X}$ is a Banach space. Prove that if $\mathcal{M}_1+\mathcal{M}_2$ are stable A-invariant subspaces and the sum $\mathcal{M}_1\dot{+}\mathcal{M}_2$ is direct, then $\mathcal{M}_1\dot{+}\mathcal{M}_2$ is also a stable A-invariant subspace.

Ex. 4.5. State and prove the result analogous to Ex. 4.4 for the case of Lipschitz stability.

Ex. 4.6. Let $S \in L(\ell_2)$ be the left shift:

$$S(x_0,x_1,\ldots) = (x_1,x_2,\ldots)$$

(a) Prove that the S-invariant subspace Ker S is Lipschitz stable.

(b) Prove that the S^*-invariant subspace Im S^* is Lipschitz stable.

Ex. 4.7. Prove that every factorization of a monic scalar polynomial is stable.

Ex. 4.8. Give an example of non-stable factorization of a 2×2 monic matrix polynomial.

Ex. 4.9. Let Z be the set of $n\times n$ monic matrix polynomials $L(\lambda)$ of degree ℓ such that all factorizations of $L(\lambda)$ into monic factors are Lipschitz stable. Prove that Z is dense in the set of all $n\times n$ monic matrix polynomials of degree ℓ.

Ex. 4.10. Let $L(\lambda)$ be a monic operator polynomial with (Lipschitz) stable factorization

$$L(\lambda) = M(\lambda)N(\lambda),$$

where

$$M(\lambda) = \begin{bmatrix} M_1(\lambda) & 0 \\ 0 & M_2(\lambda) \end{bmatrix}, \quad M(\lambda) = \begin{bmatrix} N_1(\lambda) & 0 \\ 0 & N_2(\lambda) \end{bmatrix}$$

with respect to some direct sum decomposition $\mathcal{X} = \mathcal{X}_1 \dot{+} \mathcal{X}_2$. Prove that the factorizations $L_1(\lambda) = M_1(\lambda)N_1(\lambda)$, $L_2(\lambda) = M_2(\lambda)N_2(\lambda)$ are (Lipschitz) stable as well.

Ex. 4.11. Show by example that the converse of Ex. 4.10 is false in general: The factorizations $L_1 = M_1N_1$ and $L_2 = M_2N_2$ can be stable, while $L = MN$ is not.

4.10 Notes

Theorem 4.1.1 is proved in Gohberg-Markus [1]. The material in Sections 4.1 and 4.2 are standard, see e.g., Kato [1], Gohberg-Lancaster-Rodman [2,3]. The notion of the metric space of subspaces in a Banach space first appeared in Kreĭn-Krasnoselskii-Milman [1], see also Gohberg-Kreĭn [2]. Problems concerning the global topological behavior of direct sums of subspaces were considered in Rodman [4]. For further results on the topology of the set of invariant subspaces, see Douglas-Pearcy [1]. Problem 4.5.1 is a known problem (see Apostol-Clancey [1], Herrero [1], Apostol-Fialkow-Herrero-Voiculescu [1]). One-sided resolvents are widely used in operator theory (see, e.g., Apostol-Clancey [1], Herrero [1,2], Apostol-Fialkow-

Herrero-Voiculescu [1]). Lipschitz continuous dependence of matrix polynomials on their spectral pairs and vice versa was first observed in Bart-Gohberg-Kaashoek [1]. Theorems 4.6.2 and 4.7.1 are taken from Gohberg-Lancaster-Rodman [7], and the exposition of their proofs generally follows Gohberg-Lancaster-Rodman [3].

CHAPTER 5. SELF-ADJOINT OPERATOR POLYNOMIALS

Let $\mathcal{X}$ be a Hilbert space with the scalar product $\langle x,y\rangle$; $x,y \in \mathcal{X}$. In this chapter we consider monic operator polynomials

$$L(\lambda) = \sum_{j=0}^{\ell-1} \lambda^j A_j + \lambda^\ell I$$

whose coefficients are (bounded) self-adjoint operators on $\mathcal{X}$: $A_j = A_j^*$ for $j = 0,\dots,\ell-1$.

Let (X,T,Y) be a spectral triple of $L(\lambda)$. As T is similar to the companion operator C_L of $L(\lambda)$ which acts on the Hilbert space $\mathcal{X}^\ell$, we can assume without loss of generality that T acts on $\mathcal{X}^\ell$ as well. By Theorem 2.5.2 we have

$$L(\lambda)^{-1} = X(\lambda I-T)^{-1}Y.$$

Taking adjoints in this formula and using the self-adjointness of $L(\lambda)$ we obtain

$$L(\lambda)^{-1} = Y^*(\lambda I-T^*)^{-1}X^*.$$

Actually, (Y^*,T^*,X^*) is also a spectral triple of $L(\lambda)$ (this follows from Theorem 2.5.3). Thus, these spectral triples are similar, i.e.,

$$X = Y^*S, \quad T = S^{-1}T^*S, \quad Y = S^{-1}X^* \tag{1}$$

for some invertible operator S which is uniquely determined by (X,T,Y) (see Proposition 2.5.1(a)). Upon taking adjoints in (1) one verifies also that

$$X = Y^*S^*, \quad T = S^{*-1}T^*S^*, \quad Y = S^{*-1}X,$$

and hence, by uniqueness of S we must have $S = S^*$, i.e., S is self-adjoint.

Introduce the indefinite scalar product in $\mathcal{X}^\ell$ by the formula

$$[x,y] = \langle Sx,y\rangle, \quad x,y \in \mathcal{X}^\ell.$$

(recall that $\mathcal{X}^\ell$ is considered as a Hilbert space with the scalar product

$$\left\langle \begin{bmatrix} x_1 \\ x_2 \\ \vdots \\ x_\ell \end{bmatrix}, \begin{bmatrix} y_1 \\ y_2 \\ \vdots \\ y_\ell \end{bmatrix} \right\rangle = \sum_{i=1}^{\ell} \langle x_i, y_i\rangle,$$

where $x_1,\dots,x_\ell$, $y_1,\dots,y_\ell \in \mathcal{X}$). Thus, $[\cdot,\cdot]$ satisfies all the axioms for the scalar product except that it may happen $[x,x] \leq 0$ for non-zero $x \in \mathcal{X}^\ell$. An operator $A \in L(\mathcal{X}^\ell)$ is called *self-adjoint* with respect to the indefinite scalar product $[\cdot,\cdot]$, or, in short, S-self-adjoint if

$$[Ax,y] = [x,Ay] \text{ for all } x,y \in \mathcal{X}^\ell.$$

This property is easily seen to be equivalent to the equality $SA = A^*S$. Now (1) implies, in particular, that T is S-self-adjoint. As we shall see in this chapter, this is the key property for study of factorizations of $L(\lambda)$. To make this approach work we need basic information on indefinite scalar products and operators self-adjoint with respect to these products. This information is given in the next two sections.

5.1 Indefinite scalar products and subspaces

Let $\mathcal{Y}$ be a Hilbert space with the scalar product $\langle\cdot,\cdot\rangle$, and let $J \in L(\mathcal{Y})$ be an invertible bounded self-adjoint operator. A (closed) subspace $\mathcal{M} \subset \mathcal{Y}$ is called *J-nonnegative* if $\langle Jx,x\rangle \geq 0$ for all $x \in \mathcal{M}$, and *J-nonpositive* if $\langle Jx,x\rangle \leq 0$ for all $x \in \mathcal{M}$. Replacing here the condition "$\langle Jx,x\rangle \geq 0$ for all $x \in \mathcal{M}$" by a generally stronger condition "$\langle Jx,x\rangle > 0$ for all $x \in \mathcal{M}\setminus\{0\}$," we obtain the definition of a J-positive subspace. Analogously a

J-negative subspace is defined. A subspace $\mathcal{M}$ which is one of these four types will be called *J-semidefinite*. For a J-semidefinite subspace $\mathcal{M}$ the Schwarz inequality is valid:

$$|\langle Jx, y\rangle|^2 \le \langle Jx, x\rangle \langle Jy, y\rangle, \quad x, y \in \mathcal{M}. \tag{5.1.1}$$

The proof is the same as the standard proof of the Schwarz inequality.

Of special interest are maximal J-semidefinite subspace. A subspace $\mathcal{M} \subset \mathcal{Y}$ is called *maximal J-nonnegative* if $\mathcal{M}$ is J-nonnegative and there is no J-nonnegative subspace that strictly contains $\mathcal{M}$. Analogously *maximal J-nonpositive* subspaces are defined. Observe that these classes of subspaces are not empty. Indeed, the spectral subspace of J corresponding to the positive part of $\sigma(J)$ is easily seen to be maximal J-nonnegative.

For a (closed) subspace $\mathcal{M} \subset \mathcal{Y}$, let

$$\mathcal{M}^{\perp J} = \{x \in \mathcal{Y} \mid \langle Jx, x\rangle = 0 \text{ for all } y \in \mathcal{M}\}$$

be the J-orthogonal companion of $\mathcal{M}$. Clearly $\mathcal{M}^{\perp J}$ is a (closed) subspace in $\mathcal{Y}$. Easy examples (e.g., $\mathcal{M} = \text{Span}\,\{x\}$, where $x \in \mathcal{Y}$, $x \ne 0$ is such that $\langle Jx, x\rangle = 0$) show that the J-orthogonal companion of $\mathcal{M}$ need not be a direct complement to $\mathcal{M}$.

From the definition of the J-orthogonal companion it follows easily that

$$\mathcal{M}^{\perp J} = (J\mathcal{M})^{\perp}, \tag{5.1.2}$$

where by "$\perp$" we denote the orthogonal complement.

Maximal J-nonpositive (or J-nonnegative) subspaces and their companions can be conveniently studied using angular operators. To introduce these we start with the spectral J-invariant subspaces $\mathcal{Y}_+$ and $\mathcal{Y}_-$ corresponding to the positive part and the negative part of the spectrum of J, respectively, and define $J_0 \in L(\mathcal{Y})$ by

$$J_0 x = x \text{ for all } x \in \mathcal{Y}_+; \tag{5.1.3}$$

$$J_0 x = -x \text{ for all } x \in \mathcal{Y}_-.$$

Clearly, $J_0^* = J_0$ and $J_0^2 = I$. For an operator $K \in L(\mathcal{D}_+, \mathcal{Y}_-)$ defined on a subspace $\mathcal{D}_+ \subset \mathcal{Y}_+$, let

$$\mathcal{G}(K) = \{x+Kx \mid x \in \mathcal{D}_+\}$$

be its graph.

PROPOSITION 5.1.1. *Let $\mathcal{D}_+$ be a subspace in $\mathcal{Y}_+$, and let $K \in L(\mathcal{D}_+, \mathcal{Y}_-)$ be a contraction (i.e., $\|K\| \leq 1$). Then the graph $\mathcal{G}(K)$ is J_0-nonnegative. Conversely, for every J_0-nonnegative subspace $\mathcal{M}$ there is unique $\mathcal{D}_+ \subset \mathcal{Y}_+$ and unique contraction $K \in L(\mathcal{D}_+, \mathcal{Y}_-)$ such that $\mathcal{M} = \mathcal{G}(K)$. The (closed) subspace $\mathcal{D}_+$ and K are determined by the formulas $\mathcal{D}_+ = P_+(\mathcal{M})$ and $K(P_+x) = P_-x$, where $P_+ = I-P_-$ is the (orthogonal) projection on $\mathcal{Y}_+$ along $\mathcal{Y}_-$.*

PROOF. The first statement follows from the equality

$$\langle J_0(x+Kx), x+Kx\rangle = \|x\|^2 - \|Kx\|^2, \quad x \in \mathcal{D}_+.$$

Conversely, let $\mathcal{M}$ be J_0-nonnegative. As for every $x \in \mathcal{M}$ we have

$$\langle J_0x, x\rangle = \|P_+x\|^2 - \|P_-x\|^2,$$

the J_0-nonnegativity of $\mathcal{M}$ implies

$$\|P_+x\|^2 \geq \frac{1}{2}\|x\|^2.$$

Consequently the linear set $P_+(\mathcal{M})$ is a subspace. Further,

$$\|P_-x\|^2 \leq \|P_+x\|^2, \quad x \in \mathcal{M},$$

and hence, the operator $K \in L(P_+(\mathcal{M}), \mathcal{Y}_-)$ defined by $K(P_+x) = P_-x$, $x \in \mathcal{M}$ is indeed a contraction. ■

The operator K is called the *angular operator* of a J_0-nonnegative subspace $\mathcal{M}$.

It is clear from Proposition 5.1.1 that a J_0-nonnegative subspace $\mathcal{M}$ is maximal if and only if its angular operator is defined on the whole $\mathcal{Y}_+$. Thus:

COROLLARY 5.1.2. *The correspondence $K \to \mathcal{G}(K)$ is a bijection from the set of all contractions in $L(\mathcal{Y}_+, \mathcal{Y}_-)$ onto the set of all maximal J_0-nonnegative subspaces.*

Of course, Proposition 5.1.1 and Corollary 5.1.2 have obvious counterparts concerning J_0-nonpositive subspaces.

We return now to the original operator J. The following result will be needed.

THEOREM 5.1.3. *A subspace $\mathcal{M} \subset \mathcal{Y}$ is maximal J-nonnegative (resp. J-nonpositive) if and only if its J-orthogonal companion is maximal J-nonpositve (resp. J-nonnegative).*

PROOF. First observe that (5.1.2) implies

$$(\mathcal{M}^{\perp J})^{\perp J} = \mathcal{M}$$

for every subspace $\mathcal{M} \in \mathcal{Y}$. Hence it will suffice to prove that if $\mathcal{M}$ is maximal J-nonnegative (or J-nonpositive), then $\mathcal{M}^{\perp J}$ is maximal J-nonpositive (or J-nonnegative).

Define J_0 by (5.1.3). We show first that it is sufficient to prove the theorem for the case when $J = J_0$. Indeed, there exists an invertible self-adjoint (even positive) operator S such that $J = S^*J_0S$. In particular, $S(\mathcal{M})$ is a subspace for every subspace $\mathcal{M} \subset \mathcal{Y}$. It is easy to see that $\mathcal{M}$ is J-nonnegative (or J-nonpositive) if and only if $S(\mathcal{M})$ is J_0-nonnegative (or J_0-nonpositive). Also, using (5.1.2) and the analogous equality for J_0 we verify that

$$S(\mathcal{M}^{\perp J}) = (S(\mathcal{M}))^{\perp J_0}.$$

These observations easily imply that the theorem holds provided it holds for the particular case $J = J_0$. We assume now that $J = J_0$.

Let $\mathcal{M}$ be maximal J-nonnegative. By Corollary 5.1.2 its angular operator $K \in L(\mathcal{Y}_+,\mathcal{Y}_-)$ is contractive, hence $K^* \in L(\mathcal{Y}_-,\mathcal{Y}_+)$ is contractive as well. By the J-nonpositive counterpart of Corollary 5.1.2, we conclude that the graph $\mathcal{G}(K^*)$ is maximal J-nonpositive. But one can easily verify that $\mathcal{G}(K^*)$ is the J-orthogonal companion of $\mathcal{G}(K) = \mathcal{M}$. The case of maximal J-nonpositive subspace $\mathcal{M}$ is considered analogously. ∎

5.2 J-self-adjoint and J-positizable operators

As before, let J be an invertible self-adjoint operator acting on a Hilbert space $\mathcal{Y}$. Introduce the (generally indefinite) scalar product

$$[x,y] = \langle Jx,y\rangle; \quad x,y \in \mathcal{Y}.$$

Analogously to the definition of a self-adjoint operator in $\mathcal{Y}$ the notion of a J-self-adjoint operator is introduced. An operator $A \in L(\mathcal{Y})$ is called *J-self-adjoint* if $JA = A^*J$, i.e., JA is self-adjoint. In particular, a J-self-adjoint operator A is similar to A^*, and therefore the spectrum of A is symmetric relative to the real axis if $\lambda_0 \in \sigma(A)$, then also $\overline{\lambda}_0 \in \sigma(A)$. Moreover, the qualitative properties of λ_0 and $\overline{\lambda}_0$ as spectral points for A can be easily related; for instance, if λ_0 is an isolated point of $\sigma(A)$ with finite algebraic multiplicity (i.e., the Riesz projection

$$\mathcal{R}_{\lambda_0} = \frac{1}{2\pi i}\int_{\Gamma}(\lambda I-A)^{-1}d\lambda$$

corresponding to λ_0 is a finite rank operator), then the same is true for $\overline{\lambda}_0$, and the partial multiplicities corresponding to λ_0 (i.e., the sizes of Jordan blocks in the Jordan form of A restricted to $\mathcal{R}_{\lambda_0}$) coincide with those corresponding to $\overline{\lambda}_0$.

The following simple property of J-self-adjoint operators will be used later.

PROPOSITION 5.2.1. *If* A *is J-self-adjoint and* $\mathcal{N}$ *is an A-invariant subspace, then the orthogonal companion* $\mathcal{N}^{\perp J}$ *is A-invariant as well.*

PROOF. Let $x \in \mathcal{N}^{\perp J}$, so $[x,y] = 0$ for all $y \in \mathcal{N}$. Then

$$[Ax,y] = [x,Ay] = 0$$

for $y \in \mathcal{N}$ because $\mathcal{N}$ is A-invariant. So $Ax \in \mathcal{N}^{\perp J}$. ∎

We will be interested in a particular class of J-self-adjoint operators. A J-self-adjoint operator is, by definition, *J-definitizable* if there is a polynomial $p(\lambda)$ with real coefficients such that

$$[p(A)x,x] \geq 0$$

for all $x \in \mathcal{Y}$. Replacing here $p(\lambda)$ by $-p(\lambda)$ we see that A is J-definitizable if and only if there is a polynomial $p(\lambda)$ with real coefficients such that

$$[p(A)x,x] \leq 0$$

for all $x \in \mathcal{Y}$ (that is why the term "J-definitizable" is used, and not "J-positizable"). Also, the requirement that the coefficients of $p(\lambda)$ be real is not essential (because one can consider $p(\lambda)+\overline{p}(\lambda)$ in place of $p(\lambda)$, where $\overline{p}(\lambda)$ designates the polynomial with complex conjugate coefficients).

The following theorem is the basic result on J-definitizable operators we need.

THEOREM 5.2.2. (a) *Let $A \in L(\mathcal{Y})$ be a J-definitizable operator. Then there exists an A-invariant subspace which is also maximal J-nonnegative; analogously, there exists an A-invariant maximal J-nonpositive subspace.*

(b) *Let J_1 and J_2 be two invertible self-adjoint operators such that J_1-J_2 is a finite rank operator. Let $A_1 \in L(\mathcal{Y})$ be J_1-definitizable and $A_2 \in L(\mathcal{Y})$ be J_2-self-adjoint. If A_1-A_2 is a finite rank operator, then A_2 is J_2-definitizable.*

The statement (a) of this theorem can be strengthened somewhat; namely, there is an A-invariant maximal J-nonnegative (or maximal J-nonpositive) subspace $\mathcal{M}$ such that $\sigma(A|_{\mathcal{M}})$ lies in the closed upper half plane; analogously, such subspace $\mathcal{M}$ with $\sigma(A|_{\mathcal{M}})$ in the closed lower half plane can be found.

The proof of Theorem 5.2.2 is based on rather deep properties of operators in the spaces with indefinite scalar product and is therefore beyond the scope of this book. For the proof of Theorem 5.2.2(a) see Langer [2], also Langer [3]. The proof of Theorem 5.2.2(b) is based on the proof of Theorem 1 in Jonas-Langer [1] (see Rodman [7]).

In connection with Theorem 5.2.2(a) note the following well-known open problem.

PROBLEM 5.2.1. *Does every J-self-adjoint operator* A *have an invariant maximal J-nonnegative subspace (or, for that matter, an invariant subspace)?*

We indicate an important corollary from Theorem 5.2.2.

COROLLARY 5.2.3. *If J is such that the spectral subspace of J corresponding to the positive (or negative) part of $\sigma(J)$ is finite dimensional, then every J-self-adjoint operator is J-definitizable.*

PROOF. Assuming that the spectral subspace of J corresponding to the negative part of $\sigma(J)$ is finite dimensional, write with respect to a suitable orthogonal decomposition $\mathcal{Y} = \mathcal{Y}_1 \oplus \mathcal{Y}_1^{\perp}$ of $\mathcal{Y}$: $J = J_1 \oplus (-J_2)$, where J_1 and J_2 are positive definite, and $\mathcal{Y}_1^{\perp}$ is finite dimensional. Let A be J-self-adjoint, and partition A accordingly:

$$A = \begin{bmatrix} A_{11} & A_{12} \\ A_{21} & A_{22} \end{bmatrix} .$$

Put

$$\tilde{J} = \begin{bmatrix} J_1 & 0 \\ 0 & J_2 \end{bmatrix} ; \qquad \tilde{A} = \begin{bmatrix} A_{11} & A_{12} \\ -A_{21} & A_{22} \end{bmatrix} .$$

One checks easily that $\tilde{A}$ is $\tilde{J}$-self-adjoint. As $\tilde{J}$ is positive definite, $\tilde{A}$ is also $\tilde{J}$-definitizable. It remains to note that both operators $J-\tilde{J}$ and $A-\tilde{A}$ are of finite rank, and to apply Theorem 5.2.2. ∎

5.3 Factorization and invariant semidefinite subspaces

We consider now the problem of factorization of monic operator polynomials with self-adjoint coefficients.

Let

$$L(\lambda) = \sum_{j=0}^{\ell-1} \lambda^j A_j + \lambda^{\ell} I,$$

where $A_0, \ldots, A_{\ell-1} \in L(\mathcal{X})$ are self-adjoint operators acting on a Hilbert space $\mathcal{X}$, and let (X, T, Y) be a spectral triple for $L(\lambda)$. As we have seen in the introduction to this chapter, there is unique invertible self-adjoint operator J such that

$$X = Y^* J, \quad T = J^{-1} T^* J, \quad Y = J^{-1} X^*. \tag{5.3.1}$$

We call J the operator *associated* with (X, T, Y).

THEOREM 5.3.1. *Let $\mathcal{N}$ be a T-invariant J-semidefinite subspace. If ℓ is even, then the operator* $\mathrm{col}[XT^i]_{i=0}^{k-1}|_{\mathcal{N}}$ *(where* $k = \frac{\ell}{2}$*) is left invertible. If ℓ is odd, let* $k = \frac{\ell-1}{2}$*; then the operator* $\mathrm{col}[XT^i]_{i=0}^{k-1}|_{\mathcal{N}}$ *(in case $\mathcal{N}$ is J-nonpositive) or* $\mathrm{col}[XT^i]_{i=0}^{k}|_{\mathcal{N}}$ *(in case $\mathcal{N}$ is J-nonnegative) is left invertible.*

PROOF. We verify first that it is sufficient to prove the theorem for one particular triple (X, T, Y). Indeed, let (X', T', Y') be another spectral triple of $L(\lambda)$. Then

$$X = X'S, \quad T = S^{-1} T' S, \quad Y = S^{-1} Y'$$

for some invertible operator S. It is easily seen that $J' \stackrel{\mathrm{def}}{=} S^{-1*} J S^{-1}$ is the operator associated with (X', T', Y'). A subspace $\mathcal{N}$ is T-invariant J-nonpositive (or J-nonnegative) if and only if $S\mathcal{N}$ is T'-invariant J'-nonpositive (or J'-nonnegative). As

$$\mathrm{col}[XT^i]_{i=0}^{p}|_{\mathcal{N}} = \mathrm{col}[X'T'^i]_{i=0}^{p}|_{S\mathcal{N}}$$

for $p = 0, 1, \ldots$, clearly the required properties for (X', T', Y') follow from those for (X, T, Y).

Now make a particular choice of the spectral triple and assume that

$$\begin{aligned} X &= [I \ \ 0 \ \cdots \ 0]; \\ T &= C_L \ \text{(the companion operator for } L(\lambda)); \\ Y &= \mathrm{col}[\delta_{i\ell} I]_{i=1}^{\ell}. \end{aligned} \tag{5.3.2}$$

One verifies that the associated operator is

$$(5.3.3) \qquad J_L = \begin{bmatrix} A_1 & \cdots & A_{\ell-1} & I \\ \vdots & A_{\ell-1} & I & \\ A_{\ell-1} & I & & \\ I & & & 0 \end{bmatrix}$$

(cf. formula (2.5.2)).

Assume ℓ is even and $\{x_p\}_{p=1}^{\infty}$, $x_p \in \mathcal{N}$ is a sequence such that

$$(5.3.4) \qquad \mathrm{col}[XT^i]_{i=0}^{k-1} x_p \to 0 \text{ as } p \to \infty.$$

Write

$$(5.3.5) \qquad x_p = \mathrm{col}[x_{jp}]_{j=1}^{\ell}, \quad x_{jp} \in \mathcal{X}.$$

The particular choice (5.3.2) implies that

$$\mathrm{col}[XT^i]_{i=0}^{k-1} = [I_{\mathcal{X}^k} \quad 0];$$

so in view of (5.3.4),

$$\lim_{p\to\infty} x_{jp} = 0 \quad \text{for } j = 1,\dots,k.$$

Now using (5.3.3), we obtain

$$\lim_{p\to\infty} \langle J_L x_p, x_p \rangle = 0.$$

As $\mathcal{N}$ is T-invariant and J-semidefinite, by the Cauchy-Schwarz inequality

$$|\langle J_L T x_p, x_p \rangle| \le \langle J_L T x_p, T x_p \rangle \langle J x_p, x_p \rangle \to 0$$

as $p \to \infty$. But

$$\langle J_L T x_p, x_p \rangle = \left\langle \begin{bmatrix} A_1 & & A_{\ell-1} & I \\ \vdots & & I & \\ A_{\ell-1} & & & \\ I & & 0 & \end{bmatrix} \begin{bmatrix} x_{2p} \\ \vdots \\ x_{np} \\ y \end{bmatrix}, \begin{bmatrix} x_{1p} \\ \vdots \\ x_{np} \end{bmatrix} \right\rangle,$$

where $y = -A_0x_{1p}-A_1x_{2p}-\cdots-A_{\ell-1}x_{\ell p}$; hence

$$\langle J_LTx_p,x_p\rangle = \|x_{k+1,p}\|^2+R_p,$$

where $\lim_{p\to\infty} R_p = 0$. We obtain that $\lim_{p\to\infty} x_{k+1,p} = 0$. This implies in turn that

$$\lim_{p\to\infty} \langle J_LTx_p,x_p\rangle = 0,$$

and hence, using the Cauchy-Schwarz inequality again, we have that

$$\langle J_LT^2x_p,Tx_p\rangle = \langle J_LT^3x_p,x_p\rangle \to 0 \text{ as } p \to \infty.$$

On the other hand,

$$\langle J_LT^3x_p,x_p\rangle = \|x_{k+2,p}\|^2+Q_p,$$

where $Q_p \to 0$. So $x_{k+2,p} \to 0$. Continuing by induction, we show that $\lim_{p\to\infty} x_p = 0$. This proves that the operator $\text{col}[XT^i]_{i=0}^{k-1}$ is regular and hence (because $\mathcal{X}$ is a Hilbert space) is left invertible.

Consider now the case ℓ is odd. Assume, for instance, that $\mathcal{N}$ is J-nonpositive. Let $\{x_p\}_{p=1}^{\infty}$, $x_p \in \mathcal{N}$ be a sequence such that (5.3.4) holds, where $k = \frac{\ell-1}{2}$. Write x_p in the form (5.3.5) and observe that

$$\lim_{p\to\infty} x_{jp} = 0 \quad \text{for } j = 1,\ldots,k.$$

Now

$$0 \geq \langle J_Lx_p,x_p\rangle = \|x_{k+1,p}\|^2+R_p,$$

where $R_p \to 0$ as $p \to \infty$. Hence $\lim_{p\to\infty} x_{k+1,p} = 0$. Applying this procedure to the sequence $\{Tx_p\}_{p=1}^{\infty}$ and using induction, we prove that $x_p \to 0$.

Finally, assume that ℓ is odd and $\mathcal{N}$ is J-nonnegative. Let $\{x_p\}_{p=1}^{\infty}$, $x_p \in \mathcal{N}$ be such that

$$\lim_{p\to\infty} \operatorname{col}[XT^i]_{i=0}^{k} x_p = 0.$$

Then $\langle J_L x_p, x_p\rangle \to 0$ and consequently

$$|\langle J_L Tx_p, Tx_p\rangle| = |\langle J_L T^2 x_p, x_p\rangle| \leq \langle JT^2 x_p, T^2 x_p\rangle \langle Jx_p, x_p\rangle \to 0.$$

On the other hand,

$$\langle J_L Tx_p, Tx_p\rangle = \|x_{k+2,p}\|^2 + R_p,$$

where $R_p \to 0$. It follows that $x_{k+2,p} \to 0$ as $p \to \infty$. Continuing in this way, we obtain that every coordinate of x_p tends to zero when $p \to \infty$. ■

The following result is, in a certain sense, dual to Theorem 5.3.1. As before, (X,T,Y) is a spectral triple for $L(\lambda)$ with the associated operator J.

THEOREM 5.3.2. *Let $\mathcal{N}$ be a T-invariant subspace for which $\mathcal{N}^{\perp J}$ is J-semidefinite. If ℓ is even, then the operator* $\operatorname{col}[XT^i]_{i=0}^{k-1}|_{\mathcal{N}}$, *where* $k = \frac{\ell}{2}$, *has dense range. If ℓ is odd, let* $k = \frac{\ell-1}{2}$; *then the operator* $\operatorname{col}[XT^i]_{i=0}^{k-1}|_{\mathcal{N}}$ *(in case $\mathcal{N}^{\perp J}$ is J-nonnegative) or* $\operatorname{col}[XT^i]_{i=0}^{k}|_{\mathcal{N}}$ *(in case $\mathcal{N}^{\perp J}$ is J-nonpositive) has dense range.*

PROOF. As in the proof of Theorem 5.3.1 we can assume that (X,T,Y) are given by (5.3.2). Assume first ℓ is even. Let x be orthogonal to $\operatorname{Im}[\operatorname{col}(XT^i)_{i=0}^{k-1}|_{\mathcal{N}}]$. Write $x = \operatorname{col}[x_j]_{j=1}^{k}$, $x_j \in \mathcal{X}$.

We have for every $y \in \mathcal{N}$:

$$\begin{aligned}\langle J^{-1}\begin{bmatrix} x \\ 0 \end{bmatrix}, Jy\rangle &= \langle J^{-1}\begin{bmatrix} I \\ 0 \end{bmatrix}x, Jy\rangle \\ &= \langle x, [I \; 0]y\rangle = 0\end{aligned}$$

because of our assumption on x. Consequently (formula (5.1.2)),

$$J^{-1}\begin{bmatrix} x \\ 0 \end{bmatrix} \in \mathcal{N}^{\perp J}.$$

On the other hand, J^{-1} has the form

$$J^{-1} = \begin{bmatrix} & 0 & & . & I \\ & & . & & \\ & I & & & * \\ I & & & & \end{bmatrix},$$

so the first k coordinates of $J^{-1}\begin{bmatrix} x \\ 0 \end{bmatrix}$ are zeros, and $J^{-1}\begin{bmatrix} x \\ 0 \end{bmatrix} \in \operatorname{Ker} \operatorname{col}[XT^i]_{i=0}^{k-1}$. Now $\mathcal{N}^{\perp J}$ is T-invariant (Proposition 5.2.1) and J-semidefinite. By Theorem 5.3.1

$$\operatorname{Ker}(\operatorname{col}[XT^i]_{i=0}^{k-1})\big|_{\mathcal{N}^{\perp J}} = \{0\},$$

so $x = 0$, and the theorem is proved (for ℓ even).

In the case when ℓ is odd, the proof is analogous. We omit the details. ■

COROLLARY 5.3.3. *Let $\mathcal{N}$ be T-invariant subspace which is either maximal J-nonnegative or maximal J-nonpositive. If ℓ is even, then the operator $\operatorname{col}[XT^i]_{i=0}^{k-1}\big|_{\mathcal{N}}$ is invertible $(k = \frac{\ell}{2})$. If ℓ is odd, then the operator $\operatorname{col}[XT^i]_{i=0}^{k-1}\big|_{\mathcal{N}}$ $(k = \frac{\ell-1}{2})$ is invertible in case $\mathcal{N}$ is maximal J-nonpositive, and the operator $\operatorname{col}[XT^i]_{i=0}^{k}\big|_{\mathcal{N}}$ is invertible in case $\mathcal{N}$ is maximal J-nonnegative.*

For the proof use Theorems 5.3.1 and 5.3.2 and the fact that $\mathcal{N}$ is T-invariant maximal J-nonnegative (J-nonpositive) if and only if $\mathcal{N}^{\perp J}$ is T-invariant maximal J-nonpositive (J-nonnegative), which is proved in Theorem 5.1.3.

Recalling Theorem 2.7.1, we see that the subspaces described in Corollary 5.3.3 are supporting, i.e., give rise to factorizations of $L(\lambda)$. Such factorizations will be called *special*. Thus, a factorization $L(\lambda) = L_2(\lambda)L_1(\lambda)$, where $L_1(\lambda)$ and $L_2(\lambda)$ are monic operator polynomials (not necessarily with self-adjoint coefficients), is called *positive special* if its supporting subspace $\mathcal{N}$ (with respect to some spectral triple (X,T,Y) of $L(\lambda)$) is maximal J-nonnegative, where J is the operator associated with (X,T,Y). In this case the degree $L_1(\lambda)$ is $\frac{\ell}{2}$ if ℓ is even, and $\frac{\ell+1}{2}$ if ℓ is odd. Arguing as in the beginning of the proof of Theorem 5.3.1, one verifies that the property of positive special factorization does not depend on the choice of (X,T,Y) and indeed is a property of factorization itself. If the supporting subspace $\mathcal{N}$ is maximal J-nonpositive, the factorization will be called *negative special*. In this case the degree of $L_1(\lambda)$ is $\frac{\ell}{2}$ if ℓ is even, and $\frac{\ell-1}{2}$ if ℓ is odd. Again, this is a property of the factorization itself and does not depend on the choice of spectral triple.

Corollary 5.3.3 tells us that there is one-to-one correspondence between the set of all T-invariant maximal J-nonnegative (maximal J-nonpositive) subspaces and positive special (negative special) factorizations of $L(\lambda)$. In particular, appealing to Theorem 5.2.2 we obtain:

THEOREM 5.3.4. *Let*

$$L(\lambda) = \sum_{j=0}^{\ell-1} \lambda^j A_j + \lambda^\ell I$$

be a monic operator polynomial with self-adjoint coefficients. Let

$$C_L = \begin{bmatrix} 0 & I & 0 & \cdots & 0 \\ 0 & 0 & I & & 0 \\ \vdots & \vdots & \vdots & & \vdots \\ 0 & 0 & 0 & \cdots & I \\ -A_0 & -A_1 & & & -A_{\ell-1} \end{bmatrix},$$

$$J_L = \begin{bmatrix} A_1 & & & A_{\ell-1} & I \\ \vdots & & \cdot^{\cdot^{\cdot}} & I & \\ \vdots & A_{\ell-1} & \cdot^{\cdot^{\cdot}} & & \\ A_{\ell-1} & I & & & 0 \\ I & & & & \end{bmatrix}.$$

If the J_L-self-adjoint operator C_L is J_L-definitizable, then $L(\lambda)$ admits positive special factorizations and negative special factorizations.

In the finite-dimensional case ($\dim \mathcal{X} < \infty$) positive and negative special factorizations exist always (see Theorem 5.2.2(b)).

PROBLEM 5.3.1. *Does every operator polynomial with self-adjoint coefficients of degree ≥ 2 admit a non-trivial factorization into product of monic operator polynomials?*

PROBLEM 5.3.2. *Does every monic operator polynomial with self-adjoint coefficients admit a special factorization?*

Observe that if $L(\lambda)$ admits a positive (negative) special factorization, then $L(\lambda)$ admits a negative (positive) special factorization as well (see Proposition 5.3.6 below). Observe also that the affirmative answer to Problem 5.2.1 will yield the affirmative answer to Problem 5.3.2. It is not clear whether Problems 5.2.1 and 5.3.2 are equivalent.

Consider again a monic operator polynomial $L(\lambda)$ with self-adjoint coefficients, with a spectral triple (X,T,Y) and the associated operator J. Obviously every factorization

$$L(\lambda) = L_2(\lambda)L_1(\lambda) \tag{5.3.6}$$

with monic factors $L_1(\lambda)$ and $L_2(\lambda)$ leads to a factorization

$$L(\lambda) = (L_1(\bar{\lambda}))^*(L_2(\bar{\lambda}))^*. \tag{5.3.7}$$

There is a simple connection between the supporting projections (see Section 2.9) of (5.3.6) and those of (5.3.7).

THEOREM 5.3.5. *Let* P *be a supporting projection corresponding to the factorization* (5.3.6). *Then*

$$Q = J^{-1}(I-P^*)J \tag{5.3.8}$$

is a supporting projection corresponding to (5.3.7).

PROOF. It is easy to see that Q is indeed a projection. Denoting by $\mathcal{N}$ the supporting subspace corresponding to the factorization (5.3.6), and letting (X_1,T_1) and (T_2,Y_2) be right and left spectral pairs of $L_1(\lambda)$ and $L_2(\lambda)$, respectively, we have (Theorem 2.9.2):

$$\mathcal{N} = \text{Im } P;\ X_1 = X_{|\mathcal{N}};\ T_1 = T_{|\mathcal{N}};$$

$$T_2 = (I-P)T_{|\text{Ker } P};\ Y_2 = (I-P)Y\colon \mathcal{X} \to \text{Ker } P.$$

Now

$$\mathcal{N}^{\perp J} = (J\mathcal{N})^{\perp} = (\text{Im}(JP))^{\perp} = \text{Ker}(P^*J^*)$$

$$= \text{Ker}(J^{-1}P^*J) = \text{Im}(J^{-1}(I-P^*)J) = \text{Im } Q.$$

Moreover, since $\mathcal{N}$ is T-invariant and $JT = T^*J$, the subspace $\mathcal{N}^{\perp J}$ = Im Q is easily seen to be T-invariant as well.

Consider the operators $Y_2^*\colon \text{Ker } P \to \mathcal{X}$ and $T_2^*\colon \text{Ker } P \to \text{Ker } P$. For every $x \in \text{Ker } P$ and every $y \in \mathcal{X}$ we have

$$\langle Y_2^*x, y\rangle = \langle x, Y_2y\rangle = \langle x, (I-P)Yy\rangle$$

$$= \langle Y^*(I-P^*)x, y\rangle = \langle Y^*JJ^{-1}(I-P^*)JJ^{-1}x, y\rangle$$

$$= \langle Y^*JQJ^{-1}x, y\rangle,$$

and using the equality $Y^*J = X$, it follows that

$$Y_2^* = XQJ^{-1}|_{\text{Ker } P}\ . \tag{5.3.9}$$

Analogously, for every $x, y \in \text{Ker } P$ we obtain

$$\langle T_2^*x, y\rangle = \langle x, T_2y\rangle = \langle x, (I-P)Ty\rangle$$

$$= \langle T^*(I-P^*)x, y\rangle = \langle JTJ^{-1}(I-P^*)JJ^{-1}x, y\rangle$$

$$= \langle JTQJ^{-1}x, y\rangle$$

and hence (denoting by $P_0\colon \mathcal{Y} \to \text{Ker } P$ the orthogonal projection on Ker P),

$$T_2^* = P_0JTQJ^{-1}|_{\text{Ker } P} = P_0JQTQJ^{-1}|_{\text{Ker } P}\ , \tag{5.3.10}$$

where in the last equality we have used the T-invariance of Im Q (which is equivalent to $QTQ = TQ$).

We shall need the equality

$$P_0(I-P^*)(I-P) = I-P, \tag{5.3.11}$$

which can be verified by observing that for every vector $x \in \mathcal{Y}$ the difference

$$P_0(I-P^*)(I-P)x-(I-P)x$$

belongs to Ker P and is orthogonal to every vector in Ker P.

Consider the operator Z: Ker $P \to$ Im Q defined by

$$Zx = QJ^{-1}x = J^{-1}(I-P^*)x, \quad x \in \text{Ker } P.$$

Using (5.3.11) one verifies that Z is left invertible with the left inverse P_0JQ. On the other hand, using the equality $(I-P)P_0 = P_0$ and its adjoint $P_0(I-P^*) = P_0$, together with

$$I-P^* = (I-P^*)(I-P)P_0$$

(which is obtained by taking adjoints in (5.3.11)), we see that

$$I-P^* = (I-P^*)P_0(I-P^*).$$

This implies $ZP_0JQ = Q$. In other words, Z is invertible with the inverse P_0JQ.

Return now to the formulas (5.3.9) and (5.3.10). Clearly, (Y_2^*, T_2^*) is a right spectral pair for $(L_2(\bar{\lambda}))^*$. Formulas (5.3.9) and (5.3.10), together with $Z^{-1} = P_0JQ$ imply that $(X|_{\text{Im } Q}, T|_{\text{Im } Q})$ is a right spectral pair for $(L_2(\bar{\lambda}))^*$ as well. As the supporting subspace $\mathcal{M}$ for the factorization (5.3.7) is uniquely determined by the properties that $\mathcal{M}$ is T-invariant and $(X_{|\mathcal{M}}, T_{|\mathcal{M}})$ is a right spectral pair for $(L_2(\bar{\lambda}))^*$, it follows that $\mathcal{M}$ = Im Q, and hence Q is a supporting projection for (5.3.7). ∎

In the course of the proof of Theorem 5.3.5, we have seen that if $\mathcal{N}$ is the supporting subspace for (5.3.6), then $\mathcal{N}^{\perp J}$

is the supporting subspace for (5.3.7). This leads to the following fact concerning special factorizations.

PROPOSITION 5.3.6. *If the factorization* (5.3.6) *is positive (negative) special, then the factorization* (5.3.7) *is negative (positive) special.*

For the proof combine Theorem 5.1.3 with the observation immediately preceding the proposition.

5.4 Classes of polynomials with special factorizations

We use the results of previous sections to establish existence of special factorizations for certain classes of polynomials with self-adjoint coefficients.

THEOREM 5.4.1. *Let* $L(\lambda)$ *and* $\tilde{L}(\lambda)$ *be monic operator polynomials with self-adjoint coefficients such that for every* $\lambda \in \mathbb{C}$ *the difference* $L(\lambda)-\tilde{L}(\lambda)$ *is a finite rank operator. Assume that the companion operator* C_L *of* $L(\lambda)$ *is* J_L*-definitizable, where* J_L *is given by* (5.3.3). *Then* $\tilde{L}(\lambda)$ *has positive special and negative special factorizations.*

PROOF. Let $C_{\tilde{L}}$ be the companion operator of $\tilde{L}$ and let $J_{\tilde{L}}$ be defined for $\tilde{L}$ as in (5.3.3). Then $C_L-C_{\tilde{L}}$ and $J_L-J_{\tilde{L}}$ are finite rank operators. By Theorem 5.2.2(b), $C_{\tilde{L}}$ is $J_{\tilde{L}}$-definitizable, and an application of Theorem 5.2.2(a) completes the proof. ∎

Observe that we had to assume J_L-definitizability of C_L in Theorem 5.4.1 to make it possible to use the perturbation Theorem 5.2.2. Thus, the following problem (related to Problem 5.3.2) appears naturally:

PROBLEM 5.4.1. *Is the property of having special factorizations preserved under finite rank perturbations of the coefficients of the operator polynomials?*

COROLLARY 5.4.2. *Let* $\mathcal{X} = \mathcal{Z}^n$ *for some Hilbert space* $\mathcal{Z}$, *and assume that all coefficients of a monic operator polynomial* $L(\lambda)$ *with self-adjoint coefficients are of the form* $[\alpha_{ij}I_{\mathcal{Z}}+K_{ij}]_{i,j=1}^{n}$, *where* K_{ij} *are finite rank operators (because* *of the self-adjointness we necessarily have* $\alpha_{ij} = \bar{\alpha}_{ji}$ *and* $K_{ij} = K_{ji}^{*}$). *Then* $L(\lambda)$ *admits positive special and negative special factorizations.*

This corollary follows immediately from Theorem 5.4.1 upon noticing that the companion operator of the operator polynomial $\tilde{L}(\lambda)$ obtained from $L(\lambda)$ by replacing each coefficient $[\alpha_{ij}I_{\mathcal{Z}}+K_{ij}]_{i,j=1}^{n}$ by $[\alpha_{ij}I_{\mathcal{Z}}]_{i,j=1}^{n}$ is $J_{\tilde{L}}$-definitizable.

PROBLEM 5.4.2. *Is the result of Corollary 5.4.2 still valid if the operators* K_{ij} *are assumed to be merely compact (not necessarily finite rank)?*

Another important class of polynomials that admit special factorizations is given by the following theorem.

THEOREM 5.4.3. *Let* $L(\lambda) = \lambda^{\ell}I+\sum_{j=0}^{\ell-1}\lambda^{j}A_{j}$ *with* $\ell \geq 2$ *and* A_j *self-adjoint, and assume that* $A_0,\ldots,A_{\ell-3}$ *are finite-dimensional operators while the* $A_{\ell-2}$*-invariant subspace corresponding to the positive part of* $\sigma(A_{\ell-2})$ *is finite-dimensional. Then* $L(\lambda)$ *admits positive special and negative special factorizations.*

It is convenient to prove a simple lemma which will be used in the proof of Theorem 5.4.3.

LEMMA 5.4.4. *Let* $\mathcal{Y}$ *be a Hilbert space,* $K_0 \in L(\mathcal{Y})$ *a positive semidefinite operator (i.e.,* $\langle K_0x,x\rangle \geq 0$ *for all* $x \in \mathcal{Y}$*), and* $K_1 \in L(\mathcal{Y})$ *a self-adjoint operator of finite rank* m. *Then the* (K_0+K_1)*-invariant subspace corresponding to the negative part of* $\sigma(K_0+K_1)$ *is at most* m*-dimensional.*

PROOF. Suppose not. Then the spectral theorem for the self-adjoint operator K_0+K_1 easily implies that there exists $(m+1)$-dimensional subspace $\mathcal{L} \subset \mathcal{Y}$ such that $\langle (K_0+K_1)x,x\rangle < 0$ for every $x \in \mathcal{L}$, $x \neq 0$. Denoting by $P_{\mathcal{L}}$ the orthogonal projection on $\mathcal{L}$, let $\hat{K}_j = P_{\mathcal{L}}K_jP_{\mathcal{L}}|_{\mathcal{L}} \in L(\mathcal{L})$, $j = 0,1$. Then $\hat{K}_0$ and $\hat{K}_1$ are self-adjoint operators on the $(m+1)$-dimensional space $\mathcal{L}$, $\hat{K}_0+\hat{K}_1$ is negative definite, and $\hat{K}_1$ has rank at most m. Because of the latter property, there exists $x_0 \in \mathcal{L}$, $x_0 \neq 0$ such that $x_0 \perp$ Im

$\hat{K}_1$. Now $\langle K_0x_0, x_0\rangle = \langle \hat{K}_0x_0, x_0\rangle = \langle (\hat{K}_0+\hat{K}_1)x_0, x_0\rangle < 0$, which contradicts the positive semidefiniteness of K_0. ∎

PROOF OF THEOREM 5.4.3. Let C_L be the companion operator of $L(\lambda)$, and define J_L by (5.3.3). A calculation shows that

$$J_LC_L^{\ell-1} = \begin{bmatrix} 0 & 0 & \cdots & -A_0 & 0 \\ \vdots & \vdots & . \cdot & -A_1 & 0 \\ 0 & -A_0 & \cdot^{\cdot} & \vdots & \vdots \\ -A_0 & -A_1 & \cdots & -A_{\ell-2} & 0 \\ 0 & 0 & \cdots & 0 & I \end{bmatrix}.$$

The hypotheses of the theorem imply, in view of Lemma 5.4.4, that the invariant subspace of $J_LC_L^{\ell-1}$ corresponding to the negative part of $\sigma(J_LC_L^{\ell-1})$ is finite-dimensional.

Assume first that the operator $\hat{J} = J_LC_L^{\ell-1}$ is invertible. This will be the case if $\ell = 2$ (so the hypothesis that $A_0,\ldots,A_{\ell-3}$ are finite dimensional is vacuous), and A_0 is invertible. Clearly, $\hat{J}^* = \hat{J}$ and C_L is $\hat{J}$-self-adjoint. By Corollary 5.2.3, C_L is $\hat{J}$-definitizable, i.e., $\hat{J}p(C_L)$ is positive semidefinite for some polynomial $p(\lambda)$. But then clearly, $Jq(C_L)$ is positive semidefinite, where $q(\lambda) = \lambda^{\ell-1}p(\lambda)$. Thus, C_L is J-definitizable, and it remains to apply Theorem 5.3.4.

The case when $\hat{J}$ is not invertible is more difficult. We shall not prove here Theorem 5.4.3 in this case; this can be done by using the ideas of the proof of Theorem 1 in Jonas-Langer [1]. See Rodman [7] for the full proof of Theorem 5.4.3. ∎

5.5 Positive semidefinite operator polynomials

A monic operator polynomial $L(\lambda)$ with coefficients in $L(\mathfrak{X})$ is said to be *positive semidefinite* if

$$\langle L(\lambda), x, x\rangle \geq 0$$

for every real λ and every $x \in \mathcal{X}$. Clearly, a positive definite operator polynomial has self-adjoint coefficients (here we use the fact that a bounded operator A on $\mathcal{X}$ is self-adjoint if and only if $\langle Ax, x\rangle$ is real for every $x \in \mathcal{X}$), and its degree is even. Examples of positive semidefinite operator polynomials are easily given by letting

$$L(\lambda) = (M(\bar{\lambda}))^{*}M(\lambda), \tag{5.5.1}$$

where $M(\lambda)$ is any monic operator polynomial. It turns out that (5.5.1) represents a general formula for positive semidefinite operator polynomials.

THEOREM 5.5.1. *Assume $\mathcal{X}$ is separable. Then every positive semidefinite (monic) operator polynomial $L(\lambda)$ of degree ℓ with coefficients in $L(\mathcal{X})$ admits a factorization (5.5.1), where $M(\lambda)$ is monic of degree $\frac{\ell}{2}$. Moreover, $M(\lambda)$ can be chosen so that the spectrum of M lies in the closed upper half-plane.*

The proof of Theorem 5.5.1 is beyond the scope of this book. It can be found in Markus [1] (Lemmas 34.13, 34.12), and in Rosenblum-Rovnyak [1] (Theorem 6.7).

Here we only describe factorization (5.5.1) in terms of the supporting subspaces:

THEOREM 5.5.2. *Let $L(\lambda)$ be a positive semidefinite operator polynomial, and let*

$$L(\lambda) = N(\lambda)M(\lambda), \tag{5.5.2}$$

where $N(\lambda)$ and $M(\lambda)$ are monic operator polynomials. Then (5.5.2) is of the form (5.5.1), i.e., $N(\lambda) = (M(\bar{\lambda}))^{*}$, if and only if the supporting subspace $\mathcal{N}$ for (5.5.2) with respect to a spectral triple (X,T,Y) of $L(\lambda)$ satisfies $\mathcal{N} = \mathcal{N}^{\perp J}$, where J is the self-adjoint operator associated with (X,T,Y).

PROOF. Follows from the proof of Theorem 5.3.5 according to which $\mathcal{N}^{\perp J}$ is the supporting subspace of the factorization $L(\lambda) = (M(\bar{\lambda}))^{*}N(\bar{\lambda})^{*}$. ∎

In particular, every factorization of type (5.5.1) is positive special, as well as negative special. The converse is also true, i.e., if $L(\lambda) = N(\lambda)M(\lambda)$ is a factorization of $L(\lambda)$ which is simultaneously positive special and negative special, then necessarily $N(\lambda) = (M(\bar{\lambda}))^*$. Indeed, the supporting subspace $\mathcal{N}$ of such factorization is J-neutral, that is $\langle Jx,x\rangle = 0$ for all $x \in \mathcal{N}$. The Cauchy-Schwartz inequality implies now that $\langle Jx,y\rangle = 0$ for all $x,y \in \mathcal{N}$, i.e., $\mathcal{N} \subset \mathcal{N}^{\perp J}$. As $N^{\perp J}$ is again maximal J-nonnegative and maximal J-nonpositive (Theorem 5.1.3), we must have $\mathcal{N} = \mathcal{N}^{\perp J}$.

5.6 Strongly hyperbolic operator polynomials

Let

$$L(\lambda) = \lambda^2 I+\lambda B+A \tag{5.6.1}$$

be a monic operator polynomial of second degree with $B, A \in L(\mathcal{X})$, where $\mathcal{X}$ is a Hilbert space. We say that (5.6.1) is *strongly hyperbolic* if for every $x \in \mathcal{X}$, $x \neq 0$ the two zeros of the scalar polynomial

$$\langle L(\lambda)x,x\rangle = 0 \tag{5.6.2}$$

are real and distinct. If this happens, then the coefficients $\langle Bx,x\rangle$ and $\langle Ax,x\rangle$ of (5.6.2) are all real and hence B and A are self-adjoint.

From now on we assume that (5.6.1) is a strongly hyperbolic operator polynomial. Denote the zeros of (5.6.2) by

$$p_2(x) > p_1(x).$$

Clearly, for every non-zero $\lambda \in \mathbb{C}$ we have $p_j(x) = p_j(\lambda x)$, and hence one can consider p_j as a real-valued function defined on the unit sphere S of $\mathcal{X}$. The function p_j is continuous and bounded, and hence the set

$$\Delta_j = \{\alpha \in \mathbb{R} \mid \alpha = p_j(x) \text{ for some } x \neq 0\}$$

is bounded and connected (the connectivity of Δ_j follows from the connectivity of S). So Δ_j must be a segment of the real line (with or without one or both endpoints), or possibly a one point (which can be considered a degenerated segment). The set Δ_1

(resp. Δ_2) will be called the *first* (resp. *second*) spectral zone of $L(\lambda)$.

We establish first relatively simple properties of the spectral zones.

PROPOSITION 5.6.1. *The spectrum* $\Sigma(L)$ *lies on the real line, and moreover*

$$\Sigma(L) \subset \overline{\Delta}_1 \cup \overline{\Delta}_2.$$

PROOF. For $x \in \mathcal{X}$, $\|x\| = 1$ we have

$$\|L(\lambda)x\| \geq |\langle L(\lambda)x,x\rangle| = |\lambda - p_j(x)|\ |\lambda - p_2(x)|$$

$$\geq \operatorname{dist}(\lambda,\Delta_1)\cdot\operatorname{dist}(\lambda,\Delta_2),$$

where $\operatorname{dist}(\lambda,\Delta_j)$ is the distance from the point $\lambda \in \mathbb{C}$ to Δ_j. Replacing in this inequality λ by $\overline{\lambda}$ we obtain

$$\|(L(\lambda))^* x\| \geq \operatorname{dist}(\overline{\lambda},\Delta_1)\cdot\operatorname{dist}(\overline{\lambda},\Delta_2).$$

These inequalities imply that $L(\lambda)$ is invertible for $\lambda \notin \overline{\Delta}_1 \cup \overline{\Delta}_2$, and the proposition follows. ∎

A basic property of a strongly hyperbolic polynomial is that its spectral zones do not intersect:

THEOREM 5.6.2. *We have*

$$\Delta_1 \cap \Delta_2 = \phi.$$

In the proof of Theorem 5.6.2, the following fact will be used:

LEMMA 5.6.3. *Let* $A,B \in L(\mathcal{X})$ *be self-adjoint operators and assume that for some vectors* $x,y \in \mathcal{X}$ *we have*

$$\langle Ax,x\rangle = \langle Ay,y\rangle = 0;\ \langle Bx,x\rangle < 0;\ \langle By,y\rangle > 0.$$

Then there exists $z \in \mathcal{X}$, $z \neq 0$ *such that*

$$\langle Az,z\rangle = \langle Bz,z\rangle = 0. \tag{5.6.3}$$

PROOF. The proof is based on the Toeplitz-Hausdorff theorem according to which the numerical range $\mathcal{N}(C)$ of an arbitrary operator $C \in L(\mathcal{X})$ (where $\mathcal{X}$ is a Hilbert space) is a convex set. Recall that

$$\mathcal{N}(C) = \{\langle Cx,x\rangle \mid x \in \mathcal{X},\ \|x\| = 1\}.$$

A transparent proof of the Toeplitz-Hausdorff theorem can be found in Raghavendran [1].

Returning to the proof of the lemma, we can assume without loss of generality that $\|x\| = \|y\| = 1$. Let $C = B+iA$. Then

$$\langle Cx,x\rangle < 0,\ \langle Cy,y\rangle > 0.$$

By the Toeplitz-Hausdorff theorem there is a $z \in \mathcal{X}$, $\|z\| = 1$ such that $\langle Cz,z\rangle = 0$. Separating here real and imaginary parts, we obtain (5.6.3). ∎

PROOF OF THEOREM 5.6.2. As for each $x \neq 0$ the roots of $\langle L(\lambda)x,x\rangle$ are real and distinct, the derivative of the polynomial $\langle L(\lambda)x,x\rangle$ has opposite signs at the two roots:

$$\langle L'(p_2(x))x,x\rangle < 0 < \langle L'(p_1(x))x,x\rangle. \tag{5.6.4}$$

Arguing by contradiction, assume that $\Delta_1 \cap \Delta_2 \neq \emptyset$. So for some real λ and some non-zero vectors $x,y \in \mathcal{X}$ we have

$$\langle L(\lambda)x,x\rangle = \langle L(\lambda)y,y\rangle = 0;$$

$$\lambda = p_1(x) = p_2(y).$$

By (5.6.4)

$$\langle L'(\lambda)x,x\rangle \langle L'(\lambda)y,y\rangle < 0.$$

By Lemma 5.6.3 there is $z \in \mathcal{X}$, $z \neq 0$ such that

$$\langle L(\lambda)z,z\rangle = \langle L'(\lambda)z,z\rangle = 0.$$

However, this contradicts the assumption that all roots of $\langle L(\lambda)z,z\rangle$ are simple. ∎

We state now the main theorem on factorization of strongly hyperbolic operator polynomials.

THEOREM 5.6.4. *Assume* $L(\lambda) = \lambda^2 I+\lambda B+A$ *is strongly hyperbolic. Then* $L(\lambda)$ *admits positive special factorization*

(5.6.5) $$L(\lambda) = (\lambda I - Z_2^*)(\lambda I - Z_1)$$

and negative special factorization

(5.6.6) $$L(\lambda) = (\lambda I - Z_1^*)(\lambda I - Z_2).$$

For every positive special factorization (5.6.5) *we have*

$$\sigma(Z_1) = \overline{\Delta}_1; \quad \sigma(Z_2) = \overline{\Delta}_2,$$

and

$$\Sigma(L) = \sigma(Z_1) \cup \sigma(Z_2^*).$$

We make several observations before embarking on the proof of Theorem 5.6.4. First, by Proposition 5.3.6 the factorization (5.6.6) is negative special if and only if (5.6.5) is positive special, so we will restrict our attention to (5.6.5) only. Secondly, (5.6.5) easily implies that

$$Z_2 = -B - Z_1^*.$$

The proof of Theorem 5.6.4 will be given in the next section.

5.7 Proof of Theorem 5.6.4

Without loss of generality we can assume that both B and A are invertible and positive definite. Indeed, choose $b > 0$, sufficiently big so that the operators

$$B' = 2bI + B; \ A' = b^2 I + bB + A$$

are invertible and positive definite, and define

$$M(\lambda) = \lambda^2 I + \lambda B' + A' = L(\lambda + b).$$

Let

$$C_M = \begin{bmatrix} 0 & I \\ -A' & -B' \end{bmatrix}; \quad C_L = \begin{bmatrix} 0 & I \\ -A & -B \end{bmatrix}$$

be the companion operators of M and L, respectively, and let

$$G_M = \begin{bmatrix} B' & I \\ I & 0 \end{bmatrix} ; \quad G_L = \begin{bmatrix} B & I \\ I & 0 \end{bmatrix} .$$

One checks easily that

$$(C_L - bI) \begin{bmatrix} I & 0 \\ bI & I \end{bmatrix} = \begin{bmatrix} I & 0 \\ bI & I \end{bmatrix} C_M$$

and

$$\begin{bmatrix} I & 0 \\ bI & I \end{bmatrix}^* G_L \begin{bmatrix} I & 0 \\ bI & I \end{bmatrix} = G_M .$$

Thus, if $\mathcal{M}$ is a C_M-invariant maximal G_M-nonnegative subspace, then $\mathcal{N} = \begin{bmatrix} I & 0 \\ bI & I \end{bmatrix} (\mathcal{M})$ is C_L-invariant maximal G_L-nonnegative, and it is easy to see the right divisor of $L(\lambda)$ with the supporting subspace $\mathcal{N}$ is given by $\lambda I-(Y+bI)$, where $\lambda I-Y$ is the right divisor of $M(\lambda)$ with the supporting subspace $\mathcal{M}$.

From now on we assume that both B and A are invertible and positive definite. Introduce the operators

$$H = \begin{bmatrix} 0 & A^{1/2} \\ -A^{1/2} & -B \end{bmatrix} \in L(\mathcal{X}^2)$$

(here $A^{1/2}$ is the positive definite square root of A);

$$J = \begin{bmatrix} I & 0 \\ 0 & -I \end{bmatrix} \in L(\mathcal{X}^2).$$

The operator J induces the indefinite scalar product

$$[x,y]_J = \langle Jx,y \rangle ; \qquad x,y \in \mathcal{X}^2 .$$

LEMMA 5.7.1. *If*

$$[x,x]_J = [Hx,x]_J = 0, \tag{5.7.1}$$

then $x = 0$.

PROOF. Let $x = \begin{bmatrix} y \\ z \end{bmatrix}$; $y, z \in \mathcal{X}$. Then (5.7.1) amounts to

$$\|y\| = \|z\|;\ \langle Bz, z\rangle = -2\ \mathrm{Re}\langle A^{1/2}z, y\rangle.$$

Arguing by contradiction assume that $y \neq 0$, $z \neq 0$. Then

$$\langle A^{1/2}z, A^{1/2}z\rangle \langle y, y\rangle = \langle Az, z\rangle \langle z, z\rangle$$

$$< \frac{1}{4} \langle Bz, z\rangle^2,$$

where we have used the fact that the roots of $\langle L(\lambda)z, z\rangle = 0$ are real and distinct, which is equivalent to

$$\langle Bz, z\rangle^2 > 4 \langle Az, z\rangle \|z\|^2.$$

So

$$\langle A^{1/2}z, A^{1/2}z\rangle \langle y, y\rangle < (\mathrm{Re}\langle A^{1/2}z, y\rangle)^2 \leq |\langle A^{1/2}z, y\rangle|^2,$$

a contradiction with the Cauchy-Schwartz inequality. ∎

LEMMA 5.7.2. *There is a real number* μ *such that* $H-\mu I$ *is* J*-semidefinite, i.e., the self-adjoint operator* $J(H-\mu I)$ *is either positive semidefinite or negative semidefinite.*

PROOF. Let us show first that either $[Hx, x]_J > 0$ for all non-zero x with $[x, x]_J = 0$, or $[Hx, x]_J < 0$ for all such x. Indeed, assuming the contrary and using Lemma 5.7.1, we find x and y such that

$$[x, x]_J = [y, y]_J = 0,\ [Hx, x]_J < 0 < [Hy, y]_J. \tag{5.7.2}$$

Further, there exist real numbers ξ, η and $a \neq 0$ such that

$$\mathrm{Re}(e^{i(\xi-\eta)}[x, y]_J) = 0$$

and

$$a^2[Hy, y]_J + 2a\ \mathrm{Re}(e^{i(\xi-\eta)}[Hx, y]_J) + [Hx, x]_J = 0.$$

Now for $z = e^{i\xi}x + ae^{i\eta}y$ we have

$$[Hz, z]_J = [z, z]_J = 0.$$

On the other hand, the inequalities in (5.7.2) ensure that x and y are not collinear, and hence $z \neq 0$, a contradiction with Lemma 5.7.1.

Assume, for instance, that $[Hx,x]_J > 0$ for all non-zero x with $[x,x]_J = 0$. We show that

$$\frac{[Hy,y]_J}{[y,y]_J} < \frac{[Hx,x]_J}{[x,x]_J} \tag{5.7.3}$$

for every pair x, y such that $[y,y]_J < 0 < [x,x]_J$. To verify (5.7.3), argue by contradiction again; so assume $[x_0,x_0]_J = -[y_0,y_0]_J = 1$ and

$$-[Hy_0,y_0]_J \geq [Hx_0,x_0]_J.$$

Choosing real ξ and η so that

$$\mathrm{Re}(e^{i(\xi-\eta)}[x_0,y_0]_J) = 0; \ \mathrm{Re}(e^{i(\xi-\eta)}[Hx_0,y_0]_J) \leq 0,$$

and letting

$$z_0 = e^{i\xi}x_0+e^{i\eta}y_0,$$

we see that

$$[z_0,z_0]_J = 0, \ [Hz_0,z_0]_J \leq 0, \ z_0 \neq 0.$$

This contradicts the property of H proved earlier.

Now define

$$\mu = \inf\left\{\frac{[Hx,x]_J}{[x,x]_J}\right\},$$

where the infimum is taken over all $x \in \mathcal{X}^2$ with $[x,x]_J > 0$. The inequality (5.7.3) shows that $\mu > -\infty$. The same inequality (5.7.3), together with the definition of μ and the property that $[Hx,x]_J > 0$ for all $x \neq 0$ with $[x,x]_J = 0$, show that

$$\langle J(H-\mu I)x,x\rangle = [Hx,x]_J-\mu[x,x]_J \geq 0$$

for all $x \in \mathcal{X}^2$. Consequently, $J(H-\mu I)$ is positive semidefinite. ∎

Now let

$$C_L = \begin{bmatrix} 0 & I \\ -A & -B \end{bmatrix}$$

be the companion operator for $L(\lambda)$, and let

$$G = \begin{bmatrix} B & I \\ I & 0 \end{bmatrix}.$$

One verifies that

$$S^*(-GC_L)S = J; \quad S^{-1}C_LS = H, \tag{5.7.3}$$

where

$$S = \begin{bmatrix} A^{-1/2} & 0 \\ 0 & I \end{bmatrix}.$$

Using (5.7.3), together with Lemma 5.7.2, we see that the self-adjoint operator $-GC_L(C_L-\mu I)$ is either positive semidefinite or negative semidefinite. In particular, C_L is G-definitizable, and by Theorem 5.2.2(a) there exists C_L-invariant maximal G-nonnegative subspace $\mathcal{M}$. So $\mathcal{M}$ is the supporting subspace for a special positive factorization (5.6.5).

Next, we prove that $\sigma(Z_1) \subset \bar{\Delta}_1$. Clearly, $\sigma(Z_1) \subset \bar{\Delta}_1 \cup \bar{\Delta}_2$, and because of Theorem 5.6.2 we need only to show that every real λ_0 with $\mathrm{dist}(\lambda_0,\bar{\Delta}_1) > 0$ does not belong to $\sigma(Z_1)$. Assuming the contrary, let λ_0 with

$$\mathrm{dist}(\lambda_0,\bar{\Delta}_1) = \delta > 0 \tag{5.7.4}$$

be such that λ_0 is a boundary point of $\sigma(Z_1) = \sigma(C_L|_{\mathcal{M}})$. Then there is a sequence $\{x_n\}_{n=1}^{\infty}$, $x_n \in \mathcal{M}$, such that $\|x_n\| \geq \varepsilon_0$ for some $\varepsilon_0 > 0$ independent of n, and

$$\|C_Lx_n-\lambda_0x_n\| \to 0. \tag{5.7.5}$$

Write $x_n = \begin{bmatrix} y_n \\ z_n \end{bmatrix}$; $y_n, z_n \in \mathcal{X}$. Then (5.7.5) implies

$$\|z_n - \lambda_0 y_n\| \to 0; \ \|\lambda_0^2 y_n + \lambda_0 By_n + Ay_n\| \to 0.$$

In particular, the norms $\|y_n\|$ are uniformly bounded below by a positive number; hence without loss of generality we can assume $\|y_n\| = 1$. Next,

$$\left\langle \begin{bmatrix} B & I \\ I & 0 \end{bmatrix} \begin{bmatrix} y_n \\ z_n \end{bmatrix}, \begin{bmatrix} y_n \\ z_n \end{bmatrix} \right\rangle \geq 0 \tag{5.7.6}$$

for all n. Passing to a subsequence of $\{y_n\}_{n=1}^{\infty}$ (if necessary), we deduce from (5.7.6) that the limit

$$a = \lim_{n\to\infty}\left[\langle By_n, y_n\rangle + 2\lambda_0\langle y_n, y_n\rangle\right] = \lim_{n\to\infty}\left[\langle By_n, y_n\rangle + 2\lambda_0\right]$$

is a nonnegative number.

We show that actually $a > 0$. Indeed,

$$\lambda_0^2 + \lambda_0\langle By_n, y_n\rangle + \langle Ay_n, y_n\rangle \to 0 \tag{5.7.7}$$

implies that

$$\langle By_n, y_n\rangle^2 - 4\langle Ay_n, y_n\rangle \to a^2,$$

and if a were zero, then

$$\lim_{n\to\infty}\left[p_1(y_n) + \tfrac{1}{2}\langle By_n, y_n\rangle\right] = 0$$

(here $p_1(y_n)$ is the bigger zero of the equation $\lambda^2 + \lambda \langle By_n, y_n\rangle + \langle Ay_n, y_n\rangle = 0$), and

$$\lim_{n\to\infty}\left[\lambda_0 + \frac{1}{2}\langle By_n, y_n\rangle\right] = 0,$$

a contradiction with (5.7.4). Now (5.7.7) implies

$$(5.7.8)\qquad \lambda_0^2-p_j(y_n)^2+(\lambda_0-p_j(y_n))\langle By_n,y_n\rangle$$
$$= (\lambda_0+p_j(y_n)+\langle By_n,y_n\rangle)(\lambda_0-p_j(y_n)) \to 0,\ j = 1,2.$$

As $p_1(y_n)-p_2(y_n) = a > 0$ and $|\lambda_0-p_1(y_n)| \geq \delta > 0$, we deduce from (5.7.8) that

$$(5.7.9)\qquad \lim_{n\to\infty} p_2(y_n) = \lambda_0.$$

However,

$$\langle L'(p_2(y_n))y_n,y_n\rangle = \langle By_n,y_n\rangle + 2p_2(y_n) < 0,$$

so also

$$\langle By_n,y_n\rangle + 2\lambda_0 \leq 0,$$

a contradiction with the positivity of the number a.

We have proved that $\sigma(Z_1) \subset \bar{\Delta}_1$. Analogous consideration concerning the negative special factorization (5.6.6) shows that $\sigma(Z_2) \subset \bar{\Delta}_2$ and hence $\sigma(Z_2^*) \subset \bar{\Delta}_2$. Since the spectra of Z_1 and Z_2^* have empty interiors (and consequently the operators $\lambda I-Z_1$ and $\lambda I-Z_2^*$ are never one-sided invertible for any value of λ), we have

$$\Sigma(L) = \sigma(Z_1) \cup \sigma(Z_2^*).$$

Theorem 5.6.3 is proved completely.

5.8 Invariant subspaces for unitary and self-adjoint operators in indefinite scalar products

This section is of auxiliary character. We set forth here results on invariant subspaces for unitary and self-adjoint operators in indefinite scalar product spaces that will be used in the subsequent section. The emphasis is on invariant subspaces with certain definiteness properties with respect to the indefinite scalar product, and the self-adjoint and unitary operators involved will be assumed to have certain compactness properties.

Let $\mathcal{X}$ be a separable Hilbert space with the scalar product $\langle\cdot,\cdot\rangle$ and let J be a self-adjoint and unitary operator on $\mathcal{X}$. So $\sigma(J) = \{1,-1\}$, and J has the form $\begin{bmatrix} I & 0 \\ 0 & -I \end{bmatrix}$ with respect to some orthogonal decomposition $\mathcal{X} = \mathcal{X}_+ \oplus \mathcal{X}_-$ of $\mathcal{X}$. To avoid uninteresting (in this context) cases, we assume that both $\mathcal{X}_+$ and $\mathcal{X}_-$ are non-zero.

The J-indefinite scalar product will be denoted $[\cdot,\cdot]$, so $[x,y] = \langle Jx,y\rangle$.

Consider now a J-self-adjoint operator $A \in L(\mathcal{X})$, and assume that $\operatorname{Im} A = \frac{1}{2i}(A-A^*)$, the imaginary part of A, is a compact operator. A standard theory of compact perturbations (see, e.g., Theorem I.5.2 in Gohberg-Kreĭn [1]) implies that the non-real spectrum of A consists only of isolated points with all accumulation points on the real axis (in case the non-real spectrum of A is an infinite set). Moreover, each non-real spectral point λ_0 of A is an eigenvalue and the corresponding spectral subspace

$$\operatorname{Im}\left[\frac{1}{2\pi i}\int_\Gamma (\lambda I-A)^{-1}d\lambda\right],$$

where Γ is a small circle around λ_0, is finite-dimensional. As A is J-self-adjoint, the non-real spectrum is also symmetric relative to the real axis (because $A^* = JAJ^{-1}$ is similar to A).

The main aim of this section is to prove the following theorem.

THEOREM 5.8.1. *Let* A *be J-self-adjoint with compact imaginary part. Then for every decomposition of the non-real spectrum* Λ *of* A *into two disjoint sets* Λ_1 *and* Λ_2 *such that* $\lambda \in \Lambda_1$ *if and only if* $\bar{\lambda} \in \Lambda_2$ *there is a subspace* $\mathcal{M} \subset \mathcal{X}$ *with the following properties:*

(i) $\mathcal{M}$ *is A-invariant;*

(ii) $\mathcal{M}$ *is maximal J-nonnegative;*

(iii) *the non-real spectrum of* $A|_{\mathcal{M}}$ *is precisely* Λ_1.

We remark that the condition of compactness of Im A can be easily expressed in terms of the orthogonal projections P_+ and P_- on $\mathcal{X}_+$ and $\mathcal{X}_-$ respectively. Namely, given that A is J-self-adjoint, Im A is compact if and only if $P_+AP_- \in L(\mathcal{X}_-,\mathcal{X}_+)$ is compact, or, equivalently, if and only if $P_-AP_+ \in L(\mathcal{X}_+,\mathcal{X}_-)$ is compact. We leave verification of this simple fact to the reader.

The rest of this section is geared towards the proof of Theorem 5.8.1. It is based on the following general result. It will be convenient to denote by $\mathcal{S}_+$ the set of all maximal J-nonnegative subspaces.

THEOREM 5.8.2. *Let* $A \in L(\mathcal{X})$ *satisfy the following properties:*

(a) *if* $[x,x] \geq 0$, $x \neq 0$, *then also* $[Ax,Ax] \geq 0$, $Ax \neq 0$;

(b) *the operator* P_+AP_- *is compact;*

(c) *for some* $\mathcal{L}_0 \in \mathcal{S}_+$ *the subspace* $A\mathcal{L}_0$ *is also maximal J-nonnegative.*

Then $A\mathcal{L} \in \mathcal{S}_+$ *for every* $\mathcal{L} \in \mathcal{S}_+$, *and there is* $\mathcal{L}' \in \mathcal{S}_+$ *such that* $A\mathcal{L}' = \mathcal{L}'$.

PROOF. Write

$$A = \begin{bmatrix} A_{11} & A_{12} \\ A_{21} & A_{22} \end{bmatrix}$$

with respect to the orthogonal decomposition $\mathcal{X} = \mathcal{X}_+ \oplus \mathcal{X}_-$. For any contraction $K \in L(\mathcal{X}_+,\mathcal{X}_-)$ (so $\|K\| \leq 1$) consider the operator

$$W(K) = A_{11}+A_{12}K \in L(\mathcal{X}_+).$$

We verify first that Ker $W(K) = \{0\}$. Indeed, take $x \in \mathcal{X}_+$, $x \neq 0$, and let $y = x+Kx$. Then $[y,y] \geq 0$, and by property (a) $[Ay,Ay] \geq 0$, $Ay \neq 0$. In particular, $P_+Ay \neq 0$, which means precisely that $W(K)x \neq 0$.

Consider now the angular operator K_0 for $\mathcal{L}_0$ (cf. Corollary 5.1.2). As $A\mathcal{L}_0$ is also maximal J-nonnegative, by Corollary 5.1.2 we have $P_+A\mathcal{L}_0 = \mathcal{X}_+$. So the range of $W(K_0)$ is $\mathcal{X}_+$. On the other hand, K_0 is a contraction, and we have seen before that Ker $W(K_0) = \{0\}$. Consequently, $W(K_0)$ is invertible. Now for any contraction $K \in L(\mathcal{X}_+,\mathcal{X}_-)$ the difference

$$W(K)-W(K_0) = A_{12}(K-K_0)$$

is a compact operator, because $A_{12} = P_+AP_-$ is such. Taking into account that $W(K_0)$ is invertible and Ker $W(K) = \{0\}$, we conclude that $W(K)$ is invertible for any contraction $K \in L(\mathcal{X}_+,\mathcal{X}_-)$.

Let $\mathcal{L}_K = \{x+Kx \mid x \in \mathcal{X}_+\}$ be the maximal J-nonnegative subspace with the angular operator K (Corollary 5.1.2). The subspace $A\mathcal{L}_K$ is J-nonnegative by the property (a) and $P_+A\mathcal{L}_K = \text{Im } W(K) = \mathcal{X}_+$. So by the same Corollary 5.1.2, $A\mathcal{L}_K \in \mathcal{S}_+$.

It remains to prove that $A\mathcal{L}' = \mathcal{L}'$ for some $\mathcal{L}' \in \mathcal{S}_+$. For every contraction $K \in L(\mathcal{X}_+,\mathcal{X}_-)$ define

$$\Phi(K) = (A_{21}+A_{22}K)(A_{11}+A_{12}K)^{-1} \in L(\mathcal{X}_+,\mathcal{X}_-).$$

If $\mathcal{L}_K \in \mathcal{S}_+$ has the angular operator K, then $A\mathcal{L}_K$ has the angular operator $\Phi(K)$. Since $A\mathcal{L}_K \in \mathcal{S}_+$, the operator $\Phi(K)$ is a contraction as well. We shall prove that Φ is a continuous map on the set $\mathcal{C}$ of all contractions in $L(\mathcal{X}_+,\mathcal{X}_-)$, when $\mathcal{C}$ is considered in the weak operator topology. Recall (see, e.g., Section VI.1 in Dunford-Schwartz [1]) that the basic set of neighborhoods that define the weak operator topology in $L(\mathcal{X}_+,\mathcal{X}_-)$ consists of

$$\{C \in L(\mathcal{X}_+,\mathcal{X}_-) \mid |\langle (C-C_0)x_i,y_j\rangle| < \varepsilon;$$
$$i = 1,\dots,k;\ j = 1,\dots,\ell\}$$

where $C_0 \in L(\mathcal{X}_+,\mathcal{X}_-)$; $x_1,\dots,x_k \in \mathcal{X}_+$; $y_1,\dots,y_\ell \in \mathcal{X}_-$ and $\varepsilon > 0$. Equivalently, a net (generalized sequence) $\{C_\alpha\}$ with $C_\alpha \in L(\mathcal{X}_+,\mathcal{X}_-)$ converges to $C \in L(\mathcal{X}_+,\mathcal{X}_-)$ in the weak operator topology if and only if for every $x \in \mathcal{X}_+$, $y \in \mathcal{X}_-$ the net $\{\langle C_\alpha x,y\rangle\}$ converges to $\langle Cx,y\rangle$. It is known (see, e.g., Ringrose [1]) that $L(\mathcal{X}_+,\mathcal{X}_-)$ is a locally convex topological vector space in the weak operator topology and that $\mathcal{C}$ is compact (and obviously convex) in this topology. So, as soon as we prove that Φ is continuous in the weak operator topology, by the Schauder-Tychonoff theorem (Theorem V.10.5 in Dunford-Schwartz [1]) there is $K_0 \in \mathcal{C}$ such that $\Phi(K_0) = K_0$. This means that $A\mathcal{L}'_{K_0} = \mathcal{L}'_{K_0}$.

To prove that Φ is continuous in the weak operator topology, write

$$W(K) = A_{11}(I-SK), \quad K \in \mathscr{C},$$

where $S = -A_{11}^{-1}A_{12}$ is compact (the invertibility of $A_{11} = W(0)$ is a particular case of invertibility of $W(K)$ for all $K \in \mathscr{C}$). We claim that $\|S\| < 1$. Indeed, otherwise there exists $\lambda_0 \in \mathbf{C}$, $\|\lambda_0| \geq 1$ and $x \in \mathscr{X}_+$ with $\|x\| = 1$, $y \in \mathscr{X}_-$ with $\|y\| = 1$ such that $Sy = \lambda_0 x$. Letting $K_0 \in \mathscr{C}$ be such that $K_0 x = \frac{1}{\lambda_0} y$, we obtain $W(K_0)x = 0$, a contradiction with invertibility of $W(K_0)$. Now we can write

$$\Phi(K) = (A_{21}+A_{22}K)(I-SK)^{-1}A_{11}^{-1}$$

$$= (A_{21}+A_{22}K)\sum_{p=0}^{\infty}(SK)^p A_{11}^{-1}, \qquad K \in \mathscr{C}.$$

As S is compact, it can be approximated (in the operator norm) by finite rank operators S_m:

$$\lim_{m\to\infty} \|S_m - S\| = 0.$$

Let

$$\Phi_m(K) = (A_{21}+A_{22}K)\sum_{p=0}^{m}(S_m K)^p A_{11}^{-1};$$

then clearly

$$\sup_{K\in\mathscr{C}} \|\Phi_m(K) - \Phi(K)\| \leq c_m, \tag{5.8.1}$$

where $c_m > 0$ is independent of K and $\lim_{m\to\infty} c_m = 0$. In view of (5.8.1), it is not difficult to see that Φ is continuous in the weak operator topology on $\mathscr{C}$ provided all Φ_m (regarded as maps from $L(\mathscr{X}_+,\mathscr{X}_-)$ to itself) are continuous in the weak operator topology. Writing out each S_m as finite sume of rank 1 operators, and using the fact that the set of continuous (in the weak operator topology) maps from $L(\mathscr{X}_+,\mathscr{X}_-)$ to itself is closed under addition, we have only to verify that the maps of the form

$$\Psi(K) = (A_{21}+A_{22}K)Q_1KQ_2K\cdots Q_sKA_{11}^{-1}, \quad K \in L(\mathcal{X}_+,\mathcal{X}_-)$$

are continuous (in the weak operator topology), where $Q_j \in L(\mathcal{X}_-,\mathcal{X}_+)$ are rank 1 operators. We have

$$Q_jx = \langle y_j,x\rangle z_j$$

for some $y_j \in \mathcal{X}_-$, $z_j \in \mathcal{X}_+$. Now

$$\langle\Psi(K)u,v\rangle = \langle(A_{21}+A_{22}K)z_1,v\rangle\,\langle y_1,Kz_2\rangle\cdots\langle y_{s-1},Kz_j\rangle\,\langle y_s,KA_{11}^{-1}u\rangle,$$

which implies the desired continuity of Ψ. ∎

Next, we consider the J-unitary operators. An invertible operator $U \in L(\mathcal{X})$ is called *J-unitary* if $U^*JU = J$. Consider the class $\mathcal{K}$ of all J-unitary operators U for which P_+UP_- is compact. It is not difficult to check that the set of all J-unitary operators form a group (by multiplication), and that if U is J-unitary, so is U^*. Furthermore, the class $\mathcal{K}$ forms a subgroup of J-unitary operators, and $U \in \mathcal{K}$ implies $U^* \in \mathcal{K}$. Let us verify the latter statement. Given $U \in \mathcal{K}$ write

$$U = \begin{bmatrix} U_{11} & U_{12} \\ U_{21} & U_{22} \end{bmatrix} : \mathcal{X}_+ \oplus \mathcal{X}_- \to \mathcal{X}_+ \oplus \mathcal{X}_-. \tag{5.8.2}$$

Then $U_{22}^*U_{22} = I+U_{12}^*U_{12}$ which implies that U_{22} is left invertible. As U^* is J-unitary as well, U_{22}^* is also left invertible, and hence U_{22} is invertible. Now

$$U_{12}^*U_{11}-U_{22}^*U_{21} = 0,$$

and the compactness of U_{12} implies that of U_{21}. So $U^* \in \mathcal{K}$.

It is also easy to see that for every $U \in \mathcal{K}$ there is unitary V such that U-V is compact. Indeed, let $U \in \mathcal{K}$ be given by (5.8.2) with compact U_{12} (and hence compact U_{21}). As

$$U_{22}^*U_{22} = I+U_{12}^*U_{12}; \quad U_{11}^*U_{11} = I+U_{21}^*U_{21};$$

$$U_{11}^*U_{12}-U_{21}^*U_{22} = 0; \quad U_{12}^*U_{11}-U_{22}^*U_{21} = 0;$$

it follows that $U^*U = I+S$, where S is compact. Define the self-adjoint compact operator $S_1 = (I+S)^{1/2}-I$ (the square root is well defined because $I+S$ is positive definite and invertible). Clearly, $I+S_1$ is invertible as well, and for $V = U(I+S_1)^{-1}$ we have

$$V^*V = (I+S_1)^{-1}U^*U(I+S_1)^{-1} \tag{5.8.3}$$

$$= (I+S_1)^{-1}(I+S)(I+S_1)^{-1} = I.$$

It is easy to see (using, for example, the spectral theorem for compact self-adjoint operators) that the difference $U-V = U[I-(I+S_1)^{-1}]$ is compact. As U is invertible, V is Fredholm with index zero, and together with (5.8.3) this implies that V is unitary.

It follows by the theory of perturbations by compact operators (Theorem I.5.3 in Gohberg-Kreĭn [1]) that for every $U \in \mathcal{K}$ the spectrum of U not on the unit circle consists of isolated eigenvalues with accumulation points (if any) on the unit circle, and for each non-unimodular spectral point of U the corresponding spectral U-invariant subspace is finite dimensional. Observe also that if $\lambda_0 \in \sigma(U)$, $|\lambda_0| \neq 1$, then $\bar{\lambda}_0^{-1} \in \sigma(U)$, and the multiplicities of U restricted to the finite-dimensional spectral subspace corresponding to λ_0 coincide with the multiplicities of U restricted to the spectral subspace corresponding to $\bar{\lambda}_0^{-1}$ (this follows from the similarity of U and $U^{*-1} = JUJ$).

THEOREM 5.8.3. *Let $U \in \mathcal{K}$, and let the non-unimodular spectrum of U be decomposed into two disjoint parts Λ_1 and Λ_2 such that*

$$\Lambda_2 = \{\bar{\lambda}^{-1} \mid \lambda \in \Lambda_1\}.$$

Then there exist maximal J-nonnegative subspace $\mathcal{M}_+$ *and maximal J-nonnegative subspace* $\mathcal{M}_-$ *such that* $U\mathcal{M}_+ = \mathcal{M}_+$, $U\mathcal{M}_- = \mathcal{M}_-$ *and the non-unimodular part of each spectrum* $\sigma(U|_{\mathcal{M}_+})$ *and* $\sigma(U|_{\mathcal{M}_-})$ *is precisely* Λ_1.

PROOF. As U is J-unitary, we have

$$[x,x] = [Ux,Ux], \quad x \in \mathcal{X}.$$

In particular, for every J-nonnegative subspace $\mathcal{L}$ the subspace $U\mathcal{L}$ is again J-nonnegative. Since U^{-1} is J-unitary as well, the converse is also true: if $U\mathcal{L}$ is J-nonnegative, so is $\mathcal{L}$. Hence $U\mathcal{L}$ is maximal J-nonnegative for every maximal J-nonnegative subspace $\mathcal{L}$. So Theorem 5.8.2 is applicable and there is maximal J-nonnegative subspace $\mathcal{M}_+$ such that $U\mathcal{M}_+ = \mathcal{M}_+$. Applying the already proved part of Theorem 5.8.3 with J replaced by -J, we obtain existence of maximal J-nonpositive $\mathcal{M}_-$ such that $U\mathcal{M}_- = \mathcal{M}_-$.

Next, we have to modify $\mathcal{M}_+$ and $\mathcal{M}_-$ to satisfy also the additional spectral properties. This can be done using a method from Langer [4]. We omit the details here, but Exercises 5.8-5.10 contain an outline of this procedure. ∎

Finally, we are ready to prove Theorem 5.8.1. Let A satisfy the hypotheses of Theorem 5.8.1. Choose a non-real $\zeta \notin \sigma(A)$, and define the Cayley transform

$$U = (A-\bar{\zeta}I)(A-\zeta I)^{-1}. \tag{5.8.4}$$

One verifies that U is J-unitary. Further, write

$$A = \begin{bmatrix} A_{11} & A_{12} \\ A_{21} & A_{22} \end{bmatrix}$$

with respect to the orthogonal decomposition $\mathcal{X} = \mathcal{X}_+ \oplus \mathcal{X}_-$, and choose ζ so that, in addition, $\zeta \notin \sigma(A_{11}) \cup \sigma(A_{22})$. It is not difficult to check that in this case the operator

$$Z = I-A_{21}(A_{11}-\zeta I)^{-1}A_{12}(A_{22}-\zeta T)^{-1}$$

is invertible, and

(5.8.5) $$(A-\varsigma I)^{-1} =$$

$$\begin{bmatrix} (A_{11})-\varsigma I)^{-1}-(A_{11}-\varsigma I)^{-1}A_{12}(A_{22}-\varsigma I)^{-1}X & -(A_{11}-\varsigma I)^{-1}A_{12}(A_{22}-\varsigma I)^{-1}Z^{-1} \\ (A_{22}-\varsigma I)^{-1}X & (A_{22}-\varsigma I)^{-1}Z^{-1} \end{bmatrix},$$

where $X = -Z^{-1}A_{21}(A_{11}-\varsigma I)^{-1}$. As A_{12} is compact, formulas (5.8.4) and (5.8.5) ensure that P_+UP_- is also compact, i.e., $U \in \mathcal{K}$. It remains to apply Theorem 5.8.3 and use the facts that U and A have precisely the same invariant subspaces, and that

$$\sigma(U|_{\mathcal{M}}) = \{(\lambda-\varsigma)^{-1} \mid \lambda \in \sigma(A|_{\mathcal{M}})\}$$

for any A-invariant subspace $\mathcal{M}$.

Theorem 5.8.1 is proved.

5.9 Self-adjoint operator polynomials of second degree

In this section we study operator polynomials of the form

$$L(\lambda) = \lambda^2 I+\lambda B+A$$

where B and A are self-adjoint operators in the separable Hilbert space $\mathcal{X}$, and A is assumed to be compact and positive semidefinite. Thus the class of polynomials here is different from that considered in Sections 5.6, 5.7. Namely, we do not assume here that $L(\lambda)$ is strongly hyperbolic; on the other hand we require that A is self-adjoint compact while in Sections 5.6, 5.7 A was assumed merely self-adjoint. We obtain in this section results on factorization of $L(\lambda)$ by using Theorem 5.8.1 as the main tool.

It is easy to see that $L(\lambda)$ admits factorization

$$L(\lambda) = (\lambda I-Z'')(\lambda I-Z')$$

(where $Z'',Z' \in L(\mathcal{X})$) if and only if Z' is a *right root* of the operator equation $L(Z) = 0$, i.e.,

$$Z'^2+BZ'+A = 0.$$

It will be convenient for us in this section to use the notion of right roots instead of factorizations.

Introduce the operators

$$H = \begin{bmatrix} 0 & A^{1/2} \\ -A^{1/2} & -B \end{bmatrix}, \quad J = \begin{bmatrix} I & 0 \\ 0 & -I \end{bmatrix}$$

(already encountered in Section 5.6). So H is J-self-adjoint and, because A is compact, the imaginary part of H is also compact. So by Theorem 5.8.1 for every decomposition of the non-real spectrum of H into two disjoint parts Λ and $\Lambda' = \{\bar{\lambda} \mid \lambda \in \Lambda\}$ there is H-invariant maximal J-nonnegative subspace $\mathcal{M}_\Lambda$ such that the non-real part of $\sigma(H|_{\mathcal{M}_\Lambda})$ coincides with Λ. Let $K_\Lambda \in L(\mathcal{X})$ be the corresponding angular operator

$$\mathcal{M}_\Lambda = \left\{ \begin{bmatrix} x \\ K_\Lambda x \end{bmatrix} \mid x \in \mathcal{X} \right\}$$

(in particular, $\|K_\Lambda\| \leq 1$). It is easy to see that $Z_\Lambda = K_\Lambda A^{1/2}$ is a right root of the operator equation $L(Z) = 0$. Indeed, the H-invariance of $\mathcal{M}_\Lambda$ implies that for every $x \in \mathcal{X}$ there is $y \in \mathcal{X}$ for which

$$\begin{bmatrix} 0 & A^{1/2} \\ -A^{1/2} & -B \end{bmatrix} \begin{bmatrix} x \\ K_\Lambda x \end{bmatrix} = \begin{bmatrix} y \\ K_\Lambda y \end{bmatrix}, \tag{5.9.1}$$

i.e., $y = A^{1/2}K_\Lambda x$ and

$$-A^{1/2}x-BK_\Lambda x = K_\Lambda y = K_\Lambda A^{1/2}K_\Lambda x.$$

In particular, this equality holds for every x in the range of $A^{1/2}$ which amounts to

$$K_\Lambda A^{1/2}K_\Lambda A^{1/2}+BK_\Lambda A^{1/2}+A = 0,$$

i.e.,

$$Z_\Lambda^2+BZ_\Lambda+A = 0.$$

Further, as $\|K_\Lambda\| \le 1$, we have

$$\|Z_\Lambda x\| \le \|A^{1/2}x\|$$

for all $x \in \mathcal{X}$, i.e., $A-Z_\Lambda^* Z_\Lambda$ is positive semidefinite.

We have proved a large part of the following result.

THEOREM 5.9.1. *Let $B = B^*$ and A compact and positive semidefinite ($A,B \in L(\mathcal{X})$). Then for any decomposition of the non-real part of the spectrum of the operator polynomial $L(\lambda) = \lambda^2 I+\lambda B+A$ into two disjoint parts Λ and $\Lambda' = \{\bar{\lambda} \mid \lambda \in \Lambda\}$ there is right root Z_Λ of the equation*

$$Z^2+BZ+A = 0$$

with the properties that $Z_\Lambda^* Z_\Lambda - A$ is negative semidefinite and the non-real part of $\sigma(Z_\Lambda)$ coincides with Λ.

To finish the proof of this theorem we still have to verify the following assertions:

(1) the non-real part of $\sigma(Z_\Lambda)$ coincides with the non-real part of $\sigma(H|_{\mathcal{M}_\Lambda})$;

(2) the non-real part of $\Sigma(L)$ coincides with the non-real part of $\sigma(H)$.

We start with a simple proposition.

PROPOSITION 5.9.2. *The non-real part of $\Sigma(L)$ consists of isolated points, with the accumulation points (if any) on the real line, and every non-real $\lambda_0 \in \Sigma(L)$ is an eigenvalue of $L(\lambda)$, i.e.,*

$$\operatorname{Ker} L(\lambda_0) \ne \{0\}.$$

PROOF. Let

$$C_L = \begin{bmatrix} 0 & I \\ -A & -B \end{bmatrix}$$

be the companion operator of $L(\lambda)$, and let

$$D = \begin{bmatrix} 0 & I \\ 0 & -B \end{bmatrix}.$$

As $\sigma(D)$ lies on the real line and $C_L - D$ is compact, the theory of perturbations by compact operators (Theorem I.5.3 in Gohberg-Kreĭn [1]) implies that the non-real part of $\sigma(C_L)$ has the properties described in the proposition. It remains to observe that $\mathrm{Ker}(\lambda_0 I - C_L) \neq \{0\}$ if and only if $\mathrm{Ker}\, L(\lambda_0) \neq \{0\}$. ∎

Because of Proposition 5.9.2 for the right root $Z_\Lambda = K_\Lambda A^{1/2}$ of the equation $L(Z) = 0$ the non-real part of $\sigma(Z_\Lambda)$ consists of isolated eigenvalues. As the set of non-zero eigenvalues of TS and ST is the same for any pair of operators $T, S \in L(\mathcal{X})$ (see, e.g., Theorem 3, Section 3.10 in Birman-Solomjak [1]), we see that the non-real part of $\sigma(A^{1/2}K_\Lambda)$ coincides with the non-real part of $\sigma(Z_\Lambda)$ and consists of eigenvalues only. But it is easy to check that if $\lambda_0 \neq 0$ is an eigenvalue of $H|_{\mathcal{M}_\Lambda}$ if and only if λ_0 is an eigenvalue of $A^{1/2}K_\Lambda$, and hence also Z_Λ.

Finally, we verify the assertion (2). Observe the following equality:

$$H\begin{bmatrix} A^{1/2} & 0 \\ 0 & I \end{bmatrix} = \begin{bmatrix} A^{1/2} & 0 \\ 0 & I \end{bmatrix} C_L, \tag{5.9.2}$$

where $C_L = \begin{bmatrix} 0 & I \\ -A & -B \end{bmatrix}$. Let λ_0 be a non-real eigenvalue of C_L; the corresponding eigenvector is necessarily of the form $\begin{bmatrix} x \\ \lambda_0 x \end{bmatrix}$ for some $x \neq 0$. The equality (5.9.2) shows that $\begin{bmatrix} A^{1/2}x \\ \lambda_0 x \end{bmatrix}$ is an eigenvector of H corresponding to λ_0, so $\lambda_0 \in \sigma(H)$. Conversely, let $\lambda_0 \in \sigma(H)$, λ_0 is non-real. As λ_0 is an isolated point of $\sigma(H)$, and λ_0 is an eigenvalue whose corresponding spectral subspace is finite dimensional, the point $\bar{\lambda}_0$ is an eigenvalue of

H^*. The corresponding eigenvector of H^* is necessarily of the form $\begin{bmatrix} -\lambda^{-1}A^{1/2}y \\ y \end{bmatrix}$ for some $y \neq 0$. Now take adjoints in (5.9.2):

$$\begin{bmatrix} A^{1/2} & 0 \\ 0 & I \end{bmatrix} H^* = C_L^* \begin{bmatrix} A^{1/2} & 0 \\ 0 & I \end{bmatrix}.$$

It follows that $\overline{\lambda}_0$ is an eigenvalue of C_L^* with corresponding eigenvector $\begin{bmatrix} -\lambda^{-1}A^{1/2}y \\ y \end{bmatrix}$. Consequently, $\lambda_0 \in \sigma(C_L)$. We have proved the assertion (2), and together with it Theorem 5.9.1 is proved.

5.10 Exercises

Ex. 5.1. Prove that the spectral subspace of J corresponding to the positive (resp. negative) part of $\sigma(J)$ is maximal J-nonnegative (resp. maximal J-nonpositive).

Ex. 5.2. A subspace $\mathcal{M}$ is called uniformly J-positive (resp. uniformly J-negative) if

$$[x,x] \geq \alpha\|x\|^2 \quad (\text{resp. } [x,x] \leq -\alpha\|x\|^2)$$

for all $x \in \mathcal{M}$, where α is a positive constant independent of x. Let $\mathcal{M}$ be uniformly J-positive (resp. uniformly J-negative) subspace. Prove that all subspaces sufficiently close to $\mathcal{M}$ in the gap metric are also uniformly J-positive (resp. uniformly J-negative).

Ex. 5.3. Show by example that the assertion of Ex. 5.2 is false for J-nonnegative (or J-nonpositive) subspaces $\mathcal{M}$.

Ex. 5.4. Show that the bijection established in Corollary 5.1.2 is continuous.

Ex. 5.5. Prove Theorem 5.3.2 for odd ℓ.

Ex. 5.6. A monic operator polynomial $L(\lambda)$ is called *positive definite* if $\langle L(\lambda)x,x\rangle > 0$ for every real λ and every nonzero $x \in \mathcal{X}$, and *uniformly positive definite* if there is $\alpha > 0$ such that

$$\langle L(\lambda)x,x\rangle \geq \alpha\|x\|^2 \quad \text{for all } x \in \mathcal{X}.$$

Give an example of a positive definite operator polynomial which is not uniformly positive definite.

Ex. 5.7. Prove that a monic operator with self-adjoint coefficients $L(\lambda)$ is uniformly positive definite if and only if $\Sigma(L) \cap \mathbb{R} = \emptyset$.

Ex. 5.8. In exercises 5.8-5.10,

$$J = \begin{bmatrix} I & 0 \\ 0 & -I \end{bmatrix} ; \quad P^+ = \begin{bmatrix} I & 0 \\ 0 & 0 \end{bmatrix} ; \quad P^- = \begin{bmatrix} 0 & 0 \\ 0 & I \end{bmatrix} ;$$

U is a fixed J-unitary operator, and $U \in \mathcal{K}$. Let $\mathcal{M}$ be a U-invariant maximal J-nonnegative subspace. Prove that there exists a projection E such that

$$(5.10.1) \qquad P^+E = P^+; \; EP^+ = E; \; \|E\| \le \sqrt{2}; \; \operatorname{Im} E = \mathcal{M}; \; UE = EUE.$$

Hint: Let $E = P^+ + P^-KP^+$, where $K = \operatorname{Im} P^+ \to \operatorname{Im} P^-$ is the angular operator of $\mathcal{M}$.

Ex. 5.9. Let $E' \in L(\mathcal{X})$ be given satisfying the properties:

$$P^+E' = P^+; \; E'P^+ = E'; \; \|E'\| \le \sqrt{2}; \; UE' = E'UE'.$$

(i) Show that $\operatorname{Im} E'$ is a subspace (i.e., is closed), and $\operatorname{Im} E'$ is U-invariant.

(ii) Show that for a pair λ and $\bar{\lambda}^{-1}$ of non-unimodular ($|\lambda| \ne 1$) eigenvalues of U, there exists $E'' \in L(\mathcal{X})$ such that

$$P^+E'' = P^+; \; E''P^+ = E''; \; \|E''\| \le \sqrt{2}; \; UE'' = E''UE'',$$

and, in addition, $\lambda \in \sigma(U|_{\operatorname{Im} E''})$, $\bar{\lambda} \notin \sigma(U|_{\operatorname{Im} E''})$.

Hint for part (ii): Let $\mathcal{R}_\lambda$ and $\mathcal{R}_{\bar{\lambda}}$ be the (finite dimensional) spectral U-invariant subspaces corresponding to λ and $\bar{\lambda}$, respectively. Put

$$\mathcal{M}_0 = \operatorname{Im} E' \cap (\mathcal{R}_\lambda)^{\perp J} \cap (\mathcal{R}_{\bar{\lambda}})^{\perp J}$$

and

$$\mathcal{M}_1 = \mathcal{M}_0 + \mathcal{R}_\lambda .$$

The subspace $\mathcal{M}_1$ is U-invariant and maximal J-nonnegative. Now use Ex. 5.8 with $\mathcal{M}$ replaced by $\mathcal{M}_1$.

<u>Ex. 5.10</u>. Let $\Lambda_1 = \{\lambda_1, \lambda_2, \ldots\}$. Starting with a U-invariant maximal J-nonnegative subspace $\mathcal{M}$ and the corresponding E satisfying (5.10.1), build a sequence $E_0 = E, E_1, E_2, \ldots$ where E_m is gotten from E_{m-1} in the same way E'' is gotten from E' with $\lambda = \lambda_m$ in Ex. 5.9(ii).

(i) Show that there is a weakly convergent subsequence

$$\tilde{E} = \lim E_{m_k}.$$

(ii) Prove that $\tilde{E}$ satisfies (5.10.1) (with E replaced by $\tilde{E}$).
Hint: Use ideas from the proof of Satz II.1 in Langer [6].

(iii) Prove that the non-unimodular part of $\sigma(U|_{\operatorname{Im} \tilde{E}})$ is Λ_1.

5.11 <u>Notes</u>

The method of studying operator polynomials with self-adjoint coefficients by using J-self-adjoint operators and their invariant subspaces that are maximal J-nonnegative (or maximal J-nonpositive) goes back to Kreĭn-Langer [1]. This method (used throughout this chapter) proved to be very successful, and that explains the importance of the theory of operators in indefinite scalar product spaces in this context. This theory is well developed by now. Without going into details, we just mention several monographs where various parts of this theory are exposed: Ando [1], Azizov-Iohvidov [1], Bognar [1], Iohvidov-Kreĭn-Langer [1], and review articles Langer [3], Azizov-Iohvidov [2], Iohvidov-Kreĭn [1], Kreĭn [1].

The material in Sections 5.1 and 5.2 (except for Theorem 5.2.2 and its corollary) is standard and is found in many monographs and review articles on indefinite scalar product spaces cited above. Problem 5.2.1 is well known and stated in many places, e.g., Ando [1], Azizov-Iohvidov [1]. The exposition in Section 5.3 uses Ran [1]; first results in the direction of Theorems 5.3.1 and 5.3.2 are found in Langer [1]. Explicit construction of positive and negative special factorizations in the finite-dimensional case is given in Gohberg-Lancaster-Rodman [2,6]. Theorem 5.3.5 is taken from Ran [2].

Factorization theorems for strongly hyperbolic operator polynomials of arbitrary degree have been developed in Langer [7], Marcus-Matsaev-Russu [1], and some of the main results are exposed in Markus [1]. Theorem 5.6.4 is taken from Langer [7], and the exposition of Lemma 5.7.2 follows Kühne [1]. The results in Section 5.8 are due to Kreĭn [2]; the exposition also follows that paper.

Theorem 5.9.1 was proved in Kreĭn-Langer [1]. In this paper, there are also results on factorization of weakly damped self-adjoint operator polynomials of second degree, as well as numerous connections with completeness of eigenvectors and generalized eigenvectors. In the finite dimensional case, the self-adjoint operator polynomials of second degree were studied in Duffin [1]; see also Lancaster [1], Gohberg-Lancaster-Rodman [1,6].

CHAPTER 6. SPECTRAL TRIPLES AND DIVISIBILITY OF NON-MONIC OPERATOR POLYNOMIALS

We have seen in Chapters 2-5 how the notions of spectral pairs and triples for monic operator polynomials can be used to analyze such polynomials. In this chapter we drop the requirement that polynomials are monic and develop the calculus of spectral pairs and triples in the framework of non-monic operator polynomials. In subsequent chapters, this calculus will be used to analyze non-monic operator polynomials.

6.1 Spectral triples: definition and uniqueness

Let $\mathcal{X}$ be a Banach space, and let

$$L(\lambda) = \sum_{j=0}^{p} \lambda^j A_j$$

be an operator polynomial with coefficients $A_j \in L(\mathcal{X})$. For a given open set $\Omega \subset \mathbb{C}$, a triple of operators (X,T,Y), where

$$X \in L(\mathcal{Y},\mathcal{X}),\ T \in L(\mathcal{Y}),\ Y \in L(\mathcal{X},\mathcal{Y})$$

and $\mathcal{Y}$ is a Banach space, is called a *spectral triple* of $L(\lambda)$ with respect to Ω (in short, *Ω-spectral triple*) if the following conditions are satisfied:

(P1) $$\sigma(T) \subset \Omega;$$

(P2) The operator-valued function $L(\lambda)^{-1}-X(\lambda I-T)^{-1}Y$ is analytic on Ω. More precisely, there is $\lambda_0 \in \Omega\backslash\sigma(T)$ such that $L(\lambda_0)$ is invertible, and the operator-valued function $L(\lambda)^{-1}-X(\lambda I-T)^{-1}Y$ which is analytic in a neighborhood of λ_0, admits analytic continuation to the whole of Ω;

(P3)
$$\sum_{j=0}^{p} A_j X T^j = 0;$$

(P4)
$$\bigcap_{j=0}^{\infty} \operatorname{Ker}(XT^j) = \{0\}.$$

In particular, it follows from (P2) that $L(\lambda)$ is invertible for every $\lambda \in \Omega$ which does not belong to $\sigma(T)$.

The operator T will be referred to as the *main operator* of the Ω-spectral triple (X,T,Y).

Several comments are in order concerning this notion.

First, in the above definition, the condition (P3) can be replaced by the following:

(P3′) The operator-valued function $L(\lambda)X(\lambda I-T)^{-1}$ is analytic in Ω.

Indeed, assume (P3′) and (P1) hold. Then the function $L(\lambda)X(\lambda I-T)^{-1}$ is analytic in $\mathbb{C}$, i.e., entire. For large $|\lambda|$ we have

$$L(\lambda)X(\lambda I-T)^{-1} = \left(\sum_{j=0}^{p} \lambda^j A_j\right)\left(\sum_{k=1}^{\infty} \lambda^{-k} X T^{k-1}\right) \tag{6.1.1}$$

$$= \sum_{q=-\infty}^{p-1} \lambda^q \left(\sum_{j=p}^{p} A_j X T^{j-q-1}\right),$$

where $p' = \max(0,q+1)$. As this function is entire, the coefficient of λ^{-1} must be zero, which gives precisely (P3). Conversely, if (P3) holds, then also

$$\sum_{j=0}^{p} A_j X T^{j+r} = 0$$

for $r = 1,2,\ldots$. Using formula (6.1.1), we see that the function $L(\lambda)X(\lambda I-T)^{-1}$ is entire, i.e., (P3′) holds.

This argument shows also that (P3′) in turn can be replaced by

(P3″) the operator function $L(\lambda)X(\lambda I-T)^{-1}$ is a polynomial.

Secondly, in the case when $L(\lambda)$ is monic and $\Omega = \mathbb{C}$, the $\mathbb{C}$-spectral triples of $L(\lambda)$ are precisely the spectral triples of $L(\lambda)$ introduced in Chapter 2. Indeed, in this case (P1) is trivial, (P2) follows from Theorem 2.5.2 and (P3), (P4) are consequences of the definition of right spectral pairs of monic operator polynomials. The converse statement (i.e., every $\mathbb{C}$-spectral triple of a monic operator polynomial is also a spectral triple in the sense of Chapter 2) will follow from the uniqueness (up to similarity) of Ω-spectral triples which will be proved later in this chapter.

Thirdly, it will turn out that the main operator in a Ω-spectral triple of $L(\lambda)$ is actually a linearization of $L(\lambda)$ with respect to Ω (see Chapter 1). Thus, there is a deep connection between the spectral properties of operator polynomials which manifest themselves in spectral triples, and the linearization properties of operator polynomials which manifest themselves in linearizations.

In this and the next two sections, we develop a calculus of spectral triples and prove their existence and uniqueness. An obvious necessary condition for existence of a Ω-spectral triple for $L(\lambda)$ is that the set $\Sigma(L) \cap \Omega$ is compact (indeed, in view of (P2),

$$\Sigma(L) \cap \Omega \subset \sigma(T),$$

which is compact and contained in Ω by (P1)). We shall see that this condition is also sufficient.

We start with the uniqueness property and show that Ω-spectral triples, whenever they exist, are unique up to similarity. First we derive the dual of Condition (P3′) for a spectral triple.

PROPOSITION 6.1.1. *If* (X,T,Y) *is a spectral triple for* $L(\lambda)$ *on* Ω, *then* $(\lambda I-T)^{-1}YL(\lambda)$ *has an analytic continuation to* Ω, *and consequently*

$$\sum_{j=0}^{p} T^j Y A_j = 0. \tag{6.1.2}$$

PROOF. The formula (6.1.2) follows from the analyticity of $(\lambda I-T)^{-1}YL(\lambda)$ in the same way as (P3) follows from (P3′); so it remains to prove the analyticity of $(\lambda I-T)^{-1}YL(\lambda)$.

Let Δ be a bounded Cauchy domain such that $\sigma(T) \subset \Delta \subset \overline{\Delta} \subset \Omega$. We show that for each $y \in \mathcal{X}$

$$\phi(z) \overset{\text{def}}{=} \frac{1}{2\pi i} \int_{\partial\Delta} \frac{(\lambda I-T)^{-1}YL(\lambda)}{z-\lambda}\, y d\lambda = 0 \quad (z \in \mathbb{C}\setminus\overline{\Delta}).$$

Put $H(\lambda) = L(\lambda)^{-1}-X(\lambda I-T)^{-1}Y$, $\lambda \in \Omega\setminus\Sigma(L)$. Note that $H(\lambda)$ has an analytic continuation to Ω (cf. Condition (P2)), while

$$X(\lambda I-T)^{-1}YL(\lambda) = I-H(\lambda)L(\lambda). \tag{6.1.3}$$

One proves that

$$T^n\phi(z) = \frac{1}{2\pi i} \int_{\partial\Delta} \frac{\lambda^n(\lambda I-T)^{-1}YL(\lambda)}{z-\lambda}\, y d\lambda \quad (z \in \mathbb{C}\setminus\overline{\Delta},\ n \geq 0). \tag{6.1.4}$$

Indeed,

$$T^n\phi(z) = \frac{1}{2\pi i} \int_{\partial\Delta} \frac{T^n(\lambda I-T)^{-1}YL(\lambda)}{z-\lambda}\, y d\lambda$$

$$= \frac{1}{2\pi i} \int_{\partial\Delta} \frac{(T^n-\lambda^n I)(\lambda I-T)^{-1}YL(\lambda)}{z-\lambda}\, y d\lambda + \frac{1}{2\pi i} \int_{\partial\Delta} \frac{\lambda^n(\lambda I-T)^{-1}YL(\lambda)}{z-\lambda}\, y d\lambda,$$

and the first integral is zero because of analyticity of the function $(T^n-\lambda^n I)(\lambda I-T)^{-1}$.

It follows that for each $z \in \mathbb{C}\setminus\overline{\Delta}$,

$$XT^n\phi(z) = \frac{1}{2\pi i} \int_{\partial\Delta} \frac{\lambda^n[I-H(\lambda)L(\lambda)]}{z-\lambda}\, y d\lambda = 0 \quad (n \geq 0).$$

Using (P4), one sees that $\phi(z) = 0$, $z \in \mathbb{C}\setminus\overline{\Delta}$.

The proof of Proposition 6.1.1 is now complete in view of Lemma 6.1.2 below. ∎

The following lemma is a well-known fact which is convenient to state explicitly.

LEMMA 6.1.2. *Let* $\mathcal{Z}$ *be a Banach space, and let* $T(\lambda)$ *be a* $\mathcal{Z}$*-valued function analytic in a neighborhood of* $\partial\Delta$*, where* Δ *is a bounded Cauchy domain. If*

$$\int_{\partial\Delta} \frac{T(\lambda)}{z-\lambda}\, d\lambda = 0 \tag{6.1.5}$$

for every $z \in \mathbb{C}\backslash\overline{\Delta}$*, then* $T(\lambda)$ *admits analytic continuation to the whole of* Δ.

PROOF. Let

$$F_1(z) = \frac{1}{2\pi i} \int_{\partial\Delta} \frac{T(\lambda)}{z-\lambda}\, d\lambda, \quad z \in \Delta$$

and

$$F_2(z) = \frac{1}{2\pi i} \int_{\partial\Delta} \frac{T(\lambda)}{z-\lambda}\, d\lambda, \quad z \in \mathbb{C}\backslash\overline{\Delta}.$$

Then, by the well-known fact on the boundary values of Cauchy integrals (see, e.g., Section 3.3 in Chapter 3 of Markushevich [1]), we have

$$T(z) = F_2(z)-F_1(z),$$

and the functions $F_1(z)$ and $F_2(z)$ are analytic in Δ and $\mathbb{C}\backslash\overline{\Delta}$, respectively. As $F_2(z) \equiv 0$, we are done. ■

THEOREM 6.1.3. *For* $i = 1,2$*, let* (X_i,T_i,Y_i) *be a spectral triple of* $L(\lambda)$ *on* Ω*, where*

$$X_i \in L(\mathcal{Y}_i,\mathcal{X}),\ T_i \in L(\mathcal{Y}_i),\ Y_i \in L(\mathcal{X},\mathcal{Y}_i).$$

Then there exists a unique invertible operator $S\colon \mathcal{Y}_1 \to \mathcal{Y}_2$ *with the property that*

$$X_2S = X_1,\ T_2S = ST_1,\ Y_2 = SY_1. \tag{6.1.6}$$

The similarity S *and its inverse* S^{-1} *are given by the formulas*

$$S = (2\pi i)^{-1} \int_{\partial\Delta} (\lambda-T_2)^{-1}Y_2L(\lambda)X_1(\lambda-T_1)^{-1}d\lambda, \tag{6.1.7}$$

(6.1.8) $$S^{-1} = (2\pi i)^{-1} \int_{\partial\Delta} (\lambda - T_1)^{-1} Y_1 L(\lambda) X_2 (\lambda - T_2)^{-1} d\lambda,$$

where Δ *is a bounded Cauchy domain such that* $(\sigma(T_1) \cup \sigma(T_2)) \subset \Delta \subset \overline{\Delta} \subset \Omega$.

PROOF. We use the same approach as in the proof of Theorem 1.2.1.

Let V be the operator defined by the right-hand side of (6.1.8). Note that the definition of V and S does not depend on the particular choice of the Cauchy domain Δ. Therefore, we choose a bounded Cauchy domain Δ' such that $\Delta \subset \overline{\Delta} \subset \overline{\Delta'} \subset \Omega$. Then

$$VS = (2\pi i)^{-1} \int_{\partial\Delta'} (\mu - T_1)^{-1} Y_1 L(\mu) X_2 (\mu - T_2)^{-1} S d\mu$$

$$= (2\pi i)^{-2} \int_{\partial\Delta'} \Big(\int_{\partial\Delta} (\mu - T_1)^{-1} Y_1 L(\mu) X_2 (\mu - T_2)^{-1} (\lambda - T_2)^{-1} \cdot Y_2 L(\lambda) X_1 (\lambda - T_1)^{-1} d\lambda \Big) d\mu.$$

We use the resolvent identity to rewrite the integrand as

$$(\mu - T_1)^{-1} Y_1 L(\mu) X_2 (\mu - T_2)^{-1} \frac{Y_2 L(\lambda) X_1 (\lambda - T_1)^{-1}}{\lambda - \mu} - \frac{(\mu - T_1)^{-1} Y_1 L(\mu) X_2}{\lambda - \mu} (\lambda - T_2)^{-1} Y_2 L(\lambda) X_1 (\lambda - T_1)^{-1}.$$

Observe that for a fixed $\mu \in \partial\Delta'$, the first term is analytic in λ on Δ. It follows that the double integral of the first term is zero. To integrate the second term we interchange the order of integration. By Proposition 6.1.1, the function $(\mu - T_1)^{-1} Y_1 L(\mu)$ is analytic on Ω. It follows that

$$VS = (2\pi i)^{-1} \int_{\partial\Delta} (\lambda - T_1)^{-1} Y_1 L(\lambda) X_2 (\lambda - T_2)^{-1} Y_2 L(\lambda) X_1 (\lambda - T_1)^{-1} d\lambda.$$

Now we use the formula (6.1.3), together with Condition (P3′) and Proposition 6.1.1, and get

$$VS = (2\pi i)^{-1} \int_{\partial\Delta} (\lambda - T_1)^{-1} Y_1 L(\lambda) X_1 (\lambda - T_1)^{-1} d\lambda .$$

For $n \geq 0$ we multiply by the operator $X_1 T_1^n$ from the left, apply the formula $T_1(\lambda - T_1)^{-1} = \lambda(\lambda - T_1)^{-1} - I$, and make use of Condition (P3′) n times. For $n \geq 0$ this yields

$$X_1 T_1^n VS = (2\pi i)^{-1} \int_{\partial\Delta} \lambda^n X_1 (\lambda - T_1)^{-1} Y_1 L(\lambda) X_1 (\lambda - T_1)^{-1} d\lambda .$$

Finally, we apply formula (6.1.3) and Condition (P3′) once again and get

$$X_1 T_1^n VS = X_1 \cdot (2\pi i)^{-1} \int_{\partial\Delta} \lambda^n (\lambda - T_1)^{-1} d\lambda , \quad n \geq 0.$$

As $\partial\Delta$ encloses $\sigma(T_1)$, we eventually get

$$X_1 T_1^n VS = X_1 T_1^n , \quad n \geq 0.$$

Now we use Condition (P4) and conclude that $VS = I_{\mathcal{Y}_1}$. In the same way it is shown that $SV = I_{\mathcal{Y}_2}$. Hence, S is invertible and $S^{-1} = V$.

To prove (6.1.6), we use arguments exposed previously, namely formula (6.1.3), Proposition 6.1.1, and Condition (P3′). For instance, denoting $H_2(\lambda) = L(\lambda)^{-1} - X_2(\lambda - T_2)^{-1} Y_2$ we have

$$\begin{aligned} X_2 S &= (2\pi i)^{-1} \int_{\partial\Delta} X_2 (\lambda - T_2)^{-1} Y_2 L(\lambda) X_1 (\lambda - T_1)^{-1} d\lambda \\ &= (2\pi i)^{-1} \int_{\partial\Delta} X_1 (\lambda - T_1)^{-1} d\lambda - (2\pi i)^{-1} \int_{\partial\Delta} H_2(\lambda) L(\lambda) X_1 (\lambda - T_1)^{-1} d\lambda . \end{aligned}$$

Because of (P3′), the second integral vanishes and the first one equals X_1.

Finally, to prove the uniqueness of S, assume that

$$X_2S_i = X_1, T_2S_i = S_iT_1, \quad i = 1,2 ,$$

where S_1 and S_2 are invertible operators. Then

$$X_1T_1^j = X_2T_2^jS_1 = X_2T_2^jS_2, \quad j = 0,1,\dots ,$$

hence, in view of (P4), $S_1 = S_2$. ∎

We remark that all the definitions, results, and proofs given in this section and in Sections 6.2-6.6 can be applied verbatim also to operator-valued functions $L: \Omega \to L(\mathcal{X})$ that are analytic in Ω (the only significant change is that (P3) is not applicable in this framework and one has to use (P3′) instead). For such functions $L(\lambda)$ the *spectrum* $\Sigma(L)$ is defined naturally:

$$\Sigma(L) = \{\lambda \in \Omega \mid L(\lambda) \text{ is not invertible}\}.$$

On rare occasions, it will be necessary to use some of the material of Sections 6.2, 6.3, and 6.4 for analytic (on Ω) operator functions with compact spectrum.

6.2 Calculus of spectral triples

In this section, a calculus of spectral triples is developed. In the next section, these results will be used to construct explicitly a spectral triple for a given operator polynomial starting from linear polynomials. First we derive a lemma that will play an essential role in what follows.

LEMMA 6.2.1. *Assume that the operator* $X \in L(\mathcal{Y},\mathcal{X})$, $T \in L(\mathcal{Y})$, $Y \in L(\mathcal{X},\mathcal{Y})$ *satisfy the Conditions* (P1), (P2), (P3). *Let* $x \in \mathcal{Y}$ *be such that* $X(\lambda I-T)^{-1}x = 0$ *for* $\lambda \in U$, *where* U *is a non-empty open set in the complement of* $\sigma(T)$. *Then* $X(\lambda I-T)^{-1}x = 0$ *for all* $\lambda \notin \sigma(T)$.

PROOF. Write $h(\lambda) = X(\lambda I-T)^{-1}x$, $\lambda \notin \sigma(T)$. As $L(\lambda)X(\lambda I-T)^{-1}$ is a polynomial (see the remarks after the definition of a spectral triple), so is $L(\lambda)h(\lambda)$. Since $L(\lambda)h(\lambda) = 0$ for $\lambda \in U$, we actually have $L(\lambda)h(\lambda) = 0$ for all

$\lambda \in \mathbb{C}$. Now $L(\lambda)$ is invertible for $\lambda \in \Omega\backslash\sigma(T)$, and, consequently, $h(\lambda) = 0$ for every $\lambda \in \Omega\backslash\sigma(T)$.

Let Δ be a bounded Cauchy domain such that $\sigma(T) \subset \Delta \subset \bar{\Delta} \subset \Omega$. We know already that $h(\lambda) = 0$ on the boundary of Δ, and since $h(\lambda)$ is analytic in $\mathbb{C}\backslash\Delta$, it follows that $h(\lambda) = 0$ for $\lambda \in \mathbb{C}\backslash\Delta$. ∎

THEOREM 6.2.2. *For* $i = 1,2$, *let* $L_i(\lambda)$ *be an operator polynomial with* Ω*-spectral triple* (X_i, T_i, Y_i). *Put*

$$T = \begin{bmatrix} T_1 & Y_1X_2 \\ 0 & T_2 \end{bmatrix}, \quad Y = \begin{bmatrix} R \\ Y_2 \end{bmatrix}, \quad X = [X_1 \quad Q], \tag{6.2.2}$$

where the operators R *and* Q *are defined by*

$$R = (2\pi i)^{-1} \int_{\partial\Delta} (\lambda-T_1)^{-1}Y_1\{L_2(\lambda)^{-1}-X_2(\lambda-T_2)^{-1}Y_2\}d\lambda;$$

$$Q = (2\pi i)^{-1} \int_{\partial\Delta} \{L_1(\lambda)^{-1}-X_1(\lambda-T_1)^{-1}Y_1\}X_2(\lambda-T_2)^{-1}d\lambda.$$

Here Δ *is a bounded Cauchy domain such that* $(\sigma(T_1) \cup \sigma(T_2)) \subset \Delta \subset \bar{\Delta} \subset \Omega$. *Then* (X,T,Y) *is a spectral triple for* $L = L_2L_1$ *on* Ω.

PROOF. Since $\sigma(T) \subset \sigma(T_1) \cup \sigma(T_2)$, it is clear that Property (P1) holds. For $\lambda \notin \sigma(T_1) \cup \sigma(T_2)$, we have

$$(\lambda-T)^{-1} = \begin{bmatrix} (\lambda-T_1)^{-1} & (\lambda-T_1)^{-1}Y_1X_2(\lambda-T_2)^{-1} \\ 0 & (\lambda-T_2)^{-1} \end{bmatrix}.$$

It follows that for $\lambda \notin \sigma(T_1) \cup \sigma(T_2)$

$$X(\lambda-T)^{-1}Y = X_1(\lambda-T_1)^{-1}R$$
$$+ X_1(\lambda-T_1)^{-1}Y_1X_2(\lambda-T_2)^{-1}Y_2+Q(\lambda-T_2)^{-1}Y_2.$$

Take $\omega \in \mathbb{C}\backslash\bar{\Delta}$. Using the resolvent identity and Property (P2) for L_1 and L_2, we compute that

$$X_1(\omega-T_1)^{-1}R = (2\pi i)^{-1}\int_{\partial\Delta}(\omega-\lambda)^{-1}L_1(\lambda)^{-1}H_2(\lambda)d\lambda,$$

$$Q(\omega-T_2)^{-1}Y_2 = (2\pi i)^{-1}\int_{\partial\Delta}(\omega-\lambda)^{-1}H_1(\lambda)L_2(\lambda)^{-1}d\lambda,$$

where for $i = 1,2$, $H_i(\lambda) = L_i(\lambda)^{-1}-X_i(\lambda-T_i)^{-1}Y_i$. It follows that the functions

$$L_1(\lambda)^{-1}H_2(\lambda)-X_1(\lambda-T_1)^{-1}R, \quad H_1(\lambda)L_2(\lambda)^{-1}-Q(\lambda-T_2)^{-1}Y_2$$

have an analytic continuation to Ω. Indeed,

$$\int_{\partial\Delta}(\omega-\lambda)^{-1}[L_1(\lambda)^{-1}H_2(\lambda)-X_1(\lambda-T_1)^{-1}R]d\lambda = 0$$

for all $\omega \in \mathbb{C}\backslash\overline{\Delta}$, and it remains to use Lemma 6.1.2. Now

$$L(\lambda)^{-1}-X(\lambda-T)^{-1}Y = L_1(\lambda)^{-1}L_2(\lambda)^{-1}-X(\lambda-T)^{-1}Y$$

$$= \{L_1(\lambda)^{-1}H_2(\lambda)-X_1(\lambda-T_1)^{-1}R\} + \{H_1(\lambda)L_2(\lambda)^{-1}-Q(\lambda-T_2)^{-1}Y_2\}$$

$$- H_1(\lambda)H_2(\lambda), \qquad \lambda \in \Omega\backslash(\sigma(T_1)\cup\sigma(T_2)).$$

This shows that $L(\lambda)^{-1}-X(\lambda-T)^{-1}Y$ has an analytic continuation to Ω too.

To derive (P3), we first note that for $\omega \in \mathbb{C}\backslash\overline{\Delta}$

$$Q(\omega-T_2)^{-1} = (2\pi i)^{-1}\int_{\partial\Delta}(\omega-\lambda)^{-1}H_1(\lambda)X_2(\lambda-T_2)^{-1}d\lambda.$$

This implies that $H_1(\lambda)X_2(\lambda-T_2)^{-1}-Q(\lambda-T_2)^{-1}$ has an analytic continuation to Ω. Now

$$L(\lambda)X(\lambda-T)^{-1} = [V_1(\lambda),V_2(\lambda)],$$

where

$$V_1(\lambda) = L_2(\lambda)L_1(\lambda)X_1(\lambda-T_1)^{-1};$$

$$V_2(\lambda) = L_2(\lambda)X_2(\lambda-T_2)^{-1}+L(\lambda)\{Q(\lambda-T_2)^{-1}-H_1(\lambda)X_2(\lambda-T_2)^{-1}\}.$$

This implies that $L(\lambda)X(\lambda-T)^{-1}$ has an analytic continuation to Ω, and hence (P3) holds true.

Finally, to prove (P4), take $x = (x_1,x_2) \in \bigcap_{j=0}^{\infty} \operatorname{Ker} XT^j$. Then $X(\lambda-T)^{-1}x$ vanishes on a neighborhood of infinity. Using Lemma 6.2.1, we conclude that $X(\lambda-T)^{-1}x$ is zero for $\lambda \notin \sigma(T)$. Take $\lambda \in \Omega\backslash(\sigma(T_1) \cup \sigma(T_2))$. As $X(\lambda-T)^{-1}x = 0$, we have

$$X_1(\lambda-T_1)^{-1}x_1+X_1(\lambda-T_1)^{-1}Y_1X_2(\lambda-T_2)^{-1}x_2+Q(\lambda-T_2)^{-1}x_2 = 0.$$

Multiplying from the left by $L_1(\lambda)$ and rearranging terms we get

$$X_2(\lambda-T_2)^{-1}x_2 = -L_1(\lambda)X_1(\lambda-T_1)^{-1}x_1$$
$$+ L_1(\lambda)[H_1(\lambda)X_2(\lambda-T_2)^{-1}-Q(\lambda-T_2)^{-1}]x_2.$$

In this identity the left-hand side is analytic outside $\sigma(T_2)$ and zero at infinity, whereas the right-hand side has an analytic continuation to Ω. By Liouville's theorem we have $X_2(\lambda-T_2)^{-1}x_2 = 0$ outside $\sigma(T_2)$, and therefore $x_2 = 0$. But then $X_1(\lambda-T_1)^{-1}x_1 = 0$ on a neighborhood of infinity, and therefore $x_1 = 0$. Hence $x = 0$, and Property (P4) has been established. ∎

If both $L_1(\lambda)$ and $L_2(\lambda)$ are monic operator polynomials, and $\Omega = \mathbb{C}$, then

$$L_j(\lambda)^{-1} = X_j(\lambda I-T_j)^{-1}Y_j, \quad j = 1,2,$$

(Theorem 2.5.2), and we have $R = 0$, $Q = 0$ in Theorem 6.2.2. We rediscover the result of Theorem 2.6.1.

COROLLARY 6.2.3. *Let* (X,T,Y) *be a spectral triple for* $L(\lambda)$ *on* Ω, *and let* $E(\lambda)$ *and* $F(\lambda)$ *be operator polynomials with coefficients in* $L(\mathcal{X})$ *which are invertible for all* $\lambda \in \Omega$. *Put*

$$R_E = (2\pi i)^{-1} \int_{\partial\Delta} (\lambda - T)^{-1} Y E(\lambda)^{-1} d\lambda,$$

$$Q_F = (2\pi i)^{-1} \int_{\partial\Delta} F(\lambda)^{-1} X(\lambda - T)^{-1} d\lambda,$$

where Δ *is a bounded Cauchy domain such that* $\sigma(T) \subset \Delta \subset \bar{\Delta} \subset \Omega$. *Then* (Q_F, T, R_E) *is a spectral triple for the product* $E(\lambda)L(\lambda)F(\lambda)$ *on* Ω.

For the proof observe that the trivial triple $(X = 0,\ T = 0,\ Y = 0)$, where T acts on the zero Banach space is a Ω-spectral triple for both $E(\lambda)$ and $F(\lambda)$. It remains to apply Theorem 6.2.2 twice and identify $\{0\} \oplus \mathcal{Y} \oplus \{0\}$ with $\mathcal{Y}$.

THEOREM 6.2.4. *Let* $\mathcal{X}_1, \mathcal{X}_2$ *be Banach spaces, and for* $i = 1,2$ *let* $L_i(\lambda)$ *be an operator polynomial with coefficients in* $L(\mathcal{X}_i)$ *and with* Ω*-spectral triple* (X_i, T_i, Y_i). *Then* $(X_1 \oplus X_2, T_1 \oplus T_2, Y_1 \oplus Y_2)$ *is a* Ω*-spectral triple for the operator polynomial* $L_1(\lambda) \oplus L_2(\lambda)$ *with coefficients in* $L(\mathcal{X}_1 \oplus \mathcal{X}_2)$.

The proof is obvious.

THEOREM 6.2.5. *Let* $\mathcal{X}$ *and* $\mathcal{Z}$ *be complex Banach spaces, and denote by* π *and* τ *the projection of* $\mathcal{X} \oplus \mathcal{Z}$ *onto* $\mathcal{X}$ *along* $\mathcal{Z}$ *and the natural embedding of* $\mathcal{X}$ *into* $\mathcal{X} \oplus \mathcal{Z}$ *respectively. If* (X,T,Y) *is a spectral triple for* $L(\lambda)$ *on* Ω, *where* $X \in L(\mathcal{Y},\mathcal{X})$, $T \in L(\mathcal{Y})$, $Y \in L(\mathcal{X},\mathcal{Y})$, *then* $(\tau X, T, Y\pi)$ *is a* Ω*-spectral triple for* $L(\lambda) \oplus I_{\mathcal{Z}}$. *If* $(\tilde{X},\tilde{T},\tilde{Y})$ *is a* Ω*-spectral triple for* $L(\lambda) \oplus I_{\mathcal{Z}}$, *then* $(\pi\tilde{C}, \tilde{T}, \tilde{Y}\tau)$ *is a* Ω*-spectral triple for* $L(\lambda)$.

PROOF. The first part is clear from the previous theorem, and the fact that the trivial triple $(0,0,0)$ is a spectral triple for $I_{\mathcal{Z}}$ on Ω. Let us prove the second part. Obviously, Property (P1) holds. Since $L(\lambda)\cdot\pi\tilde{C}(\lambda-\tilde{T})^{-1} = \pi\cdot(L(\lambda)\oplus I_{\mathcal{Z}})\tilde{X}(\lambda-\tilde{T})^{-1}$, and $L(\lambda)^{-1} - \pi\tilde{X}(\lambda-\tilde{T})^{-1}\tilde{Y}\tau = \pi[(L(\lambda))^{-1}\oplus I_{\mathcal{Z}}) - \tilde{X}(\lambda-\tilde{T})^{-1}\tilde{Y}]\tau$ for $\lambda \in \Omega\backslash\sigma(\tilde{T})$, it is clear that (P2) and (P3′) hold too.

To establish the final Property (P4), we assume that for some $x \in \mathcal{Y}$ the vector function $\pi\tilde{X}(\lambda-\tilde{T})^{-1}x = 0$ on a neighborhood of infinity. As (P1), (P2), and (P3) have been established already, we may apply Lemma 6.2.1 and infer that $\pi\tilde{X}(\lambda-\tilde{T})^{-1}x = 0$ for $\lambda \notin \sigma(\tilde{T})$. Let us denote by ρ the projection of $\mathcal{X} \oplus \mathcal{Z}$ onto $\mathcal{Z}$ along $\mathcal{X}$. Put

$$\tilde{H}(y) = [L(\lambda)^{-1} \oplus I_{\mathcal{Z}}]-\tilde{X}(\lambda-\tilde{T})^{-1}\tilde{Y}, \quad \lambda \in \Omega\backslash\Sigma(L).$$

Applying ρ to this identity, we see that $\rho\tilde{X}(\lambda-\tilde{T})^{-1}\tilde{Y}$ has an analytic continuation to Ω. But then

$$\rho\tilde{X}(\lambda-\tilde{T})^{-1}\tilde{Y} = 0, \quad \lambda \notin \sigma(\tilde{T}) \tag{6.2.3}$$

(by applying the Liouville theorem to the entire function $\rho\tilde{X}(\lambda-\tilde{T})^{-1}\tilde{Y}$).

Now we make use of the following equality:

$$\frac{1}{2\pi i}\int_{\partial\Delta} (\lambda-\tilde{T})^{-1}\tilde{Y}[L(\lambda) \oplus I_{\mathcal{Z}}]\tilde{X}(\lambda-\tilde{T})^{-1}d\lambda = I_{\mathcal{Y}}. \tag{6.2.4}$$

where Δ is a bounded Cauchy domain such that $\sigma(\tilde{T}) \subset \Delta \subset \bar{\Delta} \subset \Omega$. To verify (6.2.4), apply Theorem 6.1.3 with $X_i = \tilde{X}$, $T_i = \tilde{T}$, $Y_i = \tilde{Y}$ $(i = 1,2)$ and use the uniqueness of the invertible operator S satisfying (6.1.6). Premultiplying (6.2.4) by $\rho\tilde{X}$ and using (6.2.3), we obtain $\rho\tilde{X} = 0$. So, on a neighborhood of infinity we have

$$\tilde{X}(\lambda-\tilde{T})^{-1}x = \pi\tilde{X}(\lambda-\tilde{T})^{-1}x = 0,$$

and $x = 0$ because (P4) holds for $(\tilde{X},\tilde{T},\tilde{Y})$. ∎

We consider now Ω-spectral triples for hulls of operator polynomials (the definitions, notations, and facts given in the appendix to Section 3.6 will be used here).

THEOREM 6.2.6. *Let* (X,T,Y) *be a* Ω*-spectral triple for an operator polynomial* L. *Then* $(\langle X\rangle,\langle T\rangle,\langle Y\rangle)$ *is a* Ω*-spectral triple for the operator polynomial* $\langle L\rangle$ *defined by*

$$\langle L\rangle(\lambda) = \langle L(\lambda)\rangle, \quad \lambda \in \mathbb{C}.$$

PROOF. We establish Properties (P1)-(P4) for the triple $(\langle X\rangle,\langle T\rangle,\langle Y\rangle)$. Property (P1) is clear because $\sigma(\langle T\rangle) \subset \sigma(T)$. Properties (P2) and (P3′) follow from the identities

$$\langle L\rangle(\lambda)\langle X\rangle \{\lambda I_{\langle \mathcal{X}\rangle}-\langle T\rangle\}^{-1} = \langle L(\lambda)X(\lambda-T)^{-1}\rangle;$$

$$\langle L\rangle(\lambda)^{-1}-\langle X\rangle \{\lambda I_{\langle \mathcal{X}\rangle}-\langle T\rangle\}^{-1}\langle Y\rangle = \langle L(\lambda)^{-1}-X(\lambda-T)^{-1}Y\rangle,$$

which hold true for $\lambda \in \Omega\backslash\Sigma(L)$.

It remains to establish Property (P4). Choose $\langle\{x_n\}_{n=0}^{\infty}\rangle \in \langle\mathcal{Y}\rangle$, and let $\langle X\rangle\langle T\rangle^k\langle\{x_n\}_{n=0}^{\infty}\rangle = 0_{\langle\mathcal{X}\rangle}$ $(k = 0,1,2,\ldots)$. Then

$$\langle X\rangle \{\lambda I_{\langle \mathcal{Y}\rangle}-\langle T\rangle\}^{-1}\langle\{x_n\}_{n=0}^{\infty}\rangle = 0_{\langle\mathcal{X}\rangle}, \tag{6.2.5}$$

on a neighborhood of infinity. Using Lemma 6.2.1, we see that (6.2.5) holds true for all $\lambda \notin \sigma(\langle T\rangle)$, and therefore for all $\lambda \notin \sigma(T)$.

Since (X,T,Y) is a spectral triple for L on Ω, it follows from Theorem 6.1.3 (applied with $X_i = X, T_i = T, Y_i = Y$ for $i = 1,2$) that for some bounded Cauchy domain Δ such that $\sigma(T) \subset \Delta \subset \overline{\Delta} \subset \Omega$ we have

$$(2\pi i)^{-1} \int_{\partial\Delta} (\lambda-T)^{-1}YL(\lambda)X(\lambda-T)^{-1}d\lambda = I_{\mathcal{Y}}. \tag{6.2.6}$$

Observe that this integral is defined as a limit in the norm topology in $L(\mathcal{Y})$ of a sequence of Riemann sums. Since $T \to \langle T\rangle$ is a continuous algebra homomorphism from $L(\mathcal{Y})$ into $L(\langle\mathcal{Y}\rangle)$ which maps $I_{\mathcal{Y}}$ into $I_{\langle\mathcal{Y}\rangle}$, it follows from (6.2.6) that

$$(2\pi i)^{-1} \int_{\partial\Delta} (\lambda-\langle T\rangle)^{-1}\langle Y\rangle\langle L\rangle(\lambda)\langle X\rangle(\lambda-\langle T\rangle)^{-1}d\lambda = I_{\langle\mathcal{Y}\rangle}.$$

With the help of (6.2.5), we now obtain $\langle\{x_n\}_{n=0}^{\infty}\rangle = 0_{\langle\mathcal{X}\rangle}$, which establishes Property (P4). ∎

Finally, we consider spectral triples for adjoint operator polynomials. Given a Banach space $\mathcal{X}$, denote (as usual) by $\mathcal{X}^*$ the Banach space of all continuous linear functionals on $\mathcal{X}$. For $T \in L(\mathcal{X},\mathcal{Y})$ the adjoint operator $T^* \in L(\mathcal{Y}^*,\mathcal{X}^*)$ is defined by $(T^*y^*)(x) = y^*(Tx)$, $x \in \mathcal{X}$, $y^* \in \mathcal{Y}^*$.

THEOREM 6.2.7. *Let* (X,T,Y) *be a spectral triple for* $L(\lambda) = \sum_{j=0}^{p} \lambda^j A_j$ *on* Ω. *Then* (Y^*,T^*,X^*) *is a spectral triple for* L^* *on* Ω, *where*

$$L^*(\lambda) = \sum_{j=0}^{p} \lambda^j A_j^*.$$

PROOF. Property (P1) follows from the fact that $\sigma(T^*) = \sigma(T)$. If an operator-valued function $U(\lambda)$ is analytic in Ω, then so is $(U(\lambda))^*$; this proves (P2). Property (P3) follows from Proposition 6.1.1 upon taking adjoints. To prove (P4), take $g \in \bigcap_{n=0}^{\infty} \operatorname{Ker} Y^*(T^*)^n$. Then $Y^*(\lambda-T^*)^{-1}g$ vanishes on a neighborhood of infinity. Lemma 6.2.1 implies that $Y^*(\lambda-T^*)^{-1}g = 0$ for $\lambda \notin \sigma(T^*) = \sigma(T)$. So for all $x \in \mathcal{Y}$ and $\lambda \notin \sigma(T)$, we have $g((\lambda-T)^{-1}Yx) = 0$.

Take $z \in \mathcal{Y}$. For the moment we fix $\lambda \in \Omega\backslash\sigma(T)$, put $x = L(\lambda)X(\lambda-T)^{-1}z$ and conclude that $g((\lambda-T)^{-1}YL(\lambda)X(\lambda-T)^{-1}z) = 0$. But then the latter identity holds for all $z \in \mathcal{Y}$ and $\lambda \in \Omega\backslash\sigma(T)$. Now for any bounded Cauchy domain Δ such that $\sigma(T) \subset \Delta \subset \bar{\Delta} \subset \Omega$ we have in view of (6.2.6):

$$0 = \frac{1}{2\pi i}\int_{\partial\Delta} g((\lambda-T)^{-1}YL(\lambda)X(\lambda-T)^{-1}z)d\lambda$$

$$= g\left(\frac{1}{2\pi i}\int_{\partial\Delta} (\lambda-T)^{-1}YL(\lambda)X(\lambda-T)^{-1}z\, d\lambda\right) = g(z).$$

Hence $g = 0$, which establishes Property (P4). ∎

Note that Theorem 6.2.7 implies that for any spectral triple

$$\overline{\operatorname{span} \bigcup_{n=0}^{\infty} \operatorname{Im} T^n Y} = \mathcal{Y}.$$

In the definition of a spectral triple the roles of the operators Y and X are not symmetric. Of course, in (P1) and (P2) they play analogous roles, but (P3) and (P4) are conditions on the pair (X,T) only. The analogues of (P3) and (P4) for the pair (T,Y) are:

(P5) $$\sum_{j=0}^{p} T^j Y L_j = 0, \text{ where } L(\lambda) = \sum_{j=0}^{p} \lambda^j L_j;$$

(P6) $$\overline{\operatorname{span} \bigcup_{n=0}^{\infty} \operatorname{Im} T^n Y} = \mathcal{Y}.$$

Again, (P5) can be replaced here by

(P5′) the operator function $(\lambda - T)^{-1} Y L(\lambda)$ has an analytic continuation to Ω.

Note that from Proposition 6.1.1 and Theorem 6.2.7, it follows that a spectral triple (X,T,Y) for L on Ω has the properties (P5) and (P6). Conversely, if the triple (X,T,Y) satisfies the conditions (P1), (P2), (P5), and (P6) for L on Ω, then it is a spectral triple for L on Ω. To see this, observe that (Y^*, T^*, X^*) satisfies (P1) to (P4) for L^* on Ω, and hence (Y^*, T^*, X^*) is a spectral triple for L^* on Ω. But then (X^{**}, T^{**}, Y^{**}) is a spectral triple for L^{**} on Ω because of Theorem 6.2.7. It follows that

$$\sum_{j=0}^{p} L_j^{**} X^{**} T^{**j} = 0,$$

$$\bigcap_{j=0}^{\infty} \operatorname{Ker}(X^{**} T^{**j}) = \{0\}.$$

As $\mathcal{X}$ and $\mathcal{Y}$ are embedded into $\mathcal{X}^{**}$ and $\mathcal{Y}^{**}$, respectively, the pair (X,T) satisfies (P3) and (P4). Consequently, (X,T,Y) is a spectral triple for L on Ω. Hence, the notion of a spectral triple is completely symmetric with respect to Y and X.

We conclude this section with the following statement relating to spectral triples on different open sets.

THEOREM 6.2.8. *Let* $\Omega_1,\dots,\Omega_p$ *be open sets such that* $\Omega_i \cap \Omega_j = \phi$ $(i \neq j)$. *If* (X_i,T_i,Y_i) *is a spectral triple of* $L(\lambda)$ *on* Ω_i *for* $i = 1,\dots,p$, *then*

$$(6.2.7) \qquad \left([X_1X_2\cdots X_p],\ \begin{bmatrix} T_1 & 0 & \cdots & 0 \\ 0 & T_2 & \cdots & 0 \\ \vdots & \vdots & & \vdots \\ 0 & 0 & \cdots & T_p \end{bmatrix},\ \begin{bmatrix} Y_1 \\ Y_2 \\ \vdots \\ Y_p \end{bmatrix}\right)$$

is a spectral triple of $L(\lambda)$ *on* $\Omega \overset{\text{def}}{=} \bigcup_{i=1}^{p} \Omega_i$.

PROOF. We have to verify that the triple (6.2.7) satisfies all the properties required by the definition of a spectral triple for $L(\lambda)$ on Ω. The properties (P1) and (P3) are evident. For the property (P2) observe that the function

$$L(\lambda)^{-1} = \sum_{j=1}^{p} X_j(\lambda I - T_j)^{-1}Y_j$$

is analytic in each Ω_j, in view of the properties of the spectral triple (X_j,T_j,Y_j) of $L(\lambda)$ on Ω_j.

It remains to prove that

$$(6.2.8) \qquad \bigcap_{j=0}^{\infty} \operatorname{Ker}[X_1T_1^j, X_2X_2^j, \dots, X_pT_p^j] = \{0\}.$$

Let $X' = [X_2\cdots X_p]$, $T' = \operatorname{diag}[T_2,\dots,T_p]$, and assume that x and y are such that

$$X_1T_1^jx = X'T'^jy, \quad j = 0,1,\dots .$$

Then

$$(6.2.9) \qquad X_1(\lambda I - T_1)^{-1}x = X'(\lambda I - T')^{-1}y$$

for every λ with $|\lambda|$ sufficiently large. Lemma 6.2.1 ensures that (6.2.9) holds for every $\lambda \notin \sigma(T_1) \cup \sigma(T')$. As $\sigma(T_1) \cap \sigma(T') = \phi$, it follows that the function defined by (6.2.9) is analytic in $\mathbb{C}$; so by Liouville's theorem

$$X_1(\lambda_1 I - T_1)^{-1}x = 0,$$

and hence $x = 0$. Now

$$y \in \bigcap_{j=0}^{\infty} \operatorname{Ker}[X_2T_2^j, \ldots, X_pT_p^j],$$

and using induction on p, it follows that $y = 0$, thereby proving (6.2.8). ∎

6.3 Construction of spectral triples

In this section we provide explicit formulas for a Ω-standard triple of an operator polynomial $L(\lambda)$ such that $\Sigma(L) \cap \Omega$ is compact. In particular, the existence of Ω-spectral triple (which has not been proved yet) will follows.

Throughout this section we assume for simplicity that zero is inside Ω.

THEOREM 6.3.1. *Let $L(\lambda)$ be an operator polynomial with coefficients in $L(\mathcal{X})$, such that $\Sigma(L) \cap \Omega$ is compact.*

Suppose that Δ is a bounded Cauchy domain containing 0 such that $\Sigma(L) \cap \Omega \subset \Delta \subset \bar{\Delta} \cap \Omega$, and let $\mathcal{M}$ be the set of all continuous $\mathcal{X}$-valued functions f on the boundary $\partial\Delta$ which admit an analytic continuation to a $\mathcal{X}$-valued function in $(L \cup \{\infty\})\backslash(\Sigma(L) \cap \Omega)$ vanishing at ∞, while $L(\lambda)f(\lambda)$ has an analytic continuation to Ω. The set $\mathcal{M}$ endowed with the supremum norm

$$\|f\|_\infty = \sup_{\lambda\in\partial\Delta} \|f(\lambda)\|$$

is a Banach space. Put

$$T: \mathcal{M} \to \mathcal{M}, \quad (Tf)(z) = zf(z) - (2\pi i)^{-1} \int_{\partial\Delta} f(w)dw;$$

$$Y: \mathcal{X} \to \mathcal{M}, \quad (Yy)(z) = \frac{1}{2\pi i}\int_{\Gamma} \frac{L(w)^{-1}}{z-w}\, y\, dw;$$

$$X: \mathcal{M} \to \mathcal{X}, \quad Xf = (2\pi i)^{-1}\int_{\partial\Delta} f(w)dw.$$

In the definition of Y, *the contour* Γ *is the boundary of a bounded Cauchy domain* Δ' *such that* $\Sigma(L) \cap \Omega \subset \Delta' \subset \overline{\Delta'} \subset \Delta$. *Then* (X,T,Y) *is a spectral triple for* L *on* Ω.

From the proof of Theorem 6.3.1, it will be clear that this theorem remains valid if $\mathcal{M}$ is endowed with the L_2-norm (see the remark at the end of proof of Theorem 1.3.1). Hence, we may conclude that for the case when $\mathcal{X}$ is a (separable) Hilbert space, the space on which the main operator of a spectral triple acts may be taken to be a (separable) Hilbert space too.

The proof of Theorem 6.3.1 takes several steps. First we employ the calculus of spectral triples, which has been developed in the previous section, to construct a spectral triple for L on Ω assuming that a linearization of L with respect to Ω is known. This will be done in Theorem 6.3.2 below. Next, we use the fact that explicit formulas for linearizations of L may be given.

THEOREM 6.3.2. *Let* $L(\lambda)$ *be an operator polynomial with compact* $\Sigma(L) \cap \Omega$. *Consider a linearization of* $L(\lambda)$ *with respect to* Ω:

$$L(\lambda) \oplus I_{\mathcal{Z}} = E(\lambda)(\lambda - S)F(\lambda), \quad \lambda \in \Omega. \tag{6.3.1}$$

Here $\mathcal{Z}$ *is a Banach space,* S *is an operator, and* $E(\lambda)$, $F(\lambda)$ *are invertible analytic operator functions on* Ω.

Let $\pi: \mathcal{X} \oplus \mathcal{Z} \to \mathcal{X}$ *be the projection onto* $\mathcal{X}$ *along* $\mathcal{Z}$, *and let* τ *be the natural embedding of* $\mathcal{X}$ *into* $\mathcal{X} \oplus \mathcal{Z}$. *Further, let* Δ *be a bounded Cauchy domain in* Ω *such that* $\Sigma(L) \cap \Omega \subset \Delta \subset \overline{\Delta} \subset \Omega$. *Put*

$$\mathcal{Y} = \mathrm{Im}\Big[(2\pi i)^{-1}\int_{\partial\Delta} (\lambda - S)^{-1} d\lambda\Big],$$

i.e., $\mathcal{Y}$ is the spectral subspace of S *corresponding to the part of* $\sigma(S)$ *inside* Ω. *Define*

$$T: \mathcal{Y} \to \mathcal{Y}, \quad T = S|_{\mathcal{Y}};$$

$$Y: \mathcal{Y} \to \mathcal{X}, \quad Y = (2\pi i)^{-1} \int_{\partial\Delta} (\lambda - S)^{-1} E(\lambda)^{-1} \tau \, d\lambda;$$

$$X: \mathcal{X} \to \mathcal{Y}, \quad X = (2\pi i)^{-1} \int_{\partial\Delta} \pi F(\lambda)^{-1} (\lambda - S)^{-1} d\lambda.$$

Then (X,T,Y) *is a spectral triple for* L *on* Ω.

PROOF. By the linearization (6.3.1), it is clear that $\Sigma(L) \cap \Omega = \sigma(T) \cap \Omega$. Let

$$P = (2\pi i)^{-1} \int_{\partial\Delta} (\lambda - S)^{-1} d\lambda$$

be the Riesz projection of S corresponding to the part of $\sigma(S)$ inside Ω; so $\mathcal{Y} = \operatorname{Im} P$. Further, let $\mathcal{U}$ be the space on which S acts, and let $\kappa: \mathcal{Y} \to \mathcal{U}$ be the natural embedding of $\mathcal{Y}$ into $\mathcal{U}$. Then (κ, T, P) is a Ω-spectral triple for $\lambda - S$ (here P is considered as an operator from $\mathcal{U}$ into $\mathcal{Y}$).

Choose a fixed point $\lambda_0 \in \Omega$ and put $\tilde{E}(\lambda) = E(\lambda_0)^{-1} E(\lambda)$ and $\tilde{F}(\lambda) = F(\lambda) F(\lambda_0)^{-1}$. Then $\tilde{E}(\lambda)$, $\tilde{F}(\lambda)$: $\mathcal{U} \to \mathcal{U}$ are invertible and depend analytically on λ. Consider the operators $\tilde{Y}: \mathcal{U} \to \mathcal{U}$ and $\tilde{C}: \mathcal{Y} \to \mathcal{U}$ defined by

$$\tilde{Y} = (2\pi i)^{-1} \int_{\partial\Delta} (\lambda - T)^{-1} P \tilde{E}^{-1} d\lambda = (2\pi i)^{-1} \int_{\partial\Delta} (\lambda - S)^{-1} \tilde{E}(\lambda)^{-1} d\lambda,$$

$$\tilde{X} = (2\pi i)^{-1} \int_{\partial\Delta} \tilde{F}(\lambda)^{-1} (\lambda - T)^{-1} \kappa \, d\lambda = (2\pi i)^{-1} \int_{\partial\Delta} \tilde{F}(\lambda)^{-1} (\lambda - S)^{-1} d\lambda.$$

By applying Corollary 6.2.3, we get a Ω-spectral triple $(\tilde{X}, T, \tilde{Y})$ for the analytic operator function V, defined by

$$V(\lambda) = \tilde{E}(\lambda)(\lambda - S)\tilde{F}(\lambda), \quad \lambda \in \Omega.$$

(cf. the remark at the end of Section 6.1). Note that $L(\lambda) \oplus I_{\mathcal{Z}} = E(\lambda_0)V(\lambda)F(\lambda_0)$ for each $\lambda \in \Omega$. It follows that $(F(\lambda_0)^{-1}\tilde{X}, T, \tilde{Y}E(\lambda_0)^{-1})$ is a Ω-spectral triple for $L(\lambda) \oplus I_{\mathcal{Z}}$, and hence

$$(6.3.2) \qquad (\pi F(\lambda_0)^{-1}\tilde{X}, T, \tilde{Y}E(\lambda_0)^{-1}\tau)$$

is a spectral triple for L on Ω (cf. Theorem 6.2.5). Finally, observe that $Y = \tilde{Y}E(\lambda_0)^{-1}\tau$ and $X = \pi \cdot F(\lambda_0)^{-1}\tilde{X}$. ∎

We shall use Theorem 6.3.2 to derive Theorem 6.3.1. We begin with a remark. Let Ω_0 be an open set in $\mathbb{C}$ such that $\Sigma(L) \cap \Omega \subset \Omega_0 \subset \Omega$. If (X,T,Y) is a spectral triple for L on Ω_0, then trivially (X,T,Y) is a spectral triple for $L(\lambda)$ on Ω. Hence, it suffices to construct a spectral triple for $L(\lambda)$ on some open neighborhood of $\Sigma(L) \cap \Omega$. Now choose $\Omega_0 = \Delta$, where Δ is as in Theorem 6.3.1.

Let $C(\partial\Delta,\mathcal{X})$ be the Banach space of all continuous $\mathcal{X}$-valued functions on $\partial\Delta$ endowed with the supremum norm, and let S be the operator on $C(\partial\Delta,\mathcal{X})$ defined by

$$(6.3.3) \qquad (Sf)(z) = zf(z)+(2\pi i)^{-1}\int_{\partial\Delta} [L(\lambda)-I]f(\lambda)d\lambda .$$

We need certain properties of the operator S.

Firstly, in the proof of Theorem 1.3.1 (see especially formula (1.3.5)), we have shown that

$$(6.3.4) \qquad (\lambda I-S)(V-\lambda I)^{-1} = \begin{bmatrix} -I_{\mathcal{X}} & -C(\lambda) \\ 0 & -I_{\mathcal{Z}} \end{bmatrix} (L(\lambda) \oplus I_{\mathcal{Z}}), \ \lambda \in \Delta.$$

Here $\mathcal{X}$ is identified with the set of constant $\mathcal{X}$-valued functions on $\partial\Delta$; so

$$C(\partial\Delta,\mathcal{X}) = \mathcal{X} \dot{+} \mathcal{Z},$$

where

$$\mathcal{Z} = \{g \in C(\partial\Delta,\mathcal{X}) \mid \frac{1}{2\pi i}\int_{\partial\Delta} \frac{g(z)}{z}\, dz = 0\};$$

$V \in L(C(\partial\Delta,\mathcal{X}))$ is defined by

$$(Vf)(z) = zf(z), \quad z \in \partial\Delta;$$

$C(\lambda) \in L(\mathcal{Z},\mathcal{X})$ is an analytic (on Δ) operator-valued function defined by

$$(C(\lambda))f = P[V(V-\lambda)^{-1}(L(z)f(z)-f(z))], \quad f \in \mathcal{Z},$$

where P is the projection on $\mathcal{X}$ along $\mathcal{Z}$. Further, recall (see again the proof of Theorem 1.3.1) that

$$Pf = \frac{1}{2\pi i}\int_{\partial\Delta} \xi^{-1}f(\xi)d\xi, \quad f \in C(\partial\Delta,\mathcal{X})$$

and that

$$(V-\lambda I)^{-1}f(z) = (z-\lambda)^{-1}f(z), \quad z \in \partial\Delta, \ \lambda \in \Delta.$$

So (6.3.4) can be rewritten in the form

$$(6.3.5) \qquad (\lambda I-S)F(\lambda) = G(\lambda)(L(\lambda) \oplus I_{\mathcal{Z}}), \quad \lambda \in \Delta$$

where $G(\lambda) \in L(\mathcal{X}\oplus\mathcal{Z},C(\partial\Delta,\mathcal{X}))$ is given by

$$(6.3.6) \quad (G(\lambda)(y,g))(z) = -y-g(z)-C(\lambda)g$$

$$= -y-g(z) - \frac{1}{2\pi i}\int_{\partial\Delta} \frac{L(w)-I}{w-\lambda}\, g(w)dw; \quad y \in \mathcal{X}, \ g \in \mathcal{Z},$$

and $F(\lambda) \in L(\mathcal{X}\oplus\mathcal{Z},C(\partial\Delta,\mathcal{X}))$ is given by

$$(6.3.7) \qquad (F(\lambda)(y,g))(z) = (z-\lambda)^{-1}(y+g(z)), \quad y \in \mathcal{X}, \ g \in \mathcal{Z}.$$

Furthermore, the operators $G(\lambda)$, $F(\lambda)$: $\mathcal{X} \oplus \mathcal{Z} \to C(\partial\Delta,\mathcal{X})$ are invertible and depend analytically on $\lambda \in \Delta$.

It follows from (6.3.5) that $\sigma(S) \cap \Delta = \Sigma(L) \cap \Delta$ $(= \Sigma(L) \cap \Omega)$. Actually, $\sigma(S) = (\Sigma(L) \cap \Delta) \cup \partial\Delta$ (see Gohberg-Kaashoek-Lay [1]); however, we will not need this fact here.

Secondly, we identify the spectral subspace of S corresponding to the part of $\sigma(S)$ inside Δ.

LEMMA 6.3.3. *The spectral subspace $\mathcal{M}$ of* S *corresponding to the part of $\sigma(S)$ inside Δ consists of all*

$f \in C(\partial\Delta, \mathcal{X})$ *that can be extended to a* $\mathcal{X}$*-valued function analytic outside* $\Sigma(L) \cap \Omega$ *and vanishing at* ∞, *while* Lf *has an analytic continuation to* Ω. *Further, for each* $f \in \mathcal{M}$

$$(6.3.8) \qquad (Sf)(z) = zf(z) - (2\pi i)^{-1}\int_{\partial\Delta} f(w)dw.$$

PROOF. For each $f \in C(\partial\Delta, \mathcal{X})$ and $\lambda \in \Delta\backslash\Sigma(L)$ we have

$$(6.3.9) \quad [(\lambda - S)^{-1}f](z) = \frac{f(z)}{\lambda - z} - \frac{L(\lambda)^{-1}}{\lambda - z}\frac{1}{2\pi i}\int_{\partial\Delta}\frac{L(w)-I}{w-\lambda}f(w)dw, \quad z \in \partial\Delta.$$

To verify this formula, rewrite (6.3.5) in the form

$$(\lambda I - S)^{-1} = F(\lambda)(L(\lambda)^{-1} \oplus I_{\mathcal{Z}})G(\lambda)^{-1};$$

so using formulas (6.3.6) and (6.3.7), we obtain for $y \in \mathcal{X}$ and $g(z) \in \mathcal{Z}$:

$$(\lambda I - S)^{-1}(y + g(z))$$

$$= \frac{1}{z-\lambda}\begin{bmatrix} L(\lambda)^{-1} & 0 \\ 0 & I_{\mathcal{Z}}\end{bmatrix}\begin{bmatrix} -y + \frac{1}{2\pi i}\int_{\partial\Delta}\frac{L(w)-I}{w-\lambda}g(w)dw \\ -g(z)\end{bmatrix}$$

$$= \frac{1}{z-\lambda}[-y-g(z)] + \frac{L(\lambda)^{-1}}{z-\lambda}[-y+L(\lambda)y + \frac{1}{2\pi i}\int_{\partial\Delta}\frac{L(w)-I}{w-\lambda}g(w)dw]$$

$$= \frac{y+g(z)}{\lambda - z} + \frac{L(\lambda)^{-1}}{z-\lambda}[-y+L(\lambda)y$$

$$+ \frac{1}{2\pi i}\int_{\partial\Delta}\frac{L(w)-I}{w-\lambda}(y+g(w))dw - \frac{1}{2\pi i}\int_{\partial\Delta}\frac{L(w)-I}{w-\lambda}y\,dw].$$

Comparing with (6.3.9), it remains only to prove that

$$-y+L(\lambda)y = \frac{1}{2\pi i}\int_{\partial\Delta} \frac{L(w)-I}{w-\lambda}\, y\, dw,$$

which follows from the Cauchy's integral formula.

Let Δ' be a bounded Cauchy domain such that $\Sigma(L) \cap \Omega \subset \Delta' \subset \overline{\Delta'} \subset \Delta$. In view of (6.3.9), the Riesz projection P of S corresponding to the part of $\sigma(S)$ inside Δ is given by

(6.3.10)

$$(Pf)(z) = \frac{1}{2\pi i}\int_{\partial\Delta'} \frac{L(\lambda)^{-1}}{z-\lambda}\left[\frac{1}{2\pi i}\int_{\partial\Delta} \frac{L(w)-I}{w-\lambda}\, f(w)dw\right]d\lambda, \quad z \in \partial\Delta.$$

First assume that f can be extended to a $\mathcal{X}$-valued function analytic outside $\Sigma(L) \cap \Omega$ and vanishing at ∞, while Wf has an analytic continuation to Ω. Then formula (6.3.10) implies that for all $z \in \partial\Delta$ we have $(Pf)(z) = f(z)$. So $f \in \mathcal{M}$. Also, in this case,

$$\int_{\partial\Delta} L(\lambda)f(\lambda)d\lambda = 0.$$

By substituting this into the definition of S, we see that (6.3.8) holds true.

Secondly, assume that $f \in \mathcal{M}$. Then $f = Pf$, and hence we can apply formula (6.3.10) to show that f can be extended to a function analytic outside Δ' and vanishing at ∞. This extension of f will also be denoted by f. Take $z_0 \in \Delta\backslash\overline{\Delta'}$. To show that Lf admits an analytic continuation to Ω, it suffices to prove the equality

$$\frac{1}{2\pi i}\int_{\partial\Delta} \frac{L(z)f(z)}{z-z_0}\, dz = L(z_0)f(z_0). \tag{6.3.11}$$

Using $f = Pf$, we have

$$\frac{1}{2\pi i}\int_{\partial\Delta}\frac{L(z)f(z)}{z-z_0}\,dz = \frac{1}{2\pi i}\int_{\partial\Delta}\frac{L(z)}{z-z_0}\left\{\frac{1}{2\pi i}\int_{\partial\Delta'}\frac{L(\lambda)^{-1}}{z-\lambda}\right.$$

$$\left.\cdot\left[\frac{1}{2\pi i}\int_{\partial\Delta}\frac{L(w)-I}{w-\lambda}\,f(w)dw\right]d\lambda\right\}\,dz.$$

As the integrand is a continuous function in (z,λ,w) on the compact set $\partial\Delta\times\partial\Delta'\times\partial\Delta$, we may apply Fubini's theorem. At first we evaluate the integral over z and obtain

$$\frac{1}{2\pi i}\int_{\partial\Delta}\frac{L(z)}{(z-z_0)(z-\lambda)}\,dz = \frac{L(\lambda)-L(z_0)}{\lambda-z_0}\;.$$

So

$$\frac{1}{2\pi i}\int_{\partial\Delta}\frac{L(z)f(z)}{z-z_0}\,dz = \frac{1}{2\pi i}\int_{\partial\Delta'}\frac{L(\lambda)-L(z_0)}{\lambda-z_0}\,L(\lambda)^{-1}$$

$$\cdot\left[\frac{1}{2\pi i}\int_{\partial\Delta}\frac{L(w)-I}{w-\lambda}\,f(w)dw\right]\,d\lambda$$

$$= \frac{1}{2\pi i}\int_{\partial\Delta'}\frac{1}{\lambda-z_0}\left(\frac{1}{2\pi i}\int_{\partial\Delta}\frac{L(w)-I}{w-\lambda}\,f(w)dw\right)d\lambda$$

$$+\,L(z_0)\left[\frac{1}{2\pi i}\int_{\partial\Delta'}\frac{L(\lambda)^{-1}}{z_0-\lambda}\left(\frac{1}{2\pi i}\int_{\partial\Delta}\frac{L(w)-I}{w-\lambda}\,f(w)dw\right)d\lambda\right]$$

$$= \frac{1}{2\pi i}\int_{\partial\Delta}L(w)f(w)\left(\frac{1}{2\pi i}\int_{\partial\Delta'}\frac{d\lambda}{(\lambda-z_0)(w-\lambda)}\right)dw$$

$$+\,L(z_0)(Pf)(z_0) = L(z_0)f(z_0),$$

where in the last equality we have used the assumption that $f = Pf$ and the analyticity of $(\lambda-z_0)^{-1}(w-\lambda)^{-1}$, as a function of λ, in Δ'. ∎

In the case when $\mathcal{Y}$ is a separable Hilbert space, none of the previous arguments is affected if we take the space $L_2(\partial\Delta,\mathcal{Y})$ of strongly measurable $\mathcal{Y}$-valued L_2-functions on $\partial\Delta$ instead of $C(\partial\Delta,\mathcal{Y})$.

We are ready now to prove Theorem 6.3.1.

PROOF OF THEOREM 6.3.1. The fact that $\mathcal{M}$ is a Banach space follows from Lemma 6.3.3 because $\mathcal{M}$ appears as the spectral subspace of an operator S. One can prove this fact also directly, using the maximum modulus principle for analytic $\mathcal{X}$-valued functions and the property that the limit of a sequence of analytic $\mathcal{X}$-valued function which tends to the limit (in the norm) uniformly on every compact subset, is again analytic.

Let S be the operator defined by (6.3.3). Apply Theorem 6.3.2 to the formula (6.3.5). By Lemma 6.3.3, the spectral subspace of S corresponding to the part of $\sigma(S)$ inside Δ coincides with $\mathcal{M}$ and

$$T = S|_{\mathcal{M}}$$

(cf. (6.3.8)); here A is the operator introduced in Theorem 6.3.1. Let Δ' be a bounded Cauchy domain such that $\Sigma(L) \cap \Omega \subset \Delta' \subset \overline{\Delta'} \subset \Delta$. To finish the proof it suffices to show that the operators B and C introduced in the theorem satisfy the following identities:

$$Y = (2\pi i)^{-1} \int_{\partial\Delta'} (\lambda - S)^{-1} G(\lambda) \tau \, d\lambda; \tag{6.3.12}$$

$$X = (2\pi i)^{-1} \int_{\partial\Delta'} \pi F(\lambda)^{-1} (\lambda - S)^{-1} d\lambda, \tag{6.3.13}$$

where $F(\lambda)$ and $G(\lambda)$ are given by Equations (6.3.6) and (6.3.7), the map π is the projection of $\mathcal{X} \oplus \mathcal{Z}$ onto $\mathcal{X}$ along $\mathcal{Z}$ and τ is the natural embedding of $\mathcal{X}$ into $\mathcal{X} \oplus \mathcal{Z}$. Recall that $\mathcal{Z} = \{f \in C(\partial\Delta, \mathcal{X}) : (2\pi i)^{-1} \int_{\partial\Delta} z^{-1} f(z) dz = 0\}$. Then $(\tau y)(z) = y$ $(z \in \partial\Delta)$ and $\pi f = (2\pi i)^{-1} \int_{\partial\Delta} z^{-1} f(z) dz$.

To compute the right-hand side of (6.3.12), note that

$$(\lambda - T)^{-1} G(\lambda) \tau y = F(\lambda)(L(\lambda)^{-1} y, 0),$$

because of (6.3.5). Using the definition of $F(\lambda)$ (see (6.3.7)), we obtain for each $z \in \partial\Delta$ and $y \in \mathcal{X}$ the following equality:

$$\left[\frac{1}{2\pi i}\int_{\partial\Delta'}(\lambda-S)^{-1}G(\lambda)\tau y d\lambda\right]dz = \frac{1}{2\pi i}\int_{\partial\Delta'}\frac{L(\lambda)^{-1}}{z-\lambda}\,y d\lambda .$$

This proves (6.3.12).

Take $f \in \mathcal{M}$. Denote by f also the analytic continuation of f to $(\mathbb{C}\cup\infty)\setminus(\Sigma(L)\cap\Omega)$. Since Lf has an analytic continuation to Ω, we see from (6.3.9) that

$$[(\lambda-S)^{-1}f](z) = \frac{f(z)-f(\lambda)}{\lambda-z} .$$

As $(F(\lambda)^{-1}f)(z) = (z-\lambda)f(z)$ (see formula (6.3.7)), we have

$$\pi F(\lambda)^{-1}(\lambda-T)^{-1}f = \frac{1}{2\pi i}\int_{\partial\Delta}\frac{f(\lambda)-f(z)}{z}\,dz = f(\lambda),$$

and (6.3.13) holds. ■

We now state and prove an important corollary.

COROLLARY 6.3.4. *The spectrum of the main operator of a spectral triple for* L *on* Ω *coincides with* $\Sigma(L)\cap\Omega$.

PROOF. Let (X,T,Y) be a spectral triple for L on Ω. The properties (P1) and (P2) imply that

$$\Sigma(L)\cap\Omega \subset \sigma(T).$$

To prove the opposite inclusion, it is sufficient in view of Theorem 6.1.3 to consider one particular spectral triple. Without loss of generality we can assume that Ω is a bounded set (otherwise replace Ω by a smaller open bounded set which contains $\sigma(T)$). Recall (Theorem 1.3.1) that there is a linearization of L with respect to Ω, i.e., there exist a Banach space $\mathcal{Z}$ and invertible operators $E(\lambda)$ and $F(\lambda)$ depending analytically on $\lambda \in \Omega$ such that

$$L(\lambda)\oplus I_{\mathcal{Z}} = E(\lambda)(\lambda-S)F(\lambda), \quad \lambda\in\Omega. \tag{6.3.14}$$

From this equation and the compactness of $\Sigma(L)\cap\Omega$ (which is implied by existence of a Ω-spectral triple for W), it is clear that $\sigma(S)\cap\Omega = \Sigma(L)\cap\Omega$, while $\sigma(S)$ is the disjoint union of two compact sets $\sigma(S)\backslash\Omega$ and $\sigma(S) \cap \Omega$. Now by Theorem 6.3.2, the restriction of S to its spectral invariant subspace $\mathcal{M}$ corresponding to $\sigma(S) \cap \Omega$ is the main operator of a spectral triple for L on Ω. As $\sigma(S|_{\mathcal{M}}) = \sigma(S) \cap \Omega$, we are done. ∎

6.4 Spectral triples and linearizations

We have seen in the preceding section that a linearization of an operator polynomial L gives rise to a spectral triple for L on Ω. In this section we will see that the opposite is also true: the main operator in a Ω-spectral triple for L is a linearization of L with respect to Ω.

Let (X,T,Y) be a spectral triple for L on Ω. As before, $L(\lambda)$ is an operator polynomial with coefficients from $L(\mathcal{X})$, and $T \in L(\mathcal{Y})$. Use Property (P2) to define an analytic function $H\colon \Omega \to L(\mathcal{X})$ by $H(\lambda) = L(\lambda)^{-1}-X(\lambda-T)^{-1}Y$.

Define $Z\colon \Omega\backslash\sigma(T) \to L(\mathcal{Y})$ by

$$Z(\lambda) = -(\lambda-T)^{-1}+(\lambda-T)^{-1}YL(\lambda)X(\lambda-T)^{-1}. \tag{6.4.1}$$

First we show that Z has an analytic continuation to Ω.

Let Δ be a bounded Cauchy domain such that $\sigma(T) \subset \Delta \subset \overline{\Delta} \subset \Omega$, and take $\mu \in \mathbb{C}\backslash\overline{\Delta}$. With the help of the identity

$$(\mu-\lambda)^{-1}(\lambda-T)^{-1} = (\mu-T)^{-1}(\lambda-T)^{-1}+(\mu-\lambda)^{-1}(\mu-T)^{-1} \qquad (\lambda \in \partial\Delta),$$

we easily compute that

$$\frac{1}{2\pi i}\int_{\partial\Delta} \frac{Z(\lambda)}{\mu-\lambda}\, d\lambda = -(\mu-T)^{-1}+(\mu-T)^{-1}$$

$$\cdot \frac{1}{2\pi i}\int_{\partial\Delta} (\lambda-T)^{-1}YL(\lambda)X(\lambda-T)^{-1}d\lambda+(\mu-T)^{-1}$$

$$\cdot \frac{1}{2\pi i}\int_{\partial\Delta} \frac{YL(\lambda)X(\lambda-T)^{-1}}{\mu-\lambda}\, d\lambda = 0$$

in view of Property (P3), the fact that $\mu \notin \overline{\Delta}$, and the equality

$$\frac{1}{2\pi i}\int_{\partial\Delta}(\lambda-T)^{-1}YL(\lambda)X(\lambda-T)^{-1}d\lambda = I.$$

Now the analyticity of $Z(\lambda)$ follows from Lemma 6.1.2.

From Property (P3′) and Proposition 6.1.1, it is clear that

$$\begin{bmatrix} Z(\lambda) & (\lambda-T)^{-1}YL(\lambda) \\ L(\lambda)X(\lambda-T)^{-1} & L(\lambda) \end{bmatrix} \tag{6.4.2}$$

has an analytic continuation to Ω. Let

$$E(\lambda) = \begin{bmatrix} T-\lambda & Y \\ X & H(\lambda) \end{bmatrix}, \tag{6.4.3}$$

where $H(\lambda) = L(\lambda)^{-1}-X(\lambda-T)^{-1}Y$. Then $E(\lambda)$ is an analytic and invertible operator function on Ω, with $E(\lambda)^{-1}$ given by (6.4.2). Indeed, denoting (6.4.2) by $\tilde{E}(\lambda)$, the multiplication gives

$$E(\lambda)\tilde{E}(\lambda) = \begin{bmatrix} (T-\lambda)Z(\lambda)+YL(\lambda)X(\lambda-T)^{-1} & -YL(\lambda)+YL(\lambda) \\ XZ(\lambda)+H(\lambda)L(\lambda)X(\lambda-T)^{-1} & X(\lambda-T)^{-1}YL(\lambda)+H(\lambda)L(\lambda) \end{bmatrix}$$

and the definitions of $Z(\lambda)$ and $H(\lambda)$ imply that $E(\lambda)\tilde{E}(\lambda) = I$ for all $\lambda \in \Omega$. Analogously the equality $\tilde{E}(\lambda)E(\lambda) = I$, $\lambda \in \Omega$ is verified.

THEOREM 6.4.1. *If* (X,T,Y) *is a spectral triple for* L *on* Ω*, then the operator* T *is a linearization of* L *on* Ω*. In fact,*

$$E(\lambda)\begin{bmatrix} I_{\mathcal{Y}} & 0 \\ 0 & L(\lambda) \end{bmatrix} = \begin{bmatrix} \lambda-T & 0 \\ 0 & I_{\mathcal{Z}} \end{bmatrix} F(\lambda), \quad (\lambda \in \Omega), \tag{6.4.4}$$

where $E(\lambda)$ *and* $F(\lambda)$ *are invertible operators depending analytically on the parameter* $\lambda \in \Omega$ *and are given by*

$$E(\lambda) = \begin{bmatrix} T-\lambda & Y \\ X & H(\lambda) \end{bmatrix}, \quad F(\lambda) = \begin{bmatrix} -I & (\lambda-T)^{-1}YL(\lambda) \\ X & I-X(\lambda-T)^{-1}YL(\lambda) \end{bmatrix}.$$

Here $H(\lambda) = L(\lambda)^{-1}-X(\lambda-T)^{-1}Y$.

PROOF. We have seen already that $E(\lambda)$ is analytic and invertible. The inverse of $F(\lambda)$ is easy to calculate; it is given by

$$F(\lambda)^{-1} = \begin{bmatrix} -I+(\lambda-T)^{-1}YL(\lambda)X & (\lambda-T)^{-1}YL(\lambda) \\ X & I \end{bmatrix}.$$

Finally, equality (6.4.4) is established directly. ■

As an illustration of Theorem 6.4.1 and its proof, assume that $L(\lambda)$ is a monic operator polynomial of degree ℓ. Let (X,T,Y) be a spectral triple of $L(\lambda)$ with respect to $\mathbb{C}$. Then we know (Theorem 2.5.2) that $L(\lambda)^{-1} = X(\lambda-T)^{-1}Y$, so in this case

$$E(\lambda) = \begin{bmatrix} T-\lambda & Y \\ X & 0 \end{bmatrix}, \tag{6.4.5}$$

which is analytic and invertible on the whole complex plane. The invertibility of (6.4.5) can be seen directly, by taking the spectral triple

$$T = C_L \text{ (the companion operator);}$$

$$X = [I \ 0 \ \cdots \ 0], \quad Y = \mathrm{col}[\delta_{i\ell}I]_{i=1}^{\ell}$$

where $\mathcal{Y} = \mathcal{X}^{\ell}$. Then (6.4.5) takes the form

$$\begin{bmatrix} -\lambda & I & 0 & \cdots & & 0 \\ 0 & -\lambda & I & & & \vdots \\ \vdots & \vdots & & \ddots & \ddots & \vdots \\ 0 & 0 & \cdots & -\lambda & I & 0 \\ -A_0 & -A_1 & \cdots & -A_{\ell-2} & -\lambda-A_{\ell-1} & I \\ I & 0 & \cdots & 0 & 0 & 0 \end{bmatrix},$$

where A_j, $j = 0,\dots,\ell-1$, are the coefficients of L, and its invertibility for every $\lambda \in \mathbb{C}$ is evident.

Now we are in the position to assert the existence of a linearization with respect to any open set $\Omega \subset \mathbb{C}$, and not only with respect to bounded open sets as in Theorem 1.3.1.

COROLLARY 6.4.2. *Let* $L(\lambda)$ *be an operator polynomial with coefficients from* $L(\mathcal{X})$, *and let* $\Omega \subset \mathbb{C}$ *be an open set such that* $\Sigma(L) \cap \Omega$ *is compact. Then there exists a linearization* $T \in L(\mathcal{Y})$ *of* $L(\lambda)$ *with respect to* Ω. *If* $\mathcal{X}$ *is a (separable) Hilbert space, then* $\mathcal{Y}$ *can be chosen to be a (separable) Hilbert space as well.*

For the proof use the existence of a Ω-spectral triple for L (which is ensured by Theorem 6.3.2), and apply Theorem 6.4.1.

6.5 Spectral pairs and divisibility

We have seen in Chapter 2 that the right divisibility of monic operator polynomials can be expressed in terms of their right spectral pairs, while the left divisibility is studied in terms of left spectral pairs. Here we define the left and right spectral pairs of non-monic operator polynomials and prove that this description of divisibility carries over to the non-monic case. The definition is given in terms of the spectral triples introduced in Section 6.1; an equivalent definition which does not rely on the notion of spectral triples will be given in the next section.

Let $L(\lambda)$ be an operator polynomial with coefficients in $L(\mathcal{X})$ (as usual, $\mathcal{X}$ is a Banach space), and let $\Omega \subset \mathbb{C}$ be an open set such that $\Sigma(L) \cap \Omega$ is compact. Then there exists a spectral triple (X,T,Y) of $L(\lambda)$ with respect to Ω. The pair of operators (X,T) will be called *right spectral pair* of $L(\lambda)$ on Ω (in short, right Ω-spectral pair), while the pair (T,Y) will be called *left spectral pair* of $L(\lambda)$ on Ω, or left Ω-spectral pair.

From the properties of spectral triples described in Section 6.1, one obtains corresponding properties of spectral pairs. Some of them are listed below.

PROPOSITION 6.5.1. (a) *Right and left spectral paris of* L *on* Ω *are unique up to similarity: If* (X_1,T_1) *and* (X_2,T_2) *are right* Ω*-spectral pairs of* L, *then* $X_1 = X_2S$, $T_1 = S^{-1}T_2S$ *for some invertible operator* S, *which is unique. Analogously, if* (T_1,Y_1) *and* (T_2,Y_2) *are left* Ω*-spectral pairs of* L, *then* $Y_1 = VY_2$, $T_1 = VT_2V^{-1}$ *for some invertible operator* V, *which is unique.* (b) *For operator polynomials* L_1 *and* L_2, *let* (X_1,T_1) *and* (T_2,Y_2) *be a right* Ω*-spectral pair of* L_1 *and a left* Ω*-spectral pair of* L_2, *respectively. Then*

$$\left([X_1 \ Q_1], \begin{bmatrix} T_1 & Q_2 \\ 0 & T_2 \end{bmatrix}\right)$$

is a right Ω*-spectral pair for* $L = L_2L_1$, *where* Q_1 *and* Q_2 *are suitable operators, and*

$$\left(\begin{bmatrix} T_1 & Q_2 \\ 0 & T_2 \end{bmatrix}, \begin{bmatrix} Q_3 \\ Y_2 \end{bmatrix}\right)$$

is a left Ω*-spectral pair for* L, *where* Q_3 *is a suitable operator.* (c) *Let* (X,T) *and* (T,Y) *be right and left* Ω*-spectral pairs of* L, *respectively, and let*

$$\tilde{L}(\lambda) = E(\lambda)L(\lambda)F(\lambda),$$

where $E(\lambda)$ *and* $F(\lambda)$ *are operator polynomials invertible for all* $\lambda \in \Omega$. *Then*

$$\left(\frac{1}{2\pi i}\int_{\partial\Delta} F(\lambda)^{-1}X(\lambda-T)^{-1}d\lambda, T\right)$$

is a right Ω*-spectral pair of* $\tilde{L}$, *and*

$$\left(T, \frac{1}{2\pi i}\int_{\partial\Delta} (\lambda-T)^{-1}YE(\lambda)^{-1}d\lambda\right)$$

is a left Ω*-spectral pair of* L. *Here* Δ *is a bounded Cauchy domain such that* $\sigma(T) \subset \Delta \subset \overline{\Delta} \subset \Omega$. *(Warning: the triple* (X,T,Y) *is not necessarily a spectral triple for* $L(\lambda)$ *on* Ω.) (d) *Let* (X,T) *and* (T,Y) *be right and left* Ω*-spectral pairs of* L, *respectively. Then* (Y^*, T^*) *and* (T^*, X^*) *are right and left* Ω*-spectral pairs, respectively, of the operator polynomial* L^* *whose coefficients are adjoints to the corresponding coefficients of* L. *(The same warning as in part* (c) *applies.)*

The parts (a), (b), (c), (d) of this proposition follow from Theorems 6.1.3, 6.2.2, Corollary 6.2.3, and Theorem 6.2.7, respectively.

We remark that the part (c) remains valid if $E(\lambda)$ and $F(\lambda)$ are analytic operator functions invertible for all $\lambda \in \Omega$.

It is worth noticing that for a right Ω-spectral pair (X,T) of L there is *unique* operator Y such that (X,T,Y) is a Ω-spectral triple of L. The easy verification of this statement is left to the reader.

We consider divisibility of operator polynomials with respect to the open set Ω. So, given operator polynomials L and L_1 with coefficients in $L(\mathcal{X})$, we say that L_1 is a *right* (resp. *left*) *divisor of* L *with respect to* Ω, or, in short, Ω*-right* (resp. Ω*-left*) *divisor*, if there exists an analytic operator function $Q: \Omega \to L(\mathcal{X})$ such that $L(\lambda) = Q(\lambda)L_1(\lambda)$ for all $\lambda \in \Omega$ (resp. $L(\lambda) = L_1(\lambda)Q(\lambda)$ for all $\lambda \in \Omega$). In the case $\Omega = \mathbb{C}$, the notions of right (or left) divisors will be used. Thus, L_1 is called a *right* (resp. *left*) *divisor* of L if $L = QL_1$ (resp. $L = L_1Q$) for some entire operator function Q. If $\mathcal{X}$ is finite dimensional, then $Q(\lambda)$ can be chosen a polynomial (observe that because we do not put any restrictions on the invertibility of L and L_1, the operator function Q may be non-unique). In the infinite-dimensional case, it may happen that L_1 is a right (or left) divisor of L, but there is no operator polynomial Q such that $L = QL_1$ (or $L = L_1Q$). Example: Let $A \in L(\mathcal{X})$ be quasinilpotent (i.e., $\sigma(A) = \{0\}$) but not nilpotent operator, and put $L(\lambda) \equiv I$, $L_1(\lambda) = I+\lambda A$.

To describe divisibility in terms of spectral pairs we need a notion of restriction. For $i = 1,2$, let (X_i, T_i) be a pair of operators $T_i: \mathcal{Y}_i \to \mathcal{Y}_i$ and $X_i: \mathcal{Y}_i \to \mathcal{X}$. The pair (X_2, T_2) is called a *right restriction* of (X_1, T_1) if there exists a left invertible operator $S: \mathcal{Y}_2 \to \mathcal{Y}_1$ such that $X_1S = X_2$, $T_1S = ST_2$.

Analogously, let (T_i,Y_i) be a pair of operators $T_i: \mathcal{Y}_i \to \mathcal{Y}_i$ and $Y_i: \mathcal{X} \to \mathcal{Y}_i$ $(i = 1,2)$. We call (T_2,Y_2) a *left restriction* of (T_1,Y_1) if there exists a right invertible operator $S: \mathcal{Y}_1 \to \mathcal{Y}_2$ such that $SY_1 = Y_2$, $ST_1 = T_2S$.

THEOREM 6.5.2. *For* $i = 1,2$, *let* L_i *be an operator polynomial with* $\Sigma(L_i) \cap \Omega$ *compact, and let* (X_i,T_i) *be a right spectral pair for* L_i *on* Ω. *Then the pair* (X_2,T_2) *is a right restriction of the pair* (X_1,T_1) *if and only if the operator polynomial* L_2 *is a* Ω-*right divisor of the operator polynomial* L_1.

THEOREM 6.5.3. *For* $i = 1,2$, *let* L_i *be an operator polynomial with* $\Sigma(L_i) \cap \Omega$ *compact, and let* (T_i,Y_i) *be a left spectral pair for* L_i *on* Ω. *Then the pair* (T_2,Y_2) *is a left restriction of the pair* (T_2,Y_2) *if and only if* L_2 *is a* Ω-*left divisor of* L_1.

We shall prove Theorem 6.5.2 only (Theorem 6.5.3) can be proved by an analogous argument).

PROOF OF THEOREM 6.5.2. Let $Y_i \in L(\mathcal{X},\mathcal{Y}_i)$ be the (unique) operator such that (X_i,T_i,Y_i) is a spectral triple for L_i on Ω, $i = 1,2$. Suppose (X_2,T_2) is a right restriction of (X_1,T_1). Then there exists a left invertible operator $S \in L(\mathcal{Y}_2,\mathcal{Y}_1)$ such that

$$X_1S = X_2,\quad T_1S = ST_2. \tag{6.5.1}$$

So for $\lambda \in \Omega\backslash\{\Sigma(L_1) \cup \Sigma(L_2)\}$ we have:

$$L_1(\lambda)L_2(\lambda)^{-1}$$

$$= L_1(\lambda)[L_2(\lambda)^{-1}-X_2(\lambda-T_2)^{-1}Y_2]+L_1(\lambda)X_1(\lambda-T_1)^{-1}SY_2.$$

By Property (P2) for (X_2,T_2,Y_2) and Property (P3$'$) for (X_1,T_1,Y_1), it is clear that $L_1L_2^{-1}$ has an analytic continuation to Ω. So L_2 is a right divisor of L_1 on Ω.

Conversely, let L_2 be a right divisor of L_1 on Ω, and let $H = L_1L_2^{-1}$, which is an analytic operator function on Ω. Since $\Sigma(L_1)$ and $\Sigma(L_2)$ are compact subsets of Ω, clearly

$\Sigma(H) \cap \Omega = \{\lambda \in \Omega | H(\lambda)$ is not invertible$\}$ is compact too. Let (X_0, T_0, Y_0) be a spectral triple for $H(\lambda)$ on Ω (the existence of such triple is ensured by Theorem 6.3.1; see also the remark at the end of Section 6.1), where $T_0 \in L(\mathcal{Y}_0)$. By Theorem 6.2.2, we construct a Ω-spectral triple (X, T, Y) for $L_1 = HL_2$ using the Ω-spectral triples (X_0, T_0, Y_0) and (X_2, T_2, Y_2). We have

$$T = \begin{bmatrix} T_2 & Y_2X_0 \\ 0 & T_0 \end{bmatrix}, \quad X = (X_2 \quad Q),$$

where

$$Q = (2\pi i)^{-1} \int_{\partial\Delta} \{L_2(\lambda)^{-1} - X_2(\lambda - T_2)^{-1}Y_2\}X_0(\lambda - T_0)^{-1} d\lambda,$$

and Δ is a bounded Cauchy domain such that $(\Sigma(L_1) \cup \Sigma(L_2)) \cap \Omega \subset \Delta \subset \overline{\Delta} \subset \Omega$. Using the uniqueness of Ω-spectral triples for L_1 (Theorem 6.1.3), we conclude that there exists an invertible operator $\tilde{S} \in L(\mathcal{Y}_2 \oplus \mathcal{Y}_0, \mathcal{Y}_1)$ such that $ST = T_1\tilde{S}$ and $X_1\tilde{S} = X$. Define $S \in L(\mathcal{Y}_2, \mathcal{Y}_1)$ by $Sx_2 = \tilde{S}(x_2, 0)$. Then S is left invertible and satisfies the identities (6.5.1). Hence (X_2, T_2) is a right restriction of (X_1, T_1). ■

COROLLARY 6.5.4. *Let* L_1 *and* L_2 *be operator polynomials with* $\Sigma(L_j) \cap \Omega$ *compact,* $j = 1,2$.

Then L_1 *and* L_2 *have equal right (resp. left)* Ω*-spectral pairs if and only if there exists an invertible operator* $E(\lambda) \in L(\mathcal{X})$ *depending analytically on* $\lambda \in \Omega$ *such that* $L_1(\lambda) = E(\lambda)L_2(\lambda)$ *(resp.* $L_1(\lambda) = L_2(\lambda)E(\lambda)$*) for all* $\lambda \in \Omega$.

PROOF. If (X, T) is a right Ω-spectral pair for both L_1 and L_2, then, in view of Theorem 6.5.2, the functions L_1 and L_2 are Ω-right divisors of each other. Hence, the operator functions E and F, which are defined on $\Omega \backslash (\Sigma(L_1) \cup \Sigma(L_2))$ by $E = L_1L_2^{-1}$ and $F = L_2L_1^{-1}$, have an analytic continuation to Ω. But then these continuations take invertible values on all of Ω. Analogously, the assertion concerning left Ω-spectral pairs is proved. ■

Let $L(\lambda) = L_1(\lambda)L_2(\lambda)$, where L_1, L_2 are operator polynomials with compact $\Sigma\,(L_j) \cap \Omega$. The polynomial L_2 (resp.

L_1) is called a *right* (resp. *left*) *Ω-spectral divisor* of L *if* $\Sigma(L_1) \cap \Sigma(L_2) \cap \Omega = \phi$. Note that in this case $\Sigma(L) \cap \Omega$ is the union of the disjoint compact sets $\Sigma(L_1) \cap \Omega$ and $\Sigma(L_2) \cap \Omega$, and hence $\Sigma(L_2) \cap \Omega$ is a compact and relatively open subset of $\Sigma(L) \cap \Omega$.

To describe spectral divisors, we use the notion of a spectral subspace. A subspace $\mathcal{M} \subset \mathcal{Y}$ is called a *spectral subspace* for an operator $T \in L(\mathcal{Y})$ if $\mathcal{M}$ is the image of the Riesz projection for T, i.e.,

$$\mathcal{M} = \mathrm{Im}\left[\frac{1}{2\pi i} \int_{\Gamma} (\lambda - T)^{-1} d\lambda\right]$$

for some simple rectifiable contour Γ such that $\Gamma \cap \sigma(T) = \phi$ (in this case we say that the spectral subspace $\mathcal{M}$ corresponds to the part of σ(T) which is inside Γ).

THEOREM 6.5.5. *Let* L, L_2 *be operator polynomials with* $\Sigma(L) \cap \Omega$ *and* $\Sigma(L_2) \cap \Omega$ *compact and let* (X,T) *and* (X_2,T_2) *be right Ω-spectral pairs for* L *and* L_2 *respectively. Denote by* $\mathcal{Y}$ *(resp.* $\mathcal{Y}_2$*) the Banach space on which* T *(resp.* T_2*) acts. Then* L_2 *is a right Ω-spectral divisor of* L *is and only if there exists a left invertible operator* $S \in L(\mathcal{Y}_2,\mathcal{Y})$ *such that*

$$XS = X_2, \quad TS = ST_2, \tag{6.5.2}$$

and Im S *is a spectral subspace of the operator* T.

THEOREM 6.5.6 *Let* L, L_2 *be as in Theorem* 6.5.5, *and let* (T,Y) *and* (T_2,Y_2) *be left Ω-spectral pairs for* L *and* L_2, *respectively. Denote by* $\mathcal{Y}$ (resp. $\mathcal{Y}_2$) *the space on which* T *(resp.* T_2*) acts. Then* L_2 *is a left Ω-spectral divisor of* L *if and only if there exists a right invertible operator* $S \in (\mathcal{Y},\mathcal{Y}_2)$ *such that* $SY = Y_2$, $ST = T_2S$ *and* Ker S *is a spectral subspace of the operator* T.

We prove Theorem 6.5.5 only (the proof of Theorem 6.5.6 is analogous).

PROOF OF THEOREM 6.5.5. Let L_2 be a right Ω-spectral divisor of L, and let $L_1 = LL_2^{-1}$ be the quotient. Let (X_1,T_1,Y_1) be a spectral triple for L_1 on Ω. As in the proof of Theorem

6.5.2, we construct two spectral triples for L on Ω: one of the form (X,T,Y) and the other one of the form $(\tilde{X},\tilde{T},\tilde{Y})$, where

$$\tilde{T} = \begin{bmatrix} T_2 & * \\ 0 & T_1 \end{bmatrix}, \quad \tilde{X} = [X_2 \quad *]. \tag{6.5.3}$$

Then these two spectral triples are similar (Theorem 6.1.3); so

$$X\tilde{S} = \tilde{X}, \quad T\tilde{S} = \tilde{S}\tilde{T} \tag{6.5.4}$$

for some invertible operator $\tilde{S} \in L(\mathcal{Y}_2 \oplus \mathcal{Y}_1, \mathcal{Y})$; here $\mathcal{Y}$, $\mathcal{Y}_1$, and $\mathcal{Y}_2$ are the Banach spaces on which T, T_1, and T_2, respectively, act. Define $S \in (\mathcal{Y}_2, \mathcal{Y})$ by $Sx_2 = \tilde{S}(x_2, 0)$. Then S is left invertible and satisfies (6.5.2). From (6.5.3), it is clear that $\mathcal{Y}_2 \oplus (0)$ is the spectral subspace of $\tilde{T}$ corresponding to $\sigma(T_2) = \Sigma(L_2) \cap \Omega$ (because $\sigma(T_1) = \Sigma(L_1) \cap \Omega$). Hence, Im S is the spectral subspace of A corresponding to the same set $\sigma(T_2) = \Sigma(L_2) \cap \Omega$.

The converse statement is proved by reversing this argument. ∎

6.6 Characterization of spectral pairs

In the previous section we have defined spectral pairs using the notion of spectral triples. Here we present intrinsic characterizations of spectral pairs which do not refer to spectral triples. Thoughout this section

$$L(\lambda) = \sum_{j=0}^{p} \lambda^j A_j,$$

where $A_0, \ldots, A_p \in L(\mathcal{X})$, and $\Omega \subset \mathbb{C}$ is an open set such that $\Sigma(L) \cap \Omega$ is compact.

THEOREM 6.6.1. *A pair* (X,T) *of operators* $X \in L(\mathcal{Y},\mathcal{X})$ *and* $T \in L(\mathcal{Y})$ *is a right spectral pair on* Ω *for* L *if and only if the following four conditions are fulfilled:*

$$\sigma(T) \subset \Omega; \tag{Q1}$$

(Q2) $$\sum_{j=0}^{p} L_j X T^j = 0$$

(Q3) *the operator*

$$\mathrm{col}[XT^j]_{j=0}^{p-1} \in L(\mathcal{Y}, \mathcal{X}^p)$$

is left invertible;

(Q4) *every other pair of operators satisfying* (Q1), (Q2), *and* (Q3) *is a right restriction of* (X,T).

More explicitly, condition (Q4) means the following. Let (X_0, T_0) be a pair of operators $X_0 \in L(\mathcal{Y}_0, \mathcal{X})$; $T_0 \in L(\mathcal{Y}_0)$ with the following properties: (i) $\sigma(T_0) \subset \Omega$; (ii) $\sum_{j=0}^{p} L_j X_0 T_0^j = 0$; (iii) the operator $\mathrm{col}[X_0 T_0^j]_{j=0}^{p-1}$ is left invertible. Then there exists a left invertible operator $S \in L(\mathcal{Y}_0, \mathcal{Y})$ such that $XS = X_0$ and $TS = ST_0$.

PROOF. Let (X,T) be a right Ω-spectral pair of L, and let $Y \in L(\mathcal{X}, \mathcal{Y})$ be such that (X,T,Y) is a Ω-spectral triple for L. Then properties (Q1) and (Q2) follow from (P1) and (P3) of the definition of spectral triples. To derive (Q3), consider the operator-valued function $U(\lambda) = (\lambda - T)^{-1} Y L(\lambda)$. By Proposition 6.1.1, $U(\lambda)$ is analytic in $\mathbb{C}$. Hence, because $L(\lambda)$ is a polynomial of degree at most p, $U(\lambda)$ is a polynomial of degree at most p-1. Write

$$U(\lambda) = \sum_{j=0}^{p-1} \lambda^j U_j,$$

and use the fact that

$$\frac{1}{2\pi i} \int_{|\lambda|=r} U(\lambda) X (\lambda - T)^{-1} d\lambda = I \tag{6.6.1}$$

for r large enough (this fact follows from Theorem 6.1.3 by taking there $X_i = X$, $Y_i = Y$, $T_i = T$ for i = 1,2). Equality (6.6.1) means (upon developing $(\lambda - T)^{-1}$ into Laurent series in a neighborhood of infinity)

$$\sum_{j=0}^{p-1} U_j X T^j = I,$$

and (Q3) follows.

Finally, let us prove (Q4). Fix a bounded Cauchy domain Δ such that $(\Sigma(L) \cap \Omega) \cup \sigma(T_0) \subset \Delta \subset \bar{\Delta} \subset \Omega$. Without loss of generality we can assume that $0 \in \Omega$ and that X and T are those defined in Theorem 6.3.1. (so $\mathcal{Y} = \mathcal{M}$, where the Banach space $\mathcal{M}$ is also defined in Theorem 6.3.1). Let (X_0, T_0) be a pair of operators $X_0 \in L(\mathcal{Y}_0, \mathcal{X})$, $T_0 \in L(\mathcal{Y}_0)$ with the properties (i), (ii), and (iii), and define the operator $S_0 \in L(\mathcal{Y}_0, X(\partial\Delta, \mathcal{X}))$ by

$$(S_0 x)(z) = X_0(z-T_0)^{-1}x, \quad x \in \mathcal{Y}_0, \ z \in \partial\Delta.$$

Because of condition (ii) which implies that $L(\lambda)X_0(\lambda-T_0)^{-1}$ is analytic in Ω, the range of S_0 is contained in $\mathcal{M}$, so we may consider S_0 as an operator in $L(\mathcal{Y}_0, \mathcal{M})$. The next observation is that S_0 is left invertible. Indeed, let $[U_0 U_1 \cdots U_{p-1}]$ be a left inverse of $\operatorname{col}[X_0 T_0^j]_{j=0}^{p-1}$ (here $U_j \in L(\mathcal{X}, \mathcal{Y}_0)$), and define $K_0 \in L(\mathcal{M}, \mathcal{Y}_0)$ as follows: Given $f(z) \in \mathcal{M}$, write the Laurent series $f(z) = \sum_{j=-1}^{-\infty} f_j z^j$ in a neighborhood of infinity, and put $K_0 f(z) = U_0 f_{-1} + U_1 f_{-2} + \cdots + U_{p-1} f_{-p}$. The continuity of K_0 follows from the continuity of each linear transformation $V_j: \mathcal{M} \to \mathcal{X}$ defined by $V_j f(z) = f_j$, which in turn follows from the maximum modulus principle for $\mathcal{X}$-valued analytic functions defined on $(\mathbb{C} \cup \{\infty\}) \setminus \Delta$. Now the equality $K_0 S_0 = I_{\mathcal{Y}_0}$ is clear. Further, one verifies without difficulty that $XS_0 = X_0$ and $TS_0 = S_0 T_0$. So (X_0, T_0) is a right restriction of (X, T), and (Q4) follows.

Conversely, let (X, T) be a pair of operators satisfying (Q1)-(Q4), and let (X', T') be some right Ω-spectral pair of L. Then, according to the first part of the proof, (X', T') also possesses the properties (Q1)-(Q4). Hence, (X, T) and (X', T') are right restrictions of each other, which implies their similarity: $X = X'S$, $T = S^{-1}T'S$ for some invertible operator S. So (X, T) is a right Ω-spectral pair of L as well. ∎

The left counterpart of Theorem 6.6.1 runs as follows.

THEOREM 6.6.2. *A pair* (Y,T) *of operators* $Y \in L(\mathcal{X},\mathcal{Y})$, $T \in L(\mathcal{Y})$ *is a left* Ω*-spectral pair of* L *if and only if the following conditions are fulfilled:*

(Q5)
$$\sigma(T) \subset \Omega;$$

(Q6)
$$\sum_{j=0}^{p} A^{j}YL_{j} = 0;$$

(Q7) *the operator*

$$[Y, TY, \dots, T^{p-1}Y] \in L(\mathcal{X}^{p}, \mathcal{Y})$$

is right invertible;

(Q8) *every other pair of operators* (Y',T') *satisfying* (Q5), (Q6), *and* (Q7) *is a left restriction of* (Y,T): $SY = Y'$, $ST = T'S$ *for some right invertible operator* S.

We omit the proof of Theorem 6.6.2; it can be done analogously to the proof of Theorem 6.6.1.

6.7 Reduction to monic polynomials

If $L(\lambda)$ is a monic operator polynomial of degree ℓ, then its spectral triples on $\mathbb{C}$ can be described in terms of the companion operator $C_L \in L(\mathcal{X}^{\ell})$ (see Chapter 2). Namely, a triple of operators (X,T,Y) is a $\mathbb{C}$-spectral triple of L if and only if it is similar to the triple

$$([I\ 0\ \cdots\ 0], C_L, \mathrm{col}[\delta_{i\ell}I]_{i=1}^{\ell}). \tag{6.7.1}$$

Indeed, the part "if" follows from the results of Chapter 2 (Theorem 2.5.2 and the definition of right spectral pair of monic operator polynomials), while the part "only if" follows from Theorem 6.1.3. Consequently, a pair of operators is a right (resp. left) $\mathbb{C}$-spectral pair of L if and only if it is similar to the pair

$$([I\ 0\ \cdots\ 0], C_L) \ \text{(resp. } (C_L, \mathrm{col}[\delta_{i\ell}I]_{i=1}^{\ell})).$$

Starting with this observation, and using the characterizations given in Section 6.6, it is possible to explicitly construct spectral pairs for non-monic operator polynomials. Such construction is described in this section. It is somewhat more restrictive than the construction described in Section 6.3, because it applies for polynomial functions only.

We start with spectral triples for monic polynomials.

THEOREM 6.7.1. *Let $L(\lambda)$ be a monic operator polynomial, and let Ω be an open set with $\Sigma(L) \cap \Omega$ compact. Choose a bounded Cauchy domain Δ such that $\Sigma(L) \cap \Omega \subset \Delta \subset \overline{\Delta} \subset \Omega$, and denote by P_Δ the Riesz projection $(2\pi i)^{-1}\int_{\partial\Delta} (\lambda - T)^{-1} d\lambda$, where T is the main operator in a $\mathbb{C}$-spectral triple (X,T,Y) of $L(\lambda)$. Then $(X|_{\mathcal{M}}, T|_{\mathcal{M}}, P_\Delta Y)$ is a Ω-spectral triple of $L(\lambda)$, where $\mathcal{M} = \operatorname{Im} P_\Delta$ and P_Δ is considered as an operator onto $\mathcal{M}$.*

PROOF. We have to verify the properties (P1)-(P4) for the triple $(X|_{\mathcal{M}}, T|_{\mathcal{M}}, P_\Delta Y)$. (P1) is immediate, (P3) and (P4) follow because the triple (X,T,Y) satisfies (P3) and (P4). To verify (P2), let $\mathcal{N} = \operatorname{Im}(I-P_\Delta)$, and observe that $\sigma(T|_{\mathcal{N}})$ does not intersect Ω and that

$$X(\lambda-T)^{-1}Y = X|_{\mathcal{M}}(\lambda-T|_{\mathcal{M}})^{-1}P_\Delta Y + X|_{\mathcal{N}}(\lambda-T|_{\mathcal{N}})^{-1}(I-P_\Delta)Y.$$

So

$$L(\lambda)^{-1} - X|_{\mathcal{M}}(\lambda-T|_{\mathcal{M}})^{-1}P_\Delta Y = X|_{\mathcal{N}}(\lambda-T|_{\mathcal{N}})^{-1}(I-P_\Delta)Y$$

is analytic in Ω. ■

We can state now the main result of this section.

THEOREM 6.7.2. *Let $M(\lambda) = \sum_{j=0}^{\ell} \lambda^j M_j$ be an operator polynomial such that $\Sigma(M) \cap \Omega$ is compact. Then a right Ω-spectral pair for M can be constructed as follows. Let Δ be a bounded Cauchy domain such that $\Sigma(M) \cap \Omega \subset \Delta \subset \overline{\Delta} \subset \Omega$, and let $a \in \Omega\setminus\overline{\Delta}$ (in particular, this choice of a ensures that $M(a)$ is invertible). Put $\Xi = \{(\lambda-a)^{-1} \mid \lambda \in \Delta\}$, and denote by $(\tilde{X},\tilde{T})$ a right Ξ-spectral pair for the monic operator polynomial*

$L(\lambda) = (M(a))^{-1}\lambda^{\ell}M(\lambda^{-1}+a)$. *Then* $(\tilde{X},\tilde{T}^{-1}+aI)$ *is a right* Ω*-spectral pair of* $M(\lambda)$.

Observe that $(\tilde{X},\tilde{T})$ can be explicitly constructed in terms of the coefficients of $M(\lambda)$ by using Theorem 6.7.1 and the formula (6.7.1). Note also that $0 \notin \Xi$; consequently, $\tilde{T}$ is invertible.

PROOF. It is clear that the Ω-spectral pair of $M(\lambda)$ coincides with its Δ-spectral pair, so we shall prove that $(\tilde{X},\tilde{T}^{-1}+aI)$ is a right Δ-spectral pair of $M(\lambda)$.

For a number $\alpha \in \mathbb{C}$ and an integer $m \geq 1$, define the block-operator matrix

$$U_m(\alpha) = \left[\binom{m-1-j}{k-j}\alpha^{k-j}I_{\mathcal{X}}\right]_{j,k=0}^{m-1} \in L(\mathcal{X}^m),$$

where $\binom{p}{q} = 0$ for $q < 0$, and $\binom{p}{q} = \frac{p!}{(p-q)!q!}$ for $0 \leq q \leq p$. For example,

$$U_3(\alpha) = \begin{bmatrix} I & \binom{2}{1}\alpha I & \binom{2}{2}\alpha^2 I \\ 0 & I & \alpha I \\ 0 & 0 & I \end{bmatrix}$$

Note that $U_m(\alpha)$ is invertible and $(U_m(\alpha))^{-1} = U_m(-\alpha)$. Another easily verified property of $U_m(\alpha)$ that we need is that

$$(6.7.2) \qquad [U_m(\alpha)]\ \mathrm{col}[XT^i]_{i=0}^{m-1} = \mathrm{col}[X(I+\alpha T)^{m-1-i}T^i]_{i=0}^{m-1}$$

for any pair of operators $T \in L(\mathcal{Y})$, $X \in L(\mathcal{Y},\mathcal{X})$.

Return to the proof of Theorem 6.7.1. We show that $(\tilde{X},\tilde{T}^{-1}+aI)$ satisfies the conditions (Q1)-(Q4) of Theorem 6.6.1. Write

$$L(\lambda) = \sum_{j=0}^{\ell} \lambda^j A_j \quad (A_\ell = I).$$

As $(\tilde{X},\tilde{T})$ is a right Ξ-spectral pair for $L(\lambda)$, we have

(6.7.3) $$\sum_{j=0}^{\ell} A_j \tilde{X}\tilde{T}^j = 0.$$

On the other hand, the coefficients A_j can be expressed by the formulas

(6.7.4) $$A_k = [M(a)]^{-1} \sum_{j=0}^{k} \binom{\ell-j}{k-j} a^{k-j} M_{\ell-j}, \quad k = 0, \ldots, \ell.$$

Indeed, the formulas follow from the following string of equalities:

$$M(a)L(\lambda) = \lambda^{\ell} M(\lambda^{-1}+a) = \lambda^{\ell} \sum_{j=0}^{\ell} (\lambda^{-1}+a)^j M_j$$

$$= \sum_{j=0}^{\ell} (1+a\lambda)^j \lambda^{\ell-j} M_j = \sum_{k=0}^{\ell} \lambda^k \sum_{j=\ell-k}^{\ell} \binom{j}{k+j-\ell} a^{k+j-\ell} M_j$$

$$= \sum_{k=0}^{\ell} \lambda^k \sum_{j=0}^{k} \binom{\ell-j}{k-j} a^{k-j} M_{\ell-j}.$$

Substituting (6.7.4) in (6.7.3) and using (6.7.2), one obtains

$$0 = [A_0 A_1 \cdots A_{\ell}] \operatorname{col}[\tilde{X}\tilde{T}^k]_{k=0}^{\ell}$$

$$= [M_{\ell} M_{\ell-1} \cdots M_0] U_{\ell+1}(a) \operatorname{col}[\tilde{X}\tilde{T}^k]_{k=0}^{\ell}$$

$$= [M_{\ell} M_{\ell-1} \cdots M_0][\operatorname{col}[\tilde{X}(\tilde{T}^{-1}+aI)^{\ell-j}]_{j=0}^{\ell}]\tilde{T}^{\ell}.$$

Since $\tilde{T}$ is invertible, the pair $(\tilde{X}, \tilde{T}^{-1}+aI)$ satisfies condition (Q2). The condition (Q1) follows immediately from the fact that $\sigma(\tilde{T}) \subset \Xi$. To verify (Q3), observe that by Theorem 6.6.1 applied to the right Ξ-spectral pair $(\tilde{X}, \tilde{T})$ of $L(\lambda)$, we have that $\operatorname{col}[\tilde{X}\tilde{T}^i]_{i=0}^{\ell-1}$ is left invertible. It remains to note the equality

$$\operatorname{col}[\tilde{X}(\tilde{T}^{-1}+aI)^{\ell-i-1}]_{i=0}^{\ell-1} = U_{\ell}(a)\left[\operatorname{col}[\tilde{X}\tilde{T}^i]_{i=0}^{\ell-1}\right]\tilde{T}^{1-\ell},$$

which follows from (6.7.2), and the left invertibility of $\operatorname{col}[\tilde{X}(\tilde{T}^{-1}+aI)^i]_{i=0}^{\ell-1}$ is proved.

Finally, we verify the property (Q4). Let (X_1, T_1) be a pair of operators such that $\sigma(T_1) \subset \Delta$, the operator $\operatorname{col}[X_1 T_1]_{i=0}^{\ell-1}$ is left invertible, and the equality

$$\sum_{j=0}^{\ell} M_j X_1 T_1^j = 0 \tag{6.7.5}$$

holds. Then $\sigma((T_1 - aI)^{-1}) \subset \Xi$, the operator

$$\operatorname{col}[X_1 (T_1 - aI)^{-i}]_{i=0}^{\ell-1} \tag{6.7.6}$$

is left invertible, and

$$\sum_{j=0}^{\ell} A_j X_1 (T_1 - aI)^{-j} = 0. \tag{6.7.7}$$

Indeed, the left invertibility of (6.7.6) follows from the equality

$$\operatorname{col}[X_1 (T_1 - aI)^{-i}]_{i=0}^{\ell-1} = U_\ell(-a)\left[\operatorname{col}[X_1 T_1^{\ell-1-i}]_{i=0}^{\ell-1}\right](T_1 - aI)^{1-\ell},$$

and the left invertibility of $\operatorname{col}[X_1 T_1^i]_{i=0}^{\ell-1}$, while equality (6.7.7) follows from (6.7.5) taking into account the formulas (6.7.4). Applying Theorem 6.6.1 for the right Ξ-spectral pair $(\tilde{X}, \tilde{T})$ of $L(\lambda)$, we find that

$$X_1 = \tilde{X}S, \quad S(T_1 - aI)^{-1} = \tilde{T}S$$

for some left invertible operator S. But then also

$$(\tilde{T}^{-1} + aI)S = ST_1,$$

and the property (Q4) is verified. ∎

Analogously to the proof of Theorem 6.7.2, its dual statement is proved:

THEOREM 6.7.3. *Let* $M(\lambda)$, Ω, Δ, *and* Ξ *be as in Theorem* 6.7.2. *Denote by* $(\tilde{T}, \tilde{Y})$ *a left* Ξ-*spectral pair for the monic operator polynomial* $L(\lambda) = \lambda^{\ell}M(\lambda^{-1}+a)(M(a))^{-1}$. *Then* $(\tilde{T}^{-1}+aI, \tilde{Y})$ *is a left* Ω*-spectral pair for* $M(\lambda)$.

6.8 Exercises

Ex. 6.1. Prove Theorem 6.5.3 and 6.5.6.

Ex. 6.2. Prove that given a right Ω-spectral pair (X,T) of $L(\lambda)$, there is unique Y such that (X,T,Y) is a Ω-spectral triple for $L(\lambda)$.

Ex. 6.3. State and prove the result dual to Ex. 6.2.

Ex. 6.4. Let $L(\lambda)$ be an operator polynomial with self-adjoint coefficients in $L(\mathcal{X})$, where $\mathcal{X}$ is a Hilbert space, and let (X,T,Y) be a spectral triple of $L(\lambda)$ with respect to a domain Ω which is symmetric relative to the real axis. Prove that the operator T is self-adjoint in a suitable indefinite scalar produce.

Ex. 6.5. We say that an operator polynomial $L(\lambda)$ is of finite type with respect to Ω if there is a Ω-spectral triple (X,T,Y) of $L(\lambda)$ with T acting on a finite-dimensional space. Prove that $L(\lambda)$ is of finite type with respect to Ω if and only if the set $\Sigma(L) \cap \Omega$ consists only of a finite number of points, and $L(\lambda)$ is Fredholm for all $\lambda \in \Omega$ (note that the index of $L(\lambda)$ is necessarily zero for all $\lambda \in \Omega$). Hint: Use the fact that T is a linearization of $L(\lambda)$ with respect to Ω. The factorization theorems proved in Gohberg-Sigal [1,2] can be useful.

6.9 Notes

The contents of this chapter (with the exception of Section 6.7) is taken from Kaashoek-van der Mee-Rodman [1,2]. The exposition here is adapted to the framework of operator polynomials, as opposed to the exposition in Kaashoek-van der Mee-Rodman [1,2] which was done in the framework of analytic operator functions with compact spectrum. Reduction to monic polynomials described in Section 6.7 is a standard tool in the theory of matrix polynomials (see Gohberg-Rodman [1], Gohberg-Lerer-Rodman [1], Gohberg-Lancaster-Rodman [2]).

CHAPTER 7: POLYNOMIALS WITH GIVEN SPECTRAL PAIRS AND EXACTLY CONTROLLABLE SYSTEMS

In this chapter we consider the following problem: Given a pair of operators $X \in L(\mathcal{Y},\mathcal{X})$, $T \in L(\mathcal{Y})$ (here $\mathcal{X}$ and $\mathcal{Y}$ are Banach spaces), construct, if possible, an operator polynomial $L(\lambda)$ whose right spectral pair (with respect to the whole complex plane) is (X,T). By Theorem 6.6.1, a necessary condition is that $\mathrm{col}[XT^i]_{i=0}^{p-1}$ is left invertible for some p; we shall see that this condition (if $\mathcal{X},\mathcal{Y}$ are Hilbert spaces) is also sufficient. It turns out that this problem is very closely related to the problems concerning spectrum assignment in exactly controllable systems, an important topic in the modern systems theory.

We start this chapter with descriptions of such systems.

7.1 Exactly controllable systems

Let $\mathcal{X}$ and $\mathcal{Y}$ be Banach spaces, and let $A \in L(\mathcal{Y})$, $B \in L(\mathcal{X},\mathcal{Y})$. Consider the linear system

$$x_n = Ax_{n-1}+Bu_{n-1}, \quad n = 1,2,... \tag{7.1.1}$$

where $\{x_n\}_{n=0}^{\infty}$ is a sequence of vectors in $\mathcal{Y}$ and $\{u_n\}_{n=0}^{\infty}$ is a sequence of vectors in $\mathcal{X}$.

The equation (7.1.1) is often interpreted in terms of system theory. Thus, $\mathcal{X}$ is assumed to represent the states of a system, while $\mathcal{Y}$ represents controls (or inputs). The problem then becomes to choose the sequence of controls $\{u_n\}_{n=1}^{\infty}$ in a certain way to ensure desired properties of the system.

A system (7.1.1) is called exactly controllable if any state x can be reached from any initial value x_0 in finite number of steps. More precisely, this means the following: For any $x_0, x \in \mathcal{X}$ there is an integer $m(\geq 0)$ and controls $u_0,\dots,u_{m-1}$ such that if $x_1, \ldots, x_m$ are defined by (7.1.1) for $n = 1, \ldots, m$, then $x = x_m$. As the solution of (7.1.1) is given by

$$x_n = A^n x_0 + \sum_{j=0}^{n-1} A^{n-1-j} B u_j, \quad n = 1, 2, \ldots ,$$

it follows that (7.1.1) is exactly controllable if and only if

$$\mathcal{Y} = \bigcup_{n=1}^{\infty} \left(\sum_{j=0}^{n-1} \operatorname{Im} A^{n-1-j} B \right). \tag{7.1.2}$$

THEOREM 7.1.1. *The system* (7.1.1) *is exactly controllable if and only if there is a positive integer* m *such that*

$$\mathcal{Y} = \left(\sum_{j=0}^{m-1} \operatorname{Im} A^{m-1-j} B \right). \tag{7.1.3}$$

PROOF. The part "if" is obvious in view of (7.1.2).

For the part "only if," assume that (7.1.2) holds, and consider the Banach spaces $\mathcal{X}^m$, $m = 1, 2, \ldots$. Then (7.1.2) can be rewritten in the form

$$\mathcal{Y} = \bigcup_{n=1}^{\infty} \operatorname{Im} A_n,$$

where

$$A_n = [B, AB, \ldots, A^{n-1}B] \in L(\mathcal{X}^n, \mathcal{Y}).$$

Let B_m be the unit ball in $\mathcal{X}^m$. Then

$$\mathcal{Y} = \bigcup_{n=1}^{\infty} \bigcup_{k=1}^{\infty} A_n(kB_n).$$

By the Baire category theorem there are positive integers n and k such that the closure of $A_n(kB_n)$ (in the norm topology) contains an open set $\{x \in \mathcal{Y} \mid \|x - x_0\| < r\}$ for some $x_0 \in \mathcal{Y}$ and $r > 0$. (A simple proof of a version of the Baire category theorem which is adequate here is given in Gohberg-Goldberg [1].) Writing $x = x_0 - (x_0 - x)$, we see that the ball $\{x \in \mathcal{Y} \mid \|x\| < r\}$ is contained in the closure of $A_n(2kB_n)$. A standard proof of the open mapping theorem (see, e.g., II.2.1 in Dunford-Schwartz [1]) shows that

$$\{x \in \mathcal{Y} \mid \|x\| < r\} \subset A_n(4kB_n),$$

and we have $\mathcal{Y} = \operatorname{Im} A_n$. ∎

In system theoretic terms, Theorem 7.1.1 says that in an exactly controllable system any state can be reached from any other state in at most m steps, where m is a uniform bound (i.e., does not depend on the choice of the initial and terminal states).

Now consider an exactly controllable system (7.1.1), and let m be such that (7.1.3) holds, i.e.,

$$\operatorname{Im}[B,\dots,A^{m-1}B] = \mathcal{Y}. \tag{7.1.4}$$

In the sequel we need a generally stronger property than (7.1.4), namely that the operator

$$[B, AB,\dots,A^{m-1}B] \in L(\mathcal{X}^m,\mathcal{Y}) \tag{7.1.5}$$

be right invertible. Observe that if $\mathcal{X}$ is a Hilbert space, the right invertibility of (7.1.5) is actually equivalent to (7.1.4). The following characterization of this property is very useful.

THEOREM 7.1.2. *Let* $A \in L(\mathcal{Y})$ *and* $B \in L(\mathcal{X},\mathcal{Y})$ *be Banach space operators. Then* (7.1.5) *is right invertible for some integer* $m > 0$ *if and only if the operator*

$$[\lambda - A, B] \in L(\mathcal{Y}\oplus\mathcal{X},\mathcal{Y}) \tag{7.1.6}$$

is right invertible for every complex number λ.

For the proof of Theorem 7.1.2, we need the following result which we state in a more general formulation than actually needed here.

THEOREM 7.1.3. *Let* $D\colon \Omega \to L(\mathcal{X},\mathcal{Y})$ *be an operator-valued function analytic in an open set* $\Omega \subset \mathbb{C}$. *If* $D(\lambda)$ *is left (resp. right) invertible for every* $\lambda \in \Omega$, *then there is an analytic operator function* $E\colon \Omega \to L(\mathcal{Y},\mathcal{X})$ *such that*

$$E(\lambda)D(\lambda) = I_{\mathcal{X}}$$

for all $\lambda \in \Omega$ *(resp.*

$$D(\lambda)E(\lambda) = I_{\mathcal{Y}}$$

for all $\lambda \in \Omega$).

The proof of Theorem 7.1.3 is beyond the scope of this book (see Allan [1] or Zaidenberg-Kreĭn-Kuchment-Pankov [1] for the proof).

PROOF OF THEOREM 7.1.2. Assume that (7.1.5) is right invertible with a right inverse $\text{col}[Z_i]_{i=0}^{m-1} \in L(\mathcal{Y}, \mathcal{X}^m)$. Put

$$R(\lambda) = \sum_{j=0}^{m-1} \lambda^j Z_j$$

and

$$Q(\lambda) = \sum_{j=0}^{m-2} \lambda^j Q_j,$$

where

$$Q_j = -(BZ_{j+1}+ABZ_{j+2}+\cdots+A^{m-j-2}BZ_{m-1}), \quad j = 0,\ldots,m-2.$$

From

$$\sum_{j=0}^{m-1} A^j BZ_j = I$$

one easily deduces that for $|\lambda|$ sufficiently large

$$Q(\lambda)+(\lambda^{-1}I+\lambda^{-2}A+\lambda^{-3}A^2+\cdots)BR(\lambda) = \lambda^{-1}I+\lambda^{-2}A+\lambda^{-3}A^2+\cdots .$$

In other words,

$$Q(\lambda)+(\lambda-A)^{-1}BR(\lambda) = (\lambda-A)^{-1},$$

or

$$(\lambda-A)Q(\lambda)+BR(\lambda) = I. \tag{7.1.7}$$

As both sides of (7.1.7) are polynomials, the equality (7.1.7) holds for all $\lambda \in \mathbb{C}$, not only for $|\lambda|$ sufficiently large. Thus, $\begin{bmatrix} Q(\lambda) \\ R(\lambda) \end{bmatrix}$ is a right inverse of $[\lambda-A, B]$.

Conversely, assume (7.1.6) is right invertible for every complex number λ. By Theorem 7.1.3 there exist entire operator-valued functions $V(\lambda)$ and $U(\lambda)$ such that

$$(\lambda-A)V(\lambda)+BU(\lambda) = I, \quad \lambda \in \mathbb{C}.$$

In particular,

$$V(\lambda)+(\lambda-A)^{-1}BU(\lambda) = (\lambda-A)^{-1}, \quad |\lambda| > \|A\|. \tag{7.1.8}$$

Write $U(\lambda) = \sum_{j=0}^{\infty} \lambda^j U_j$. By integrating both left- and right-hand sides of (7.1.8) over a circle with center zero and radius $r > \|A\|$, we obtain

$$I = \frac{1}{2\pi i}\int_{|\lambda|=r} (\lambda-A)^{-1}d\lambda = \frac{1}{2\pi i}\int_{|\lambda|=r} (\lambda-A)^{-1}BU(\lambda)d\lambda$$

$$= \sum_{j=0}^{\infty} A^j BU_j.$$

Consequently, for m large enough (namely, such that $\|\sum_{j=m}^{\infty} A^j BU_j\| < 1$), the operator $\sum_{j=0}^{m-1} A^j BU_j$ is invertible, and (7.1.5) is right invertible. ∎

We mention also a result dual to Theorem 7.1.2.

THEOREM 7.1.4. *Let* $A \in L(\mathcal{Y})$ *and* $C \in L(\mathcal{Y},\mathcal{X})$ *be Banach space operators. Then the operator* $\mathrm{col}[CA^i]_{i=0}^{m-1} \in L(\mathcal{Y},\mathcal{X}^m)$ *is left invertible for some* m *if and only if the operator*

$$\begin{bmatrix} \lambda I-A \\ C \end{bmatrix} \in L(\mathcal{Y},\mathcal{Y}\oplus\mathcal{X})$$

is left invertible for all $\lambda \in \mathbb{C}$.

The proof of Theorem 7.1.4 is completely analogous to the proof of Theorem 7.1.2 and therefore is omitted.

7.2 Spectrum assignment theorems

Consider the system

$$x_n = Ax_{n-1}+Bu_{n-1}, \quad n = 1,2,\dots , \tag{7.2.1}$$

where $A \in L(\mathcal{Y})$, $B \in L(\mathcal{X},\mathcal{Y})$. An important problem in control is to bring about the desired behavior of the system by using state feedback, that is, by putting $u_n = Fx_n$, $n = 1,2,\dots$, where $F \in L(\mathcal{Y},\mathcal{X})$ is a suitable operator. The operator will be called the *feedback operator*. In particular, one is interested to find if there is a feedback operator F such that the system

(7.2.2) $x_n = Ax_{n-1}+Bu_{n-1}; \; u_{n-1} = Fx_{n-1}; \; n = 1,2,\dots$

is *stable*, i.e., the spectrum of $A+BF \in L(\mathcal{Y})$ lies in the open unit disc $\{\lambda \in \mathbb{C} \mid |\lambda| < 1\}$. This condition ensures that

$$\|(A+BF)^n\| \le \rho^n, \quad n = 1,2,\dots ,$$

where $0 < \rho < 1$ is independent of n, and hence the solution

$$x_n = (A+BF)^n x_0, \quad n = 1,2,\dots$$

of (7.2.2) tends to zero (as $n \to \infty$) with at least geometric rate ρ^n. More generally, it is of interest to find a feedback operator F such that A+BF has its spectrum in a prescribed set in the complex plane. Also, it is important to have some control on the behavior of F; for instance, it is desirable to keep the norm of F moderate.

It turns out that if the system (7.1.2) is exactly controllable and the spaces $\mathcal{X}$ and $\mathcal{Y}$ are Hilbert spaces, then by using suitable F one can make the spectrum of A+BF to be any prescribed non-empty compact set in the complex plane. Moreover, F can be chosen to depend continuously on A, B and the prescribed compact set. We make precise and prove this (and other related statements) in this section. Everywhere in this section it will be assumed that $\mathcal{X}$ and $\mathcal{Y}$ are Hilbert spaces.

Consider the set $\mathbb{C}_n$ of all n-tuples $\{\lambda_1,\dots,\lambda_n\}$ of complex numbers with repetitions allowed and two n-tuples obtained from each other by a permutation are considered the same element in $\mathbb{C}_n$. One defines naturally a metric on $\mathbb{C}_n$:

$$d_n(\{\lambda_1,\dots,\lambda_n\},\{\mu_1,\dots,\mu_n\}) = \inf_{\sigma} \sup_{1\le i\le n} |\lambda_i-\mu_{\sigma(i)}|,$$

where the infimum is taken over all permutations σ of the set $\{1,\dots,n\}$. Consider also the set $C(\mathbb{C})$ of all non-empty compact subsets of the complex plane with the usual Hausdorff metric (here $M, \Lambda \in C(\mathbb{C})$):

$$d(M,\Lambda) = \max\left\{\max_{y\in\Lambda}\min_{x\in M} |x-y|,\ \max_{y\in M}\min_{x\in\Lambda} |x-y|\right\}.$$

For $\lambda = \{\lambda_1,\dots,\lambda_n\} \in \mathbb{C}_n$ denote by $\|\lambda\|$ the quantity $\max_{1\le j\le n} |\lambda_j|$, and for $\Lambda \in C(\mathbb{C})$ put

$$\|\Lambda\| = \max_{z\in\Lambda} |z|.$$

THEOREM 7.2.1. *Let* $A \in L(\mathcal{Y})$, $B \in L(\mathcal{X},\mathcal{Y})$ *be such that*

$$\sum_{k=0}^{n-1} (A^k B)(\mathcal{X}) = \mathcal{Y}. \tag{7.2.3}$$

(a) *Let* $\lambda = \{\lambda_1,\dots,\lambda_n\} \in \mathbb{C}_n$. *Then there exist positive constants* ε *(depending on* A,B *only) and* K *(depending on* A,B *and* $\|\lambda\|$*) with the following property: For every pair* $A' \in L(\mathcal{Y})$, $B' \in L(\mathcal{X},\mathcal{Y})$ *such that*

$$\|A-A'\|+\|B-B'\| < \varepsilon, \tag{7.2.4}$$

and for every $\mu = \{\mu_1,\dots,\mu_n\} \in \mathbb{C}_n$, *there is a feedback* $F = F(A',B',\mu)$ *such that*

$$\prod_{j=1}^{n} (A'+B'F-\mu_j I) = 0$$

and the inequalities

$$\|F(A,B,\lambda)\| \le K \tag{7.2.5}$$

and

$$\|F(A',B',\mu)-F(A,B,\lambda)\| \le K(\|A'-A\|+\|B'-B\|+d_n(\lambda,\mu)) \tag{7.2.6}$$

hold for any $\mu \in \mathbb{C}_n$ *and any* A',B' *satisfying (7.2.4).*

(b) *Assume that* $\mathcal{Y}$ *is infinite-dimensional, and let* $\Lambda \in C(\mathbb{C})$. *Then there exist positive constants* ε *(depending on* A,B *only) and* K *(depending on* A,B *and* $\|\Lambda\|$*) with the following property: For every pair* A',B' *of operators satisfying (7.2.4) and for every* $M \in C(\mathbb{C})$, *there is a feedback* $F = F(A',B',M)$ *such that*

$$\sigma(A-BF) = M$$

and

(7.2.7) $$\|F(A,B,\Lambda)\| \le K,$$

(7.2.8) $$\|F(A',B',M)-F(A,B,\Lambda)\| \le K(\|A'-A\|+\|B'-B\|+d(\Lambda,M)).$$

The following simple proposition will be used in the proof of Theorem 7.2.1.

PROPOSITION 7.2.2. (a) *Assume* $\dim \mathcal{Y} \ge n$. *Given* $\lambda \in \mathbf{C}_n$, *for every* $\mu = \{\mu_1,\ldots,\mu_n\} \in \mathbf{C}_n$, *there exists a normal operator* $N(\mu) \in L(\mathcal{Y})$ *such that* $\sigma(N(\mu)) = \mu$ *and*

$$\|N(\mu)-N(\lambda)\| \le d_n(\lambda,\mu)$$

for any $\mu \in \mathbf{C}_n$.

(b) *Assume* $\dim \mathcal{Y} = \infty$. *Given* $\Lambda \in C(\mathbf{C})$, *for every* $M \in C(\mathbf{C})$ *there exists a normal operator* $N(M) \in L(\mathcal{Y})$ *such that* $\sigma(N(M)) = M$ *and*

$$\|N(M)-N(\Lambda)\| \le 2d(\Lambda,M)$$

for every $M \in C(\mathbf{C})$.

PROOF. We omit the proof of statement (a) (it can be done, assuming for simplicity that $\mathcal{Y}$ is finite dimensional, by taking $N(\mu)$ to be a diagonal matrix in a fixed orthonormal basis in $\mathcal{Y}$).

We prove part (b). Without loss of generality, we can assume that $\mathcal{Y}$ is separable (otherwise write $\mathcal{Y} = \mathcal{Y}_0 \oplus \mathcal{Y}_0^\perp$, where $\mathcal{Y}_0$ is separable, and take $N(M)$ in such a way that $\mathcal{Y}_0$ and $\mathcal{Y}_0^\perp$ are $N(M)$-invariant and $N(M)|_{\mathcal{Y}_0^\perp} = 0$). Let $\{\lambda_j\}_{j=1}^\infty$ be a sequence of (not necessarily different) complex numbers from Λ such that the closure of the set $\{\lambda_1,\lambda_2,\ldots\}$ is Λ itself.

We can assume that

$$\mathcal{Y} = \bigoplus_{i=0}^{\infty} \mathcal{Y}_i,$$

where $\mathcal{Y}_i$ is a Hilbert space with an orthonormal basis $\{e_j^{(i)}\}_{j=1}^\infty$. Define $N(\Lambda) \in L(\mathcal{Y})$ by $N(\Lambda)e_j^{(i)} = \lambda_j e_j^{(i)}$. Consider now $M \in C(\mathbf{C})$, which is different from Λ, so $d(M,\Lambda) > 0$. Let $y_1, y_2, \ldots$ be a countable dense subset in M. For each λ_j, let $x_j \in M$ be such that $|\lambda_j - x_j| = \min_{x\in M}|\lambda_j - x|$. For each y_i, let λ_{k_i} be such that

$$|y_i - \lambda_{k_i}| \le \min_{x\in\Lambda}|y_i - x| + d(\Lambda, M).$$

Form the operator $N(M)$ as follows:

$$N(M)e_j^{(0)} = x_j e_j^{(0)}, \quad j = 1,2,\ldots ;$$

$$N(M)e_{k_i}^{(i)} = y_i e_{k_i}^{(i)}, \quad i = 1,2,\ldots ;$$

$$N(M)e_j^{(i)} = x_j e_j^{(i)}, \quad \text{for } j \ne k_i \text{ and } i = 1,2,\ldots .$$

Clearly, $N(M)$ is normal, the spectrum of $N(M)$ is M and

$$\begin{aligned} \|N(M)-N(\Lambda)\| &\le \sup\{|\lambda_j - x_j|,\ j = 1,2,\ldots;\ |y_i - \lambda_{k_i}|,\ i = 1,2,\ldots\} \\ &\le 2d(M,\Lambda). \quad \blacksquare \end{aligned}$$

PROOF OF THEOREM 7.2.1. Apply induction on n. If $n = 1$, then (7.2.3) becomes $B\mathcal{X} = \mathcal{Y}$, i.e., B is right invertible. Choose ε so small that B' is right invertible for every $B': \mathcal{X} \to \mathcal{Y}$ satisfying $\|B'-B\| < 2\varepsilon$, and put

$$F(A',B',\mu) = B'^{-1}(N(\mu)-A'), \quad \mu \in \mathbb{C}_1$$

where $N(\mu)$ is taken from Proposition 7.2.2 and B'^{-1} is some right inverse of B' chosen so that $\|B'^{-1}-B^{-1}\| \le K_0\|B'-B\|$, where the constant K_0 depends on B only. The estimates (7.2.5) and (7.2.6) are easily verified. Indeed,

$$\|F(A,B,\lambda)\| \le \|B^{-1}\|(\|\lambda\|+\|A\|)$$

and

$$\begin{aligned}\|F(A',B',\mu)-F(A,B,\lambda)\| &\le \|B'^{-1}\|\cdot\|N(\mu)-N(\lambda)\|+\|B'^{-1}-B^{-1}\|\cdot\|N(\lambda)\| \\ &+ \|B'^{-1}\|\cdot\|A'-A\|+\|B'^{-1}-B^{-1}\|\cdot\|A\| \\ &\le (K_0\|B'-B\|+\|B^{-1}\|)d_n(\lambda,\mu)+K_0\|B'-B\|\cdot\|\lambda\| \\ &+ (K_0\|B'-B\|+\|B^{-1}\|)\|A'-A\|+K_0\|B'-B\|\cdot\|A\|,\end{aligned}$$

which proves (7.2.6).

Analogously one proves part (b) in case $n = 1$.

Assume now Theorem 7.2.1 is proved with n replaced by $n-1$. Let A,B be operators for which (7.2.3) holds. We show that there is a (closed) subspace $\mathcal{M}_1 \subset \mathcal{Y}$ such that

$$\sum_{k=0}^{n-1} A^k(\mathcal{M}_1) = \mathcal{Y} \tag{7.2.9}$$

and

$$BC = I_{\mathcal{M}_1} \tag{7.2.10}$$

for some operator $C \in L(\mathcal{M}_1,\mathcal{X})$.

Let

$$B^*B = \int_0^\infty t\, dE(t) \in L(\mathcal{X})$$

be the spectral resolution of the positive semidefinite operator B^*B, and let

$$B_\varepsilon = B(I-E(\varepsilon)),\ \varepsilon > 0.$$

By the property of spectral resolutions we have

$$\langle B^*BE(\varepsilon)x,E(\varepsilon)x\rangle \le \varepsilon\|E(\varepsilon)x\|^2 \le \varepsilon\|x\|^2,\ x \in \mathcal{X}$$

so

$$\|BE(\varepsilon)x\|^2 = \langle BE(\varepsilon)x,BE(\varepsilon)x\rangle \le \varepsilon\|x\|^2,\ x \in \mathcal{X},$$

and hence

$$\|B-B_\varepsilon\| = \|BE(\varepsilon)\| \le \sqrt{\varepsilon}.$$

Since in view of (7.2.3) the operator $[B, AB, \ldots, A^{n-1}B]$ is right invertible, for ε sufficiently close to zero the operator $[B_\varepsilon, AB_\varepsilon, \ldots, A^{n-1}B_\varepsilon]$ is right invertible as well, hence

$$\sum_{k=0}^{n-1} A^k B_\varepsilon(\mathcal{X}) = \mathcal{Y}.$$

Put $\mathcal{M}_1 = B_\varepsilon(\mathcal{X})$. As

$$\|B_\varepsilon x\|^2 = \langle B(I-E(\varepsilon))x, B(I-E(\varepsilon))x\rangle \geq \varepsilon\|x\|^2, \quad x \in \mathcal{X},$$

it follows that $\mathcal{M}_1$ is a closed subspace in $\mathcal{Y}$. Hence B_ε (considered as an operator $\mathcal{X} \to \mathcal{M}_1$) is invertible, so there is $\widetilde{C} \in L(\mathcal{M}_1, \mathcal{X})$ such that $B_\varepsilon\widetilde{C} = I_{\mathcal{M}_1}$. Putting $C = (I-E(\varepsilon))\widetilde{C}$, the equality (7.2.10) is satisfied.

Without loss of generality, we can assume $\mathcal{M}_1 \neq \mathcal{Y}$ (otherwise B is right invertible and we can repeat the proof given above for the case $n = 1$).

Choose $\varepsilon > 0$ so small that for every A', B' satisfying (7.2.4) we have

$$\sum_{k=0}^{n-1} A'^k(\mathcal{M}_1) = \mathcal{Y}$$

and

$$B'C' = I_{\mathcal{M}_1} \tag{7.2.11}$$

for some operator $C'\colon \mathcal{M}_1 \to \mathcal{X}$; and, moreover, the estimate

$$\|C-C'\| \leq K_0\|B'-B\|$$

holds, where the positive constant K_0 depends on B and $\mathcal{M}_1$ only.

With respect to the orthogonal decomposition $\mathcal{Y} = \mathcal{M}_1 \oplus \mathcal{M}_1^\perp$ write

$$A = \begin{bmatrix} A_{11} & A_{12} \\ A_{21} & A_{22} \end{bmatrix}.$$

Then (7.2.9) implies

$$\sum_{k=0}^{n-2} A_{22}^{k} A_{21}(\mathcal{M}_1) = \mathcal{M}_1^{\perp}.$$

So, applying the induction hypothesis, we find $K_1, \varepsilon_1 > 0$ such that for any operators A'_{22}, A'_{21} with

$$\|A'_{22}-A_{22}\|+\|A'_{21}-A_{21}\| < \varepsilon_1,$$

and any $\tilde{\mu} = \{\mu_1, \ldots, \mu_{n-1}\} \in \mathbb{C}_{n-1}$ there is an operator $D = D(A'_{22}, A'_{21}, \tilde{\mu}) \in L(\mathcal{M}_1^{\perp}, \mathcal{M}_1)$ such that

$$\prod_{j=1}^{n-1} (A'_{22}+A'_{21}D-\mu_j I) = 0 \tag{7.2.12}$$

and

$$\|D(A_{22}, A_{21}, \tilde{\lambda})\| \le K_1, \tag{7.2.13}$$

$$\|D(A'_{22}, A'_{21}, \tilde{\mu})-D(A_{22}, A_{21}, \tilde{\lambda})\| \le K_1(\|A'_{22}-A_{22}\|+\|A'_{21}-A_{21}\|+d_{n-1}(\tilde{\mu}, \tilde{\lambda})). \tag{7.2.14}$$

Here $\tilde{\lambda} = \{\lambda_1, \ldots, \lambda_{n-1}\}$ and the entries $\lambda_1, \ldots, \lambda_n$ are enumerated so that $\|\tilde{\lambda}\| = \|\lambda\|$.

Now, given A', B' satisfying (7.2.4), write

$$A' = \begin{bmatrix} A'_{11} & A'_{12} \\ A'_{21} & A'_{22} \end{bmatrix}$$

with respect to the orthogonal decomposition $\mathcal{Y} = \mathcal{M}_1 \oplus \mathcal{M}_1^{\perp}$, and for $\mu = \{\mu_1, \ldots, \mu_n\} \in \mathbb{C}_n$ put

$$F(A', B', \mu) = [C'(DA'_{21}+\mu_{\sigma(n)} I_{\mathcal{M}_1}-A'_{11}), C'(DA'_{22}-\mu_{\sigma(n)} D-A'_{12})]:$$

$$\mathcal{M}_1 \oplus \mathcal{M}_1^{\perp} \to \mathcal{X},$$

where C' is taken from (7.2.11), $D = D(A'_{22}, A'_{21}, \tilde{\mu})$, $\tilde{\mu} = \{\mu_{\sigma(1)}, \ldots, \mu_{\sigma(n-1)}\}$, and the permutation σ of $\{1,\ldots,n\}$ is chosen so that

$$d_n(\lambda,\mu) = \sup_{1\le i\le n} |\lambda_i - \mu_{\sigma(i)}|.$$

This choice of σ ensures that

$$d_{n-1}(\tilde{\lambda},\tilde{\mu}) \le d_n(\lambda,\mu) \text{ and } d_1(\lambda_n, \mu_{\sigma(n)}) \le d_n(\lambda,\mu).$$

Now we have

$$(7.2.15) \qquad \prod_{j=1}^{n} (A'+B'F(A',B',\mu)-\mu_{\sigma(j)}I] = 0.$$

Indeed, using the definition of $F(A',B',\mu)$, the left-hand side of (7.2.15) is

$$\left\{\prod_{j=1}^{n-1}\begin{bmatrix} DA'_{21}+(\mu_{\sigma(n)}-\mu_{\sigma(j)})I & DA'_{22}-\mu_{\sigma(n)}D \\ A'_{21} & A'_{22}-\mu_{\sigma(j)}I \end{bmatrix}\right\} \begin{bmatrix} DA'_{21} & DA'_{22}-\mu_{\sigma(n)}D \\ A'_{21} & A'_{22}-\mu_{\sigma(n)}I \end{bmatrix}$$

$$= \begin{bmatrix} I & D \\ 0 & I \end{bmatrix} \left\{\prod_{j=1}^{n-1}\begin{bmatrix} (\mu_{\sigma(n)}-\mu_{\sigma(j)})I & 0 \\ A'_{21} & A'_{22}+A'_{21}D-\mu_{\sigma(j)}I \end{bmatrix}\right\}$$

$$\cdot \begin{bmatrix} 0 & 0 \\ A'_{21} & A'_{22}+A'_{21}D-\mu_{\sigma(n)}I \end{bmatrix} \begin{bmatrix} I & -D \\ 0 & I \end{bmatrix}$$

which is easily seen to be zero in view of (7.2.12). Finally, using (7.2.13) and (7.2.14), one verifies the estimates (7.2.5) and (7.2.6). For example:

$$\|C'[D(A'_{22},A'_{21},\tilde{\mu})A'_{21}-\mu_{\sigma(n)}I_{\mathcal{M}_1}-A'_{11}]-C[D(A_{22},A_{21},\tilde{\lambda})A_{21}-\lambda_n I_{\mathcal{M}_1}-A_{11}]\|$$

$$\le \|C'\| \; \|D(A'_{22},A'_{21},\tilde{\mu})-D(A_{22},A_{21},\tilde{\lambda})\| \; \|A'_{21}\|$$

$$+ \|C'-C\| \; \|D(A_{22},A_{21},\tilde{\lambda})\| \; \|A'_{21}\| + \|C\| \; \|D(A_{22},A_{21},\tilde{\lambda})\| \; \|A'_{21}-A_{21}\|$$

$$+ \|C'\|\cdot|\mu_{\sigma(n)}-\lambda_n| + \|C'-C\|\cdot|\lambda_n| + \|C'\|\cdot\|A'_{11}-A_{11}\| + \|C'-C\| \; \|A_{11}\|.$$

This proves the part (a).

For the part (b), assume that $\mathcal{Y}$ is infinite dimensional. Then in view of (7.2.9), $\mathcal{M}_1$ is infinite dimensional as well. By Proposition 7.2.2(b), for any $M \in C(\mathbb{C})$ choose a normal operator $N(M)$ such that $\sigma(N(M)) = M$ and

$$\|N(M)-N(\Lambda)\| \le 2d(\Lambda,M).$$

Put

$$F(A',B',M) = [C'(DA'_{21}+N(M)-A'_{11}),C'(DA'_{22}-N(M)D-A'_{12}]:$$

$$\mathcal{M}_1 \oplus \mathcal{M}_1^{\perp} \to \mathcal{X},$$

where $D = D(A'_{22},A'_{21},M)$ is an operator which exists by the induction hypothesis. Then $A'+B'F(A',B',M)$ is similar to

$$T = \begin{bmatrix} N(M) & 0 \\ A'_{21} & A'_{22}+A'_{21}D \end{bmatrix}$$

(cf. the proof of (7.2.15)). As $\sigma(N(M)) = M$, $\sigma(A'_{22}+A'_{21}D) = M$ and $N(M)$ is normal, we have that $\sigma(T) = M$. The estimates (7.2.5) and (7.2.6) are proved in the same way as in the part (a).

Theorem 7.2.1 is proved. ∎

It turns out that the converse of Theorem 7.2.1 is true as well: spectrum assignability by feedback implies exact controllability. Actually, rather weak assumptions in terms of spectrum assignability allow one to deduce that the system is exactly controllable:

THEOREM 7.2.3. *Let $A \in L(\mathcal{Y})$ and $B \in L(\mathcal{X},\mathcal{Y})$ be Hilbert space operators. The following statements are equivalent:*

(i) *the system*

$$x_n = Ax_{n-1}+Bu_{n-1}; \quad n = 1,2,\ldots$$

is exactly controllable;

(ii) *there exist $F',F'' \in L(\mathcal{Y},\mathcal{X})$ such that*

$$\sigma(A+BF') \cap \sigma(A+BF'') = \phi;$$

(iii) *for each $\lambda \in \mathbb{C}$ there is $F_\lambda \in L(\mathcal{Y},\mathcal{X})$ such that $\lambda \notin \sigma(A+BF_\lambda)$.*

PROOF. The implications (i) → (ii) and (i) → (iii) follow from Theorem 7.2.1.

(ii) → (iii) is obvious, because one can take F_λ to be either F' or F''.

Assume now (iii) holds. Then for a fixed $\lambda \in \mathbb{C}$:

$$\mathcal{Y} = (A+BF_\lambda-\lambda I)(\mathcal{Y}) \subset (A-\lambda I)(\mathcal{Y})+BF_\lambda(\mathcal{Y}) \subset (A-\lambda I)(\mathcal{Y})+B(\mathcal{X}).$$

Thus

$$\operatorname{Im}[A-\lambda I,B] = \mathcal{Y}$$

for all $\lambda \in \mathbb{C}$, and it remains to appeal to Theorem 7.1.2. ■

A result dual to Theorem 7.2.1 holds for pairs of Hilbert space operators (C,A), where $A \in L(\mathcal{Y})$ and $C \in L(\mathcal{Y},\mathcal{X})$, such that the operator $\operatorname{col}[CA^i]_{i=0}^{n-1} \in L(\mathcal{Y},\mathcal{X}^n)$ is left invertible. We will not state this result explicitly; it can be obtained by applying Theorem 7.2.1 with A and B replaced by A^* and C^*, respectively. We do state a result dual to Theorem 7.2.3.

THEOREM 7.2.4. *The following statements are equivalent for Hilbert space operators* $A \in L(\mathcal{Y})$, $C \in L(\mathcal{Y},\mathcal{X})$.

(i) *the operator* $\operatorname{col}[CA^i]_{i=0}^{n-1}$ *is left invertible for some* n;

(ii) *there exist* $F',F'' \in L(\mathcal{X},\mathcal{Y})$ *such that*

$$\sigma(A+F'C) \cap \sigma(A+F''C) = \phi;$$

(iii) *for every* $\lambda \in \mathbb{C}$ *there is* $F_\lambda \in L(\mathcal{X},\mathcal{Y})$ *such that* $\lambda \notin \sigma(A+F_\lambda C)$.

The proof is analogous to the proof of Theorem 7.2.3.

We pose three open problems related to Theorem 7.2.1.

PROBLEM 7.2.1. *Is Theorem 7.2.1 valid in case both* $\mathcal{X}$ *and* $\mathcal{Y}$ *are Banach spaces*? In this framework one should replace (7.2.3) by the right invertibility of the operator

$$[B,AB,\ldots,A^{n-1}B].$$

PROBLEM 7.2.2. *Obtain more detailed information on the possible spectral properties of the operator* A+BF. A good model is provided by the following theorem in the finite-dimensional case (Rosenbrock's theorem): Let A and B be matrices of sizes $n\times n$ and $n\times p$, respectively, and assume that (A,B) is exactly controllable. Let $\lambda_1 \le \cdots \le \lambda_m$ be the minimal column indices of the rectangular matrix $[\lambda I-A,B]$ (in the sense of the Kronecker canonical form; e.g., Gantmacher [1], Gohberg-Lancaster-Rodman [3]). Then the following inequalities are necessary and sufficient for existence of a $p \times n$ matrix F with the invariant polynomials $\phi_1(\lambda), \ldots, \phi_n(\lambda)$ of A-BF:

$$\phi_1(\lambda) \equiv \cdots \equiv \phi_{n-m}(\lambda) \equiv 1;$$

$$\sum_{i=1}^{k} \text{degree}(\phi_{n-m+i}(\lambda)) \leq \sum_{i=1}^{k} \lambda_i; \quad k = 1, \ldots, m;$$

$$\sum_{i=1}^{m} \text{degree}(\phi_{n-m+i}(\lambda)) = \sum_{i=1}^{m} \lambda_i .$$

Recall that the invariant polynomials $\phi_1(\lambda), \ldots, \phi_n(\lambda)$ of an $n \times n$ matrix X are defined by the formulas $\phi_i(\lambda) = D_i(\lambda)/D_{i-1}(\lambda)$, $i = 1, \ldots, n$, where $D_0(\lambda) \equiv 1$ and for $j = 1, \ldots, n$ $D_j(\lambda)$ is the greatest monic common divisor of all $j \times j$ minors of $\lambda I - X$.

PROBLEM 7.2.3. *Study spectrum assignment problems with the hypothesis of exact controllability removed.* Here the set

$$\Lambda_0 = \{\lambda \in \mathbf{C} \mid [\lambda - A, B] \text{ is not right invertible}\}$$

should be important (cf. Theorem 7.1.2). For instance, is the following true: For every compact set M such that $M \supset \Lambda_0$ there is $F \in L(\mathcal{X}, \mathcal{Y})$ with $\sigma(A+BF) = M$?

7.3 Analytic dependence of the feedback

We have seen in the preceding section that an exactly controllable system admits spectrum assignability by feedback, and, moreover, the required feedback can be chosen to depend continuously on the system and on the prescribed spectrum. Here this problem is studied in the presence of analytic behavior of the exactly controllable system.

We set up the framework. As in Section 7.2, $\mathcal{X}$ and $\mathcal{Y}$ are assumed to be Hilbert spaces. Let $\Omega \subset \mathbf{C}$ be an open set, and let $A: \Omega \to L(\mathcal{Y})$, $B: \Omega \to L(\mathcal{X}, \mathcal{Y})$ be operator-valued functions which are analytic in Ω. Everywhere in this section it will be assumed that the system

$$x_n = A(z)x_{n-1}+B(z)u_{n-1}, \quad n = 1,2,\dots$$

is exactly controllable for every $z \in \Omega$.

THEOREM 7.3.1. *Let* $z_0 \in \Omega$, *and let* n *be such that*

$$\sum_{k=0}^{n-1} (A(z_0))^k B(z_0)(\mathcal{X}) = \mathcal{Y}.$$

Then for every n*-triple of scalar analytic functions* $\lambda_1(z),\dots,\lambda_n(z)$ *defined on an open neighborhood* U *of* z_0, *there is an analytic operator function* $F: V \to L(\mathcal{Y},\mathcal{X})$, *where* $V \subset U$ *is an open neighborhood of* z_0 *such that*

$$\prod_{j=1}^{n} (A(z)+B(z)F(z)-\lambda_j(z)I) = 0$$

for all $z \in V$.

PROOF. We mimic the ideas used in the proof of Theorem 7.2.1. Proceed by induction on n, and consider the case $n = 1$. Then $B(z)$ is right invertible for all $z \in U_1$, where $U_1 \subset U$ is some open neighborhood of z_0. Let $D(z_0)$ be some right inverse of $B(z_0)$ and define

$$V = \{z \in U_1 \mid \|B(z)-B(z_0)\| \le \tfrac{1}{2}\|D(z_0)\|^{-1}\}.$$

For every $z \in V$ let

$$D(z) = [\sum_{n=0}^{\infty} (-1)^n (D(z_0)(B(z)-B(z_0))^n]D(z_0).$$

It is easily seen that $D(z)$ is a right inverse of $B(z)$ for every $z \in V$, and that $D(z)$ is analytic on V (the latter statement follows because $D(z)$ is the limit of a sequence of analytic operator functions on V in the topology of uniform convergence on every compact subset in V). Put

$$F(z) = D(z)(\lambda_1(z)I-A(z)), \quad z \in V$$

to satisfy the requirements of Theorem 7.3.1.

Assume now Theorem 7.3.1 is proved with n replaced by n-1. As in the proof of Theorem 7.2.1, find a subspace $\mathcal{M} \subset \mathcal{Y}$ such that

$$\sum_{k=0}^{n-1} (A(z_0))^k (\mathcal{M}_1) = \mathcal{Y} \tag{7.3.1}$$

and

$$B(z_0)C = I_{\mathcal{M}_1} \tag{7.3.2}$$

for some $C \in L(\mathcal{M}_1, \mathcal{X})$. Let us examine the equality (7.3.2). As C is left invertible, Im C is a (closed) subspace. Write B(z) in the 2×2 block operator matrix form:

$$B(z) = \begin{bmatrix} B_{11}(z) & B_{12}(z) \\ B_{21}(z) & B_{22}(z) \end{bmatrix} : (\text{Im } C) \oplus (\text{Im } C)^{\perp} \to \mathcal{M}_1 \oplus \mathcal{M}_1^{\perp}$$

(so, for instance, $B_{11}(z) \in L(\text{Im } C, \mathcal{M}_1)$). The equality (7.3.2) implies that $B_{21}(z_0) = 0$ and that $B_{11}(z_0)$ is invertible (its inverse is C considered as an operator from $\mathcal{X}$ onto Im C). So $B_{11}(z)$ is invertible for all $z \in V_1$, where V_1 is a sufficiently small neighborhood of z_0. Let

$$S(z) = \begin{bmatrix} I_{\mathcal{M}_1} & 0 \\ -B_{21}(z)B_{11}(z)^{-1} & I_{\mathcal{M}_1^{\perp}} \end{bmatrix} : \mathcal{M}_1 \oplus \mathcal{M}_1^{\perp} \to \mathcal{M}_1 \oplus \mathcal{M}_1^{\perp}, \quad z \in V_1.$$

Obviously, S(z) is invertible, and $S(z)B(z) = \begin{bmatrix} * & * \\ 0 & * \end{bmatrix}$. Replacing A(z) by $S(z)A(z)S(z)^{-1}$ and B(z) by S(z)B(z), we can assume that $B_{21}(z)$ is zero for $z \in V_1$. This means

$$B(z)C = I_{\mathcal{M}_1}, \quad z \in V_1.$$

Taking V_1 smaller if necessary, we can ensure that also

$$\sum_{k=0}^{n-1} (A(z))^k (\mathcal{M}_1) = \mathcal{Y}, \quad z \in V_1.$$

Now we repeat the construction given in the proof of Theorem 7.2.1. ■

An analogous result holds, of course, for analytic operator functions $A(z): \Omega \to L(\mathcal{Y})$ and $C(z): \Omega \to L(\mathcal{Y},\mathcal{X})$ such that the operator $\mathrm{col}[C(z_0)A(z_0)^i]_{i=0}^{n-1} \in L(\mathcal{Y},\mathcal{X}^n)$ is left invertible for some $z_0 \in \Omega$.

7.4 Polynomials with given spectral pairs

We come back now to the problem of existence of operator polynomials with given right spectral pairs. Again, it will be assumed in this section that $\mathcal{X}$ and $\mathcal{Y}$ are Hilbert spaces.

THEOREM 7.4.1. *Let* $A \in L(\mathcal{Y})$ *and* $C \in L(\mathcal{Y},\mathcal{X})$. *There exists an operator polynomial* $L(\lambda)$ *with coefficients in* $L(\mathcal{X})$ *and of degree* $\leq \ell$ *such that* (C,A) *is a right spectral pair of* $L(\lambda)$ *(on the whole complex plane) if and only if the operator*

$$\mathrm{col}[CA^i]_{i=0}^{\ell-1} \in L(\mathcal{Y},\mathcal{X}^\ell) \tag{7.4.1}$$

is left invertible. In fact, if (7.4.1) is left invertible and assuming (without loss of generality) that A *is invertible,* $L(\lambda)$ *can be taken in the form*

$$L(\lambda) = I - CA^{-1}(\lambda^{-1} - (A^{-1} - BCA^{-1}))^{-1}B, \tag{7.4.2}$$

where $B \in L(\mathcal{X},\mathcal{Y})$ *is any operator such that* $(A^{-1}-BCA^{-1})^\ell = 0$.

PROOF. If (C,A) is a right $\mathbb{C}$-spectral pair of an operator polynomial of degree $\leq \ell$, then (7.4.1) is left invertible by Theorem 6.6.1.

Assume now that (7.4.1) is left invertible. Let us explain why we can assume invertibility of A without loss of generality. This follows from the following easily verified observation: If $\alpha \in \mathbb{C}$ and $(C,A-\alpha I,B)$ is a $\mathbb{C}$-spectral triple of an operator polynomial $L(\lambda)$, then the triple (C,A,B) is a $\mathbb{C}$-spectral triple of an operator polynomial $L(\lambda-\alpha)$. So it will be assumed that A is invertible.

Let $B \in L(\mathcal{X},\mathcal{Y})$ be any operator for which

$$(A^{-1}-BCA^{-1})^{\ell} = 0$$

(the existence of B is ensured by the dual to Theorem 7.2.1; observe that

$$\mathrm{col}[CA^{-1}\cdot(A^{-1})^{i}]_{i=0}^{\ell-1}$$

is left invertible together with (7.4.1)). Define $L(\lambda)$ by the formula (7.4.2). First of all, observe that $L(\lambda)$ is a polynomial of degree $\leq \ell$. Indeed, denoting $F = A^{-1}-BCA^{-1}$, we have

$$(\lambda^{-1}-F)^{-1} = \lambda(I-\lambda F)^{-1} = \lambda(I+\lambda F+\cdots+\lambda^{\ell-1}F^{\ell-1}),$$

and our assertion follows. Next we verify that $(C,A,-AB)$ is a spectral triple for L on $\mathbb{C}$. The property (P1) holds trivially while (P4) is ensured by the left invertibility of (7.4.1). Further,

$$\begin{aligned} L(\lambda)C(\lambda-A)^{-1} &= C(\lambda-A)^{-1}-CA^{-1}(\lambda^{-1}-F)^{-1}BC(\lambda-A)^{-1} \\ &= CA^{-1}(\lambda^{-1}-F)^{-1}[(\lambda^{-1}-F)A-BC](\lambda-A)^{-1} \\ &= CA^{-1}(\lambda^{-1}-F)^{-1}[\lambda^{-1}A-(A^{-1}-BCA^{-1})A-BC](\lambda-A^{-1}) \\ &= CA^{-1}(\lambda^{-1}-F)^{-1}(\lambda^{-1}A-I)(\lambda-A^{-1}) \\ &= -\lambda^{-1}CA^{-1}(\lambda^{-1}-F)^{-1}, \end{aligned}$$

where in the last but one equality we have used the definition of $F = A^{-1}-BCA^{-1}$. Clearly, $L(\lambda)C(\lambda-A)^{-1}$ is analytic in $\mathbb{C}$, and (P3′) holds true. To verify (P2), compute first $L(\lambda)^{-1}$:

$$L(\lambda)^{-1} = I+CA^{-1}(\lambda^{-1}-A^{-1})^{-1}B, \tag{7.4.3}$$

where $\lambda \neq 0$ is such that $\lambda^{-1} \notin \sigma(A^{-1})$. This formula is verified by multiplying out (7.4.2) and (7.4.3). Now

$$L(\lambda)^{-1}+C(\lambda-A)^{-1}AB = I+\lambda CA^{-1}(A-\lambda)^{-1}AB+C(\lambda-A)^{-1}AB$$

$$= I+C(I-\lambda A^{-1})(\lambda-A)^{-1}AB = I-CB,$$

which is analytic (even constant) in $\mathbb{C}$. Hence (P2) is satisfied. ∎

The proof of Theorem 7.4.1 reveals additional properties of the operator polynomial (7.4.2): namely, $L(\lambda)^{-1}$ is analytic at infinity and $[L(\lambda)^{-1}]_{\lambda=\infty} = I-CB$.

We consider now monic operator polynomials. As we have seen in Chapter 2, a necessary and sufficient condition for existence of a monic operator polynomial of degree ℓ with given right spectral pair (C,A) is that the operator

$$\operatorname{col}[CA^i]_{i=0}^{\ell-1} \tag{7.4.4}$$

is invertible. Failing that, and assuming only one-sided invertibility of (7.4.4), a procedure was given in Section 2.4 to construct monic polynomials using the one-sided inverses. However, using this procedure, it is difficult to keep track of the spectral properties of the obtained polynomial. An alternative way to construct a monic operator polynomial for which (C,A) is a part of its right spectral pair, where the operators C and A are such that (7.4.4) is left invertible, is based on the following theorem. This approach allows one to obtain much more information about the spectral structure of the constructed polynomial, as we shall see shortly.

THEOREM 7.4.2. *Let* $A \in L(\mathcal{Y})$ *and* $C \in L(\mathcal{Y},\mathcal{X})$ *be Hilbert space operators such that* $\operatorname{col}[CA^i]_{i=0}^{\ell-1}$ *is left invertible. Then for every* $\alpha \notin \sigma(A)$ *there exists a Hilbert space* $\mathcal{Y}_0$ *and* $A_0 \in L(\mathcal{Y}_0)$, $C_0 \in L(\mathcal{Y}_0,\mathcal{X})$ *such that* $\sigma(A_0) = \{\alpha\}$ *and the operator*

$$\begin{bmatrix} C & C_0 \\ CA & C_0A_0 \\ \vdots & \vdots \\ CA^{\ell-1} & C_0A_0^{\ell-1} \end{bmatrix} \in L(\mathcal{Y}\oplus\mathcal{Y}_0,\mathcal{X}^\ell) \tag{7.4.5}$$

is invertible. If, in addition, $\mathcal{X}$ is a separable Hilbert space, then one can take A_0 *with the additional property that* $(A_0-\alpha I)^p = 0$ *for some integer* $p > 0$.

PROOF. Without loss of generality, we can assume that A is invertible and $\alpha = 0$. Let

$$L(\lambda) = I-CA(\lambda^{-1}-(A-BCA))^{-1}B,$$

where B is such that $(A-BCA)^{\ell} = 0$ (cf. formula (7.4.2)). By Theorem 7.4.1, (C,A^{-1}) is a right $\mathbb{C}$-spectral pair for $L(\lambda)$. Denote

$$\tilde{L}(\lambda) = \lambda^{\ell}L(\lambda^{-1}).$$

The operator polynomial $\tilde{L}(\lambda)$ is monic of degree ℓ, and by Theorem 6.7.2 (C,A) is a right spectral pair of $\tilde{L}$ on the open set $\mathbb{C}\setminus\{0\}$. Let (C_0,A_0) be a right spectral pair of $\tilde{L}$ on a sufficiently small neighborhood U of zero (so that $\Sigma(\tilde{L}) \cap U = \{0\}$; we exclude the trivial case when $\tilde{L}(0)$ is invertible, because in this case one can take $\mathcal{Y}_0$ to be the zero space). By Theorem 6.2.8,

$$\left([C \quad C_0], \begin{bmatrix} A & 0 \\ 0 & A_0 \end{bmatrix}\right)$$

is a right $\mathbb{C}$-spectral pair of $\tilde{L}(\lambda)$. It remains to use the definition of a right spectral pair for monic polynomials given in Section 2.1.

Suppose now, in addition, that $\mathcal{X}$ is a separable Hilbert space. As for some C_0,A_0 the operator (7.4.5) is invertible, $\mathcal{Y}_0$ is a separable as well. We use now the following fact: *If* A_0 *is an operator acting on a separable Hilbert space and* $\sigma(A_0) = \{0\}$, *then there is a sequence* $\{T_n\}_{n=1}^{\infty}$ *of nilpotent operators such that* $\lim_{n\to\infty} \|T_n-A_0\| = 0$. Proof of this fact is beyond the scope of this book, and the reader is referred to Apostol-Foias-Pearcy [1] and Apostol-Voiculescu [1] for the proof (see also Halmos [1]). Using this fact, one obtains invertibility of the operator

$\mathrm{col}[CA^i, C_0T_n^i]_{i=0}^{\ell-1}$ for n large enough, and Theorem 7.4.2 is proved. ∎

The proof of Theorem 7.4.2 shows that given a pair (C,A) with left invertible $\mathrm{col}[CA^i]_{i=0}^{\ell-1}$, for any $\alpha \notin \sigma(A)$ one can construct a monic operator polynomial $L(\lambda)$ whose right spectral pair on $\mathbb{C}\backslash\{\alpha\}$ coincides with (C,A). Thus, the spectral structure of $L(\lambda)$ that is additional to the pair (C,A) is concentrated at one single point α.

Using the description of divisibility in terms of spectral pairs given in Section 6.5, we derive from Theorem 7.4.2 the following interesting fact.

THEOREM 7.4.3. *Let $L(\lambda)$ be an operator polynomial with coefficients from $L(\mathcal{X})$ and of degree $\leq \ell$, with compact spectrum $\Sigma(L)$. Then for every $\alpha \notin \Sigma(L)$ there exist a monic operator polynomial $M(\lambda)$ of degree ℓ and an entire operator-valued function $L_1(\lambda)$ with $\Sigma(L_1) = \{\alpha\}$ such that $M(\lambda) = L_1(\lambda)L(\lambda)$. If $\mathcal{X}$ is separable, then $L_1(\lambda)$ can be chosen with the additional property that α is a pole for $L_1(\lambda)^{-1}$, i.e., $(\lambda-\alpha)^pL_1(\lambda)^{-1}$ is analytic at α for some integer p.*

PROOF. Let (C,A) be a right $\mathbb{C}$-spectral pair of $L(\lambda)$. Then $\mathrm{col}[CA^i]_{i=0}^{\ell-1}$ is left invertible by Theorem 6.6.1. Using Theorem 7.4.2, let $M(\lambda)$ be a monic operator polynomial of degree ℓ whose right $\mathbb{C}$-spectral pair is $([C \;\; C_0], A \oplus A_0)$, and (C_0, A_0) satisfy the requirements of Theorem 7.4.2. By Theorem 6.5.2, $L(\lambda)$ is a right divisor of $M(\lambda)$ on the whole complex plane: $M(\lambda) = L_1(\lambda)L(\lambda)$, where $L_1(\lambda)$ is entire operator-valued function. It is easy to see that (C,A) is a right spectral pair of $M(\lambda)$, as well as of $L(\lambda)$, on $\mathbb{C}\backslash\{\alpha\}$. Now Corollary 6.5.4 implies that $L_1(\lambda)$ is invertible for $\lambda \neq \alpha$.

Assume, in addition, that $\mathcal{X}$ is separable, and choose A_0 in Theorem 7.4.2 so that $(A_0-\alpha I)^p = 0$ for some p. Write

$$L_1(\lambda)^{-1} = L(\lambda)M(\lambda)^{-1} = L(\lambda)[C \quad C_0]\begin{bmatrix}\lambda-A & 0\\ 0 & \lambda-A_0\end{bmatrix}^{-1}\begin{bmatrix}B\\ B_0\end{bmatrix},$$

where $\begin{bmatrix} B \\ B_0 \end{bmatrix}$ is such that

$$\left([C \quad C_0], \begin{bmatrix} A & 0 \\ 0 & A_0 \end{bmatrix}, \begin{bmatrix} B \\ B_0 \end{bmatrix} \right)$$

is a $\mathbb{C}$-spectral triple for $M(\lambda)$, and we have used Theorem 2.5.2. So

$$L_1(\lambda)^{-1} = L(\lambda)C(\lambda-A)^{-1}B+L(\lambda)C_0((\lambda-\alpha)-(A_0-\alpha I))^{-1}B_0$$

$$= L(\lambda)C(\lambda-A)^{-1}B+L(\lambda)C_0 \sum_{j=0}^{p-1} (\lambda-\alpha)^{-j-1}(A_0-\alpha I)^j B_0,$$

and α is indeed a pole for $L_1(\lambda)^{-1}$. ∎

In general, the function $L_1(\lambda)$ from Theorem 7.4.3 need not be a polynomial.

All the results of this section have dual statements applicable for left spectral pairs (A,B) (in place of the right ones). We omit these statements, as they are gotten by applying the results proved in this section to the pair (B^*,A^*).

7.5 Invariant subspaces and divisors

We have seen in Section 6.5 that divisibility of operator polynomials can be described in terms of restrictions of spectral pairs. Here we carry further this correspondence and describe divisibility in terms of invariant subspaces of a certain operator. As in the preceding sections, it will be assumed here that $\mathcal{X}$ is a Hilbert space.

Let $L(\lambda)$ be an operator polynomial with coefficients in $L(\mathcal{X})$ and such that the spectrum

$$\Sigma(L) = \{\lambda \in \mathbb{C} \mid L(\lambda) \text{ is not invertible}\}$$

is compact. Let $A \in L(\mathcal{Y})$, where $\mathcal{Y}$ is a suitable Hilbert space, be a global linearization of $L(\lambda)$. That is,

$$\begin{bmatrix} L(\lambda) & 0 \\ 0 & I_{\mathcal{Z}_1} \end{bmatrix} = E(\lambda) \begin{bmatrix} \lambda I - A & 0 \\ 0 & I_{\mathcal{Z}_2} \end{bmatrix} F(\lambda), \ \lambda \in \mathbb{C},$$

where $\mathcal{Z}_1$ and $\mathcal{Z}_2$ are some Hilbert spaces, and $F(\lambda)$ and $E(\lambda)$ are entire operator functions whose values are invertible operators for every $\lambda \in \mathbf{C}$. The existence of a global linearization follows from Corollary 6.4.2, and its uniqueness (up to similarity) is ensured by Theorem 1.2.1.

Recall that an operator polynomial $L_1(\lambda)$ is called a *right divisor* of $L(\lambda)$ (on $\mathbf{C}$) if $L(\lambda) = L_2(\lambda)L_1(\lambda)$ for some entire operator function $L_2(\lambda)$. We say that two right divisors $L_1(\lambda)$ and $\widetilde{L}_1(\lambda)$ of $L(\lambda)$ *belong to the same class* if $L_1\widetilde{L}_1^{-1}$ and $\widetilde{L}_1L_1^{-1}$ are both entire operator functions. Clearly, the relation of belonging to the same class is symmetric, reflexive and transitive; so it is an equivalence relation. There is also a natural inclusion relation between the classes of right divisors: Given classes A_1 and A_2, we say that $A_1 \subset A_2$ if the operator function $L_2L_1^{-1}$ is entire, where L_1 is a representative of A_1 and L_2 is a representative of A_2 (clearly the definition of $A_1 \subset A_2$ does not depend on choice of the representatives L_1 and L_2). Thus, the set of all classes of right divisors of $L(\lambda)$ becomes a lattice. This lattice has the supremal element which is the class containing L itself; it has the infimal element which the class containing the constant polynomial I.

THEOREM 7.5.1. *The lattice of the classes of right divisors of* $L(\lambda)$ *is isomorphic to the lattice of all* A*-invariant subspaces, where* **A** *is a global linearization of* $L(\lambda)$.

The isomorphism is established as follows: Fix $C \in L(\mathcal{Y},\mathcal{X})$ *such that* (C,A) *is a right* $\mathbf{C}$*-spectral pair of* $L(\lambda)$. *Let* $\mathcal{M}$ *be an* A*-invariant subspace. Then a right divisor* $M_0(\lambda)$ *of* $L(\lambda)$ *that belongs to the class of divisors which corresponds to* $\mathcal{M}$ *can be found by the formula*

$$(7.5.1) \qquad M_0(\lambda) = I-\widetilde{C}(\widetilde{A}-\alpha I)^{-1}((\lambda-\alpha)^{-1}-((\widetilde{A}-\alpha I)^{-1}-\widetilde{B}\widetilde{C}(\widetilde{A}-\alpha I)^{-1})^{-1}\widetilde{B}.$$

Here $\widetilde{A} = A_{|\mathcal{M}}$; $\widetilde{C} = C_{|\mathcal{M}}$; α *is any complex number not in the spectrum of* $\widetilde{A}$; $\widetilde{B}$ *is any operator such that*

$$((\tilde{A}-\alpha I)^{-1}-\tilde{B}\tilde{C}(\tilde{A}-\alpha I)^{-1})^{\ell} = 0,$$

and ℓ *is the degree of* L.

Conversely, let $M(\lambda)$ *be a right divisor of* $L(\lambda)$*, so* $L(\lambda) = N(\lambda)M(\lambda)$ *for some entire operator-valued function* $N(\lambda)$*, and fix* $\alpha \in \mathbb{C}$ *in the unbounded component of the resolvent set of* A *(so that* α *is in the resolvent of the restriction of* A *to any invariant subspace). Then there is unique* A*-invariant subspace* $\mathcal{M}$ *for which the operator polynomial* $M_0(\lambda)$ *given by (7.5.1) is a right divisor of* $L(\lambda)$ *belonging to the same class as* $M(\lambda)$.

The proof is obtained by combining Theorems 7.4.1 and 6.5.2.

There is a dual version of Theorem 7.5.1 concerning the classes of left divisors of $L(\lambda)$. It can be obtained from Theorem 7.5.1 by passing to the operator polynomial whose coefficients are adjoints of the coefficients of $L(\lambda)$. We leave it to the reader the statement and proof of the result dual to Theorem 7.5.1.

7.6 Exercises

Ex. 7.1. Consider the system

$$\frac{dx}{dt} = Ax(t)+Bu(t);\ x(0) = x_0; \tag{1}$$

where $A \in L(\mathcal{Y})$, $B \in L(\mathcal{Y},\mathcal{X})$, and $\mathcal{X},\mathcal{Y}$ are Banach spaces. The system (1) is called exactly controllable if for every pair of vectors $x_0, x_1 \in \mathcal{X}$ there exist $T > 0$ and a continuous function $u(t) \in \mathcal{Y}$, $0 \leq t \leq T$ such that the solution $x(t)$ of (1) satisfies $x(T) = x_1$. Prove that (1) is exactly controllable if and only if (7.1.3) holds for some m. Hint: Use the formula

$$x(t) = e^{tA}x_0 + \int_0^t e^{(t-s)A}Bs(s)ds.$$

Ex. 7.2. Assume that (1) is exactly controllable. Prove that (1) is exactly controllable in uniform time, i.e., there is $T_0 > 0$ such that any state $x_1 \in \mathcal{X}$ can be reached from any other state $x_0 \in \mathcal{X}$ in time less than or equal to T_0 (with a suitable choice of the continuous control $u(t)$).

Ex. 7.3. Prove Theorem 7.1.4.

Ex. 7.4. State and prove the dual analogue of Theorem 7.2.1.

Ex. 7.5. Prove that the polynomial $L(\lambda)$ from Theorem 7.4.1 can be chosen to depend continuously on C, A. More precisely, given $C \in L(\mathcal{Y},\mathcal{X})$, $A \in L(\mathcal{Y})$ with left invertible $\mathrm{col}[CA^i]_{i=0}^{\ell-1}$, there exist constants K, $\varepsilon > 0$ with the following property. For every pair of operators $C' \in L(\mathcal{Y},\mathcal{X})$, $A' \in L(\mathcal{Y})$ such that

$$\|C'-C\|+\|A'-A\| < \varepsilon$$

there is an operator polynomial

$$L(\lambda;C',A') = \sum_{j=0}^{\ell} \lambda^j L_j(C',A')$$

of degree $\leq \ell$ (here $L_j(C',A')$ are operators in $L(\mathcal{X})$ that depend on C' and A') for which (C',A') is a right spectral pair with respect to $\mathbb{C}$ and the inequality

$$\sum_{j=0}^{\ell} \|L_j(C',A')-L_j(C,A)\| \leq K(\|C'-C\|+\|A'-A\|)$$

holds. Moreover, $L(\lambda;C',A')$ can be chosen so that $L(\lambda;C',A')^{-1}$ exists and is analytic in a neighborhood of infinity. Hint: Use Theorem 7.2.1 and formula (7.4.2).

Ex. 7.6. State and prove results on analytic dependence of $L(\lambda)$ and C and A analogous to Ex. 7.5. Hint: Use Theorem 7.3.1.

Ex. 7.7. Prove that the isomorphism between the classes of right divisors of $L(\lambda)$ and the A-invariant subspaces established in Theorem 7.5.1 is continuous in an appropriate sense.

Ex. 7.8. Give an example of an operator polynomial $L(\lambda)$ satisfying the hypotheses of Theorem 7.4.3 for which the entire operator function $L_1(\lambda)$ is not a polynomial.

Ex. 7.9. We say that an operator polynomial $L(\lambda)$ is *p.g.i.* (short for polynomial growth at infinity of the inverse) if $\Sigma(L)$ is compact and

$$\|L(\lambda)^{-1}\| \leq C|\lambda|^p$$

for $|\lambda|$ large enough, where the positive constants C and p do not depend on λ. Prove that if $L(\lambda)$ from Theorem 7.4.3 is p.g.i., then $L_1(\lambda)$ must be a polynomial.

7.7 Notes

Exact controllability in infinite-dimensional linear systems was studied by many authors, especially in the setting of C_0-semigroups (Sourour [1], Fuhrmann [1], Louis-Wexler [1], Curtain-Pritchard [1], Eckstein [1]). Exposition of Theorem 7.1.1 is based on Sourour [1]. Theorem 7.1.2 appeared in Kaashoek-van der Mee-Rodman [3], also in Takahashi [1]. Theorem 7.2.1 is taken from Rodman [6]; less exact version appears in Takahashi [2]. See also Eckstein [1] for other spectrum assignment theorems. Theorem 7.2.3 is due to Takahashi [2]. Exposition of Section 7.2 as well as the material of Section 7.3 is based on Rodman [6]. A weaker version of Theorem 7.4.1 appeared in Kaashoek-van der Mee-Rodman [3].

CHAPTER 8: COMMON DIVISORS AND COMMON MULTIPLES

In Chapters 5 and 6 we obtained a description of divisibility of (not necessarily monic) operator polynomials in terms of spectral triples and invariant subspaces of the linearization. Here we apply these results to study common divisors and common multiples of operator polynomials with compact spectrum.

8.1 Common divisors

Let $L_1(\lambda),\dots,L_r(\lambda)$ be operator polynomials with coefficients in $L(\mathcal{X})$ where $\mathcal{X}$ is a Banach space. An operator polynomial $L(\lambda)$ (also with coefficients in $L(\mathcal{X})$) is called common *Ω-right divisor* of $L_1,\dots,L_r$ if $L_1 = Q_1L,\dots,L_r = Q_rL$ for some operator functions $Q_1,\dots,Q_r$ which are analytic in Ω (here, as usual, in this book, Ω is an open set in $\mathbb{C}$). This notion is a natural spin-off of the notion of an Ω-right divisor introduced and studied in Section 6.5.

We define now the notion of greatest common Ω-right divisor. To this end, we need to assume that $\Sigma(L_j) \cap \Omega$ is compact for $j = 1,\dots,r$. Thus, an operator polynomial L with $\Sigma(L) \cap \Omega$ compact is called *greatest common Ω-right divisor* of $L_1,\dots,L_r$ if L is a common Ω-right divisor of $L_1,\dots,L_r$ and every common Ω-right divisor $\tilde{L}$ of $L_1,\dots,L_r$ such that $\Sigma(\tilde{L}) \cap \Omega$ is compact has the property that $\tilde{L}$ is itself an Ω-right divisor of L, i.e., $L = \tilde{Q}\tilde{L}$ for some operator function $\tilde{Q}$ which is analytic in Ω.

A greatest common Ω-right divisor (if it exists) is unique up to multiplication on the left by an analytic and invertible (on Ω) operator-valued function. Indeed, let L and $\tilde{L}$ be two greatest common Ω-right divisors of $L_1,\dots,L_r$. Then

$L = Q\tilde{L}$, $\tilde{L} = \tilde{Q}L$, where Q and $\tilde{Q}$ are analytic (in Ω) operator functions. Consequently,

$$Q(\lambda)\tilde{Q}(\lambda) = I, \qquad \lambda \in \Omega\setminus(\Sigma(L) \cap \Omega);$$

$$\tilde{Q}(\lambda)Q(\lambda) = I, \qquad \lambda \in \Omega\setminus(\Sigma(\tilde{L}) \cap \Omega).$$

As the left-hand sides of these equalities are analytic in Ω, by the uniqueness of analytic continuation, we obtain that $Q(\lambda)\tilde{Q}(\lambda) = \tilde{Q}(\lambda)Q(\lambda) = I$ for all $\lambda \in \Omega$, and our claim follows.

Here is the main result on existence of greatest common divisors.

THEOREM 8.1.1. *Let $\mathcal{X}$ be a Hilbert space, and let $\Omega \subset \mathbf{C}$ be an open and simply connected set. Then for any finite set of operator polynomials $L_1,\dots,L_r$ with coefficients in $L(\mathcal{X})$ and such that $\Sigma(L_j) \cap \Omega$ is compact for $j = 1,\dots,r$ there exists a greatest common Ω-right divisor L of $L_1,\dots,L_r$. Moreover, this divisor L can be chosen in such a way that the spectrum $\Sigma(L)$ is compact and contained in Ω, the degree of L does not exceed the minimum of the degrees of $L_1,\dots,L_r$, and the function $L(\lambda)^{-1}$ is analytic at infinity (in particular, all functions $L_j(\lambda)L(\lambda)^{-1}$, $j = 1,\dots,r$, are actually polynomials). If L' is another greatest common Ω-right divisor of $L_1,\dots,L_r$, then $L'(\lambda) = Q(\lambda)L(\lambda)$ for some operator polynomial $Q(\lambda)$ which is invertible for all $\lambda \in \mathbf{C}$.*

PROOF. Without loss of generality we assume that zero is a point of Ω (otherwise we replace $L_i(\lambda)$ by $L_i(\lambda+\alpha)$ for a suitable $\alpha \in \mathbf{C}$). Choose a bounded Cauchy domain Δ containing zero such that $\bigcup_{i=1}^{r} (\Sigma(L_i) \cap \Omega) \subset \Delta \subset \overline{\Delta} \subset \Omega$, and let $\tilde{V}: L_2(\partial\Delta,\mathcal{X}) \to L_2(\partial\Delta,\mathcal{X})$ and $\tilde{Q}: L_2(\partial\Delta,\mathcal{X}) \to \mathcal{X}$ be the operators defined by the following formulas:

$$(8.1.1) \qquad \tilde{V}f(z) = zf(z)-(2\pi i)^{-1}\int_{\partial\Delta} f(w)dw, \quad z \in \partial\Delta;$$

$$(8.1.2) \qquad \tilde{Q}f = (2\pi i)^{-1}\int_{\partial\Delta} f(w)dw.$$

Here $\partial\Delta$ is the boundary of Δ. For $i = 1,\ldots,r$, let $\mathcal{M}_i$ be the subspace of $L_2(\partial\Delta,\mathcal{X})$ consisting of all functions $f \in L_2(\partial\Delta,\mathcal{X})$ that admit an analytic continuation to $(\mathbb{C} \cup \infty)\setminus(\Sigma(L_i) \cap \Omega)$ vanishing at infinity, while $L_i(\lambda)f(\lambda)$ has an analytic continuation to Ω. By Theorem 6.3.1 (see also the remark after the statement of this theorem), the restriction $(\tilde{Q}_{|\mathcal{M}_i},\tilde{V}_{|\mathcal{M}_i})$ of $(\tilde{Q},\tilde{V})$ is a right Ω-spectral pair for L_i $(i = 1,\ldots,r)$. By Theorem 6.6.1, the operator

$$\begin{bmatrix} \tilde{Q} \\ \tilde{Q}\tilde{V} \\ \vdots \\ \tilde{Q}\tilde{V}^{m_i-1} \end{bmatrix}_{|\mathcal{M}_i}$$

is left invertible, where m_i is the degree of L_i. Put $\mathcal{M} = \bigcap_{i=1}^{r} \mathcal{M}_i$; it is easy to see that the operator $\mathrm{col}[\tilde{Q}(\tilde{V}_{|\mathcal{M}})^i]_{i=0}^{m-1}$ is left invertible as well, where $m = \min(m_1,\ldots,m_r)$.

As Ω is simply connected and $\sigma(\tilde{V}_{|\mathcal{M}_i}) \subset \Omega$ for $i = 1,\ldots,r$, we also have $\sigma(\tilde{V}_{|\mathcal{M}}) \subset \Omega$. By Theorem 7.4.1 there exists an operator polynomial $L(\lambda)$ of degree not bigger than m with right spectral pair $(\tilde{Q}_{|\mathcal{M}},\tilde{V}_{|\mathcal{M}})$ (with respect to the whole complex plane). Moreover, formula (7.4.2) shows that $L(\lambda)^{-1}$ is analytic at infinity. By Theorem 6.5.2, L is an Ω-right divisor of each of $L_1,\ldots,L_r$ and the same theorem shows that actually L is a greatest common Ω-right divisor of $L_1,\ldots,L_r$.

Let now L' be another greatest common Ω-right divisor of $L_1,\ldots,L_r$. Then $L' = QL$ for some analytic (in Ω) operator function Q. Now the properties of L imply easily that Q is actually an everywhere invertible operator polynomial. ∎

As the proof of Theorem 8.1.1 shows the hypothesis of simply connectedness of Ω can be relaxed somewhat; namely, it is sufficient to require that for $i = 1,\ldots,r$ the open set Ω

contains any bounded connected component of $\mathbb{C}\backslash\Sigma(L_i)$ with which Ω has non-empty intersection.

PROBLEM 8.1.1. *Relax (if possible) the hypotheses of Theorem* 8.1.1; *for instance, can one assume that $\mathcal{X}$ is merely a Banach space and Ω is an open set (not necessarily simply connected)?*

The operator polynomials $L_1,\dots,L_r$ will be called *right Ω-prime* if the constant polynomial I is their greatest common Ω-right divisor (it is assumed implicitly that $\Sigma(L_i) \cap \Omega$ are compact sets for $i = 1,\dots,r$). Clearly, if $\Sigma(L_i) \cap \Sigma(L_j) \cap \Omega = \emptyset$ for $i \neq j$, then $L_1,\dots,L_r$ are right Ω-coprime; however, this spectral condition is far from being necessary for the right coprimeness. We shall study later in this chapter the coprime operator polynomials more closely.

Analogously one can define and study greatest common Ω-left divisors of operator polynomials. For the greatest common Ω-left divisors, a statement analogous to Theorem 8.1.1 holds true.

8.2 Common multiples

Let $\Omega \subset \mathbb{C}$ be an open set. We consider here operator polynomials L with coefficients in $L(\mathcal{X})$ ($\mathcal{X}$ being some Banach space) such that $\Sigma(L) \cap \Omega$ is compact. *This condition will be implicitly assumed for all operator polynomials throughout this section.*

An operator polynomial $L(\lambda)$ is called a *left common multiple* (with respect to Ω) of operator polynomials $L_1,\dots,L_r$ if $L = U_1L_1 = \dots = U_rL_r$ for some analytic (in Ω) operator-valued functions $U_1,\dots,U_r$.

A left common multiple L on Ω of $L_1,\dots,L_r$ is called a *least left common multiple*, if every other common multiple $\tilde{L}$ of $L_1,\dots,L_r$ on Ω is divisible by L on the right, i.e., the function $\tilde{L}(\lambda)L(\lambda)^{-1}$ is analytic on Ω. Clearly, if a least common multiple exists, it is determined uniquely up to multiplication from the left by an analytic (on Ω) $L(\mathcal{X})$-valued function whose values are invertible operators.

It is a well-known and widely used fact that in the finite-dimensional case ($\dim \mathcal{X} < \infty$), a least left common multiple always exists. See, e.g., Gohberg-Lancaster-Rodman [2] for the study of common multiples and least common multiples in the finite-dimensional case. In the infinite-dimensional case this is not true: common multiples (let along least common multiples) do not always exist (see the example given below). A necessary condition is given by the following proposition.

In this proposition and throughout the chapter, we denote by $K_m(X,T)$ the operator column $\mathrm{col}[XT^i]_{i=0}^{m-1}$.

PROPOSITION 8.2.1. *Let* $L_1,\ldots,L_r$ *be operator polynomials with right* Ω*-spectral pairs* $(C_1,A_1),\ldots,(C_r,A_r)$, *respectively. Assume that there is a left common multiple* L *of* $L_1,\ldots,L_r$ *on* Ω. *Then*

$$\operatorname{Ker} K_\ell([C_1C_2\cdots C_r],A_1\oplus\cdots\oplus A_r) = \bigcap_{i=0}^{\infty} \operatorname{Ker}[C_1A_1^i,C_2A_2^i,\ldots,C_rA_r^i] \tag{8.2.1}$$

for some integer ℓ.

PROOF. Let (C,A) be a right spectral pair for L on Ω; so $K_\ell(C,A)$ is left invertible for some ℓ. By Theorem 2.4.1 there is a Banach space G_0 and a pair of operators $A_0: G_0 \to G_0$, $C_0: G_0 \to \mathcal{X}$ such that (C,A) is a right restriction of (C_0,A_0) and the operator $K_\ell(C_0,A_0)$ is invertible. Then

$$C_0A_0^\ell[K_\ell(C_0,A_0)]^{-1}K_\ell(C,A) = CA^\ell. \tag{8.2.2}$$

By Theorem 6.5.2, for $i = 1,\ldots,r$ the pair (C_i,A_i) is a right restriction of (C,A). Now (8.2.2) implies that

$$C_0A_0^\ell[K_\ell(C_0,A_0)]^{-1}K_\ell([C_1\cdots C_r],A_1\oplus\cdots\oplus A_r) = [C_1A_1^\ell\cdots C_rA_r^\ell],$$

and (8.2.1) follows. ∎

Proposition 8.2.1 allows us to produce examples of operator polynomials without a common multiple. For example, define

$$T_k = \begin{bmatrix} 0 & 0 & \cdots & 0 & 0 \\ 1 & 0 & \cdots & 0 & 0 \\ 0 & 1 & \cdots & 0 & 0 \\ \cdot & \cdot & \cdot & \cdot & \cdot \\ 0 & 0 & \cdots & 1 & 0 \end{bmatrix}, \quad S_k = \begin{bmatrix} 0 & 0 & \cdots & 0 & 1 \\ 1 & 0 & \cdots & 0 & 0 \\ 0 & 1 & \cdots & 0 & 0 \\ \cdot & \cdot & \cdot & \cdot & \cdot \\ 0 & 0 & \cdots & 1 & 0 \end{bmatrix}$$

as operators from $\mathbf{C}^k \to \mathbf{C}^k$. Let $\mathcal{X} = \mathbf{C}^1 \oplus \mathbf{C}^2 \oplus \mathbf{C}^3 \oplus \cdots$, and put $T = T_1 \oplus T_2 \oplus T_3 \oplus \cdots$; $S = S_1 \oplus S_2 \oplus S_3 \oplus \cdots$. Then for every open set Ω in $\mathbf{C}$ containing $\sigma(T) \cup \sigma(S)$, the operator polynomials $\lambda I-T$, $\lambda I-S$ do not have a common multiple once (condition (8.2.1) is violated).

The following result reduces the existence problem for a least common multiple to the existence problem of a common multiple (in the Hilbert space case and assuming Ω is simply connected).

THEOREM 8.2.2. *Assume $\mathcal{X}$ is a Hilbert space. Let Ω be a simply connected open set in $\mathbf{C}$ and let $L_1,\dots,L_r$ be operator polynomials. Then there exists a least left common multiple of $L_1,\dots,L_r$ on Ω if and only if there exists a left common multiple of $L_1,\dots,L_r$ on Ω.*

PROOF. Assume L_0 is a common multiple of $L_1,\dots,L_r$ on Ω. Without loss of generality we may assume that $0 \in \Omega$. Choose a bounded Cauchy domain Δ containing zero such that $\bigcup_{i=0}^{r} \Sigma(L_i) \subset \Delta \subset \overline{\Delta} \subset \Omega$. Let $\mathcal{M}_i$ be the set of all $f \in L_2(\partial\Delta,\mathcal{X})$ that are analytic on $(\mathbf{C} \cup \{\infty\})\backslash(\Sigma(L_i) \cap \Omega)$, are zero at ∞ such that $L_i f$ is analytic in Ω $(i = 0,1,\dots,r)$. By Theorem 6.3.1, the restriction $(\tilde{Q}_{|\mathcal{M}_i}, \tilde{V}_{|\mathcal{M}_i})$ is a right spectral pair for L_i on Ω, where $\tilde{V}$ and $\tilde{Q}$ are given by (8.1.1) and (8.1.2), respectively. Since L_0 is a common multiple of $L_1,\dots,L_r$, we have by Theorem 6.5.2, $\mathcal{M}_0 \supset \mathcal{M}_1+\cdots+\mathcal{M}_r$. Put $\mathcal{N} = \overline{\mathcal{M}_1+\cdots+\mathcal{M}_r}$. Clearly, $\mathcal{N}$ is a $\tilde{V}$-invariant subspace and $\mathcal{N} \subset \mathcal{M}_0$. Further, by Theorem 6.6.1, $K_m(\tilde{Q}_{|\mathcal{M}_0}, \tilde{V}_{|\mathcal{M}_0})$ is left invertible for some m; so $K_m(\tilde{Q}_{|\mathcal{N}}, \tilde{V}_{|\mathcal{N}})$ is left invertible as well. Since $\sigma(\tilde{V}_{|\mathcal{M}_0}) \subset \Omega$ and Ω is simply connected, also $\sigma(\tilde{V}_{|\mathcal{N}}) \subset \Omega$. By Theorem 7.4.1 there exists an operator polynomial with right Ω-spectral pair $(\tilde{Q}_{|\mathcal{N}}, \tilde{V}_{|\mathcal{N}})$. Using (divisibility) Theorem 6.5.2, it is easily seen that $\tilde{L}$ is a least common multiple of $L_1, \ldots, L_r$ on Ω. ∎

We point out one case when the existence of a least common multiple is ensured.

THEOREM 8.2.3. *Let $\Omega \subset \mathbb{C}$ be an open set and $\mathcal{X}$ a Hilbert space. Let $L_1, \ldots, L_r$ be operator polynomials such that $\Sigma(L_i) \cap \Omega \cap \Sigma(L_j) = \emptyset$ for $i \neq j$. Then there exists a least common multiple L of $L_1, \ldots, L_r$ on Ω, which can be chosen so that* $\Sigma(L) = \bigcup_{i=1}^{r} (\Sigma(L_i) \cap \Omega)$.

PROOF. We shall consider the case $r = 2$ (the general case can be obtained easily by induction on r). For $i = 1,2$, let (C_i, A_i) be a right spectral pair for L_i on Ω, where $A_i\colon \mathcal{G}_i \to \mathcal{G}_i$, $C_i\colon \mathcal{G}_i \to \mathcal{X}$ ($\mathcal{G}_i$ is a Hilbert space). Without loss of generality we may assume that the spectral radius of A_i is less than 1, $i = 1,2$ (otherwise replace $L(\lambda)$ by $L(\alpha\lambda)$ for a suitable fixed positive number α). Consider

$$\mathcal{M} = \{(x_1, x_2) \in \mathcal{G}_1 \oplus \mathcal{G}_2 \mid C_1 A_1^j x_1 + C_2 A_2^j x_2 = 0;\ j = 0,1,\ldots\}.$$

Note that $\mathcal{M}$ is a closed subspace of $\mathcal{G}_1 \oplus \mathcal{G}_2$. We shall prove that $\mathcal{M}$ is, in fact, the zero subspace. Define

$$\mathcal{N}_1 = \{x_1 \in \mathcal{G}_1 \mid \text{there is } x_2 \in \mathcal{G}_2 \text{ such that } (x_1, x_2) \in \mathcal{M}\};$$

$$\mathcal{N}_2 = \{x_2 \in \mathcal{G}_2 \mid \text{there is } x_1 \in \mathcal{G}_1 \text{ such that } (x_1, x_2) \in \mathcal{M}\}.$$

Clearly, $\mathcal{N}_1$ and $\mathcal{N}_2$ are linear sets. Given $x_1 \in \mathcal{N}_1$, there is a unique $x_2 \in \mathcal{G}_2$ such that $(x_1, x_2) \in \mathcal{M}$ (this follows from the fact that $\operatorname{Ker} K_m(C_2, A_2) = \{0\}$ for some m); write $x_2 = S_1 x_1$. Thus $S_1\colon \mathcal{N}_1 \to \mathcal{N}_2$ is a linear map. Similarly, given $x_2 \in \mathcal{N}_2$, there is a unique $x_1 \in \mathcal{G}_1$ such that $(x_1, x_2) \in \mathcal{M}$; write $x_1 = S_2 x_2$. So $S_2\colon \mathcal{N}_2 \to \mathcal{N}_1$ is a linear map and $S_2 = S_1^{-1}$.

Choose an integer $m > 0$ such that $K_m(C_2, A_2)$ and $K_m(C_1, A_1)$ are left invertible. Let

$$\Lambda_1 = -[K_m(C_2,A_2)]^+K_m(C_1,A_1): \mathcal{G}_1 \to \mathcal{G}_2;$$

$$\Lambda_2 = -[K_m(C_1,A_1)]^+K_m(C_2,A_2): \mathcal{G}_2 \to \mathcal{G}_1,$$

where the superscript + denotes a left inverse. Then $S_i = \Lambda_i|\mathcal{N}_i$, $i = 1,2$. As for $i = 1,2$, the map S_i is a restriction of a (bounded linear) operator Λ_i, and $\mathcal{M}$ is closed, it is easily seen that $\mathcal{N}_1$ and $\mathcal{N}_2$ are closed as well. Observe that $\mathcal{N}_i$ is A_i-invariant, $i = 1,2$, and the restrictions $A_1|\mathcal{N}_1$ and $A_2|\mathcal{N}_2$ are similar. So for the boundaries of their spectra we have $\partial\sigma(A_1|\mathcal{N}_1) = \partial\sigma(A_2|\mathcal{N}_2)$, which in view of $\partial\sigma(A_i|\mathcal{N}_i) \subset \sigma(A_i)$, $i = 1,2$, and $\sigma(A_1) \cap \sigma(A_2) = \emptyset$ leads to a contradiction, unless $\mathcal{N}_i = \{0\}$, $i = 1,2$. So $\mathcal{M} = \{0\}$. Put

$$K_\infty = \begin{bmatrix} C_1 & C_2 \\ C_1A_1 & C_2A_2 \\ C_1A_1^2 & C_2A_2^2 \\ \vdots & \vdots \end{bmatrix} : \mathcal{G}_1 \oplus \mathcal{G}_2 \to \ell_2(\mathcal{X}). \tag{8.2.3}$$

Since $\sigma(A_1) \cup \sigma(A_2) \subset \{\lambda \mid |\lambda| < 1\}$, formula (8.2.3) defines a (bounded linear) operator K_∞. We have shown above that $\operatorname{Ker} K_\infty = \{0\}$. It turns out that K_∞ is moreover, left invertible. To see this, let $\langle\mathcal{G}_i\rangle = \ell_\infty(\mathcal{G}_i)/c_0(\mathcal{G}_i)$, $i = 1,2$, and $\langle\mathcal{X}\rangle = \ell_\infty(\mathcal{X})/c_0(\mathcal{X})$. Now C_i and A_i induce operators $\langle A_i\rangle: \langle\mathcal{G}_i\rangle \to \langle\mathcal{G}_i\rangle$ and $\langle C_i\rangle: \langle\mathcal{G}_i\rangle \to \langle\mathcal{X}\rangle$ (cf. the appendix to Section 3.8). Using the property $\sigma(A_i) = \sigma(\langle A_i\rangle)$, $i = 1,2$, we see that K_∞ induces an operator $\langle K_\infty\rangle: \langle\mathcal{G}_1\rangle \oplus \langle\mathcal{G}_2\rangle \to \ell_2(\langle\mathcal{X}\rangle)$. As above we prove that $\operatorname{Ker}\langle K_\infty\rangle = \{0\}$, which means that there exists $\gamma > 0$ such that $\|K_\infty x\| \geq \gamma\|x\|$ for all $x \in \mathcal{G}_1 \oplus \mathcal{G}_2$. Thus K is left invertible. Consequently, a finite column $K_\infty([C_1C_2],A_1\oplus A_2)$ is left invertible for some integer m. By Theorem 7.4.1, there is an operator polynomial L whose right spectral pair (with respect to the whole complex plane) is $([C_1C_2],A_1\oplus A_2)$. Clearly, L is a common multiple of L_1 and L_2. It is not difficult to check (using, for instance, the particular spectral pairs described in Theorem 6.3.1) that L is, in fact, a least common multiple of L_1

and L_2. Since $A_1 \oplus A_2$ is a linearization of L on Ω, clearly $\Sigma(L) = \Sigma(L_1) \cup \Sigma(L_2) \cap \Omega$. ■

It is possible to relax somewhat the hypotheses of Theorem 8.2.3, for the case that Ω is the whole complex plane. Given an operator polynomial L, define

$$\Sigma_\ell(L) = \{\lambda_0 \in \mathbb{C} \mid L(\lambda_0) \text{ has no (bounded) left inverse}\}.$$

THEOREM 8.2.4. *Assume $\mathcal{X}$ is a Hilbert space, and let $L_1, \ldots, L_r$ be operator polynomials with $\Sigma(L_1), \ldots, \Sigma(L_r)$ compact. If*

$$\Sigma_\ell(L_i) \cap \Sigma_\ell(L_j) = \emptyset$$

for $i \neq j$, then there exists a least left common multiple of $L_1, \ldots, L_r$ on $\mathbb{C}$.

PROOF. By Theorem 7.4.3 there exist monic operator polynomials $\tilde{L}_1, \ldots, \tilde{L}_r$ such that $\tilde{L}_i = Q_i L_i$, $i = 1, \ldots, r$ for some entire operator functions Q_i, and $\Sigma(\tilde{L}_i) = \Sigma(L_i) \cup \{\alpha_i\}$, where $\alpha_1, \ldots, \alpha_r$ are distinct complex numbers not belonging to $\bigcup_{i=1}^{r} \Sigma(L_i)$. As

$$\Sigma_\ell(L_i) \subset \Sigma_\ell(\tilde{L}_i) \subset \Sigma_\ell(\tilde{L}_i) \cup \{\alpha_i\},$$

we have

$$\Sigma_\ell(\tilde{L}_i) \cap \Sigma_\ell(\tilde{L}_j) = \emptyset, \quad i \neq j.$$

By Theorem 3.2.5 there is a monic operator polynomial which is a left common multiple of $\tilde{L}_1, \ldots, \tilde{L}_r$, and hence also of $L_1, \ldots, L_r$. It remains to use Theorem 8.2.2. ■

The following problem arises naturally by comparing Theorems 8.2.3 and 8.2.4.

PROBLEM 8.2.1. *Can one replace $\Sigma(L_i)$ by $\Sigma_\ell(L_i)$ in Theorem* 8.2.3?

PROBLEM 8.2.2. *Are Theorems* 8.2.2–8.2.4 *still true when $\mathcal{X}$ is merely a Banach space?*

8.3 Coprimeness and Bezout equation

Recall from Section 8.1 that operator polynomials $L_1,\dots,L_r$ with compact spectra $\Sigma(L_1),\dots,\Sigma(L_r)$ are called right coprime (with respect to $\mathbb{C}$) if I is their greatest right common divisor with respect to $\mathbb{C}$. Closely related to this notion is the Bezout equation

$$V_1(\lambda)L_1(\lambda) + V_2(\lambda)L_2(\lambda) + \cdots + V_r(\lambda)L_r(\lambda) \equiv I, \quad \lambda \in \mathbb{C}, \tag{8.3.1}$$

where $V_1,\dots,V_r$ are some operator polynomials (not necessarily with compact spectrum).

A necessary condition for existence of $V_1,\dots,V_r$ such that (8.3.1) holds is that $L_1,\dots,L_r$ are right coprime (it is assumed implicitly that the spectra of $L_1,\dots,L_r$ are compact). Indeed, if (8.3.1) holds and $L(\lambda)$ is a right common divisor of $L_1,\dots,L_r$ with respect to $\mathbb{C}$, then

$$\left(\sum_{i=1}^{r} V_i(L_iL^{-1})\right)L = I, \quad \lambda \in \mathbb{C},$$

and consequently, $L(\lambda)$ is left invertible for all $\lambda \in \mathbb{C}$. On the other hand, $\Sigma(L)$ is compact, and hence $L(\lambda)$ is (two-sided) invertible for $|\lambda|$ sufficiently large. Now the following lemma ensures that $L(\lambda)$ is invertible for all $\lambda \in \mathbb{C}$ and hence $L_1,\dots,L_r$ are right coprime.

LEMMA 8.3.1. *Let* $A(\lambda)$ *be an analytic function defined in a (connected) domain* $\Omega \subset \mathbb{C}$ *and with values in* $L(\mathcal{X})$, *where* $\mathcal{X}$ *is a Banach space. If* $A(\lambda)$ *is one-sided (either left or right) invertible for all* $\lambda \in \Omega$, *and* $A(\lambda_0)$ *is (two-sided) invertible for some* $\lambda_0 \in \Omega$, *then* $A(\lambda)$ *is invertible for all* $\lambda \in \Omega$.

PROOF. Arguing by contradiction, assume that $A(\lambda_1)$ is one-sided invertible but not (two-sided) invertible for some $\lambda_1 \in \Omega$. Let $\lambda(t)$, $0 \le t \le 1$ be a continuous path in Ω such that $\lambda(0) = \lambda_0$, $\lambda(1) = \lambda_1$, and let

$$t_0 = \sup_{0\le t\le 1} \{t \mid A(\lambda(t)) \text{ is invertible}\}.$$

By assumption, $A(\lambda(t_0))$ is either one-sided or (two-sided) invertible. Using the fact that the set of all one-sided invertible but not invertible operators is open, we obtain that $A(\lambda(t_0))$ must be invertible. It follows now in view of the assumption on λ_1 that $t_0 < 1$. But the set of all invertible operators is open as well, which contradicts the choice of t_0. ■

It $\mathcal{X}$ is finite dimensional, then the right coprimeness of $L_1,\ldots,L_r$ is sufficient for the existence of operator polynomials $V_1,\ldots,V_r$ with the property (8.3.1). (This fact is well known and widely used in mathematics as well as in modern engineering; see, e.g., Kailath [1]). In the infinite-dimensional case the right coprimeness of $L_1,\ldots,L_r$ is not sufficient in general.

EXAMPLE 8.3.1. Let $L_1(\lambda) = L_2(\lambda) = I+\lambda A$, where A is a quasinilpotent operator (i.e., $\sigma(A) = \{0\}$) on an infinite-dimensional Banach space such that $A^n \neq 0$ for $n = 1,2,\ldots$. Clearly, $L_1(\lambda)$ and $L_2(\lambda)$ are right coprime, but there is no operator polynomials $X_1(\lambda)$ and $X_2(\lambda)$ such that

$$X_1(\lambda)L_1(\lambda)+X_2(\lambda)L_2(\lambda) = I, \quad \lambda \in \mathbb{C}.$$ ■

In this section we establish necessary and sufficient conditions for solvability of the Bezout equation (8.3.1) in terms of left invertibility of certain operators. It will be convenient to consider a slightly more general equation, as follows.

Let $\mathcal{X}$ and $\mathcal{Y}$ be Banach spaces, and let $A_1(\lambda)$ and $A_2(\lambda)$ be operator polynomials with coefficients in $L(\mathcal{X})$ and $L(\mathcal{X},\mathcal{Y})$, respectively. We consider the equation

$$X_1(\lambda)A_1(\lambda)+X_2(\lambda)A_2(\lambda) = I_{\mathcal{X}}, \quad \lambda \in \mathbb{C} \tag{8.3.2}$$

where $X_1(\lambda)$ and $X_2(\lambda)$ are operator polynomials to be found, with coefficients in $L(\mathcal{X})$ and $L(\mathcal{Y},\mathcal{X})$, respectively. The equation (8.3.1) is obtained from (8.3.2) in the particular case when $\mathcal{Y} = \mathcal{X}^{r-1}$.

THEOREM 8.3.2. *Let* $A_1(\lambda)$ *be an operator polynomial with compact spectrum and such that*

$$\|A_1(\lambda)^{-1}\| \le |\lambda|^r \tag{8.3.3}$$

for $|\lambda|$ *sufficiently large and for some constant* $r > 0$ *independent of* λ. *Let* $A_2(\lambda) = \sum_{j=0}^{m} \lambda^j A_{2j}$ *with* $A_{2j} \in L(\mathcal{X},\mathcal{Y})$. *Then the following statements are equivalent:*

(i) *There exist operator polynomials (not necessarily with compact spectrum)* $X_1(\lambda): \mathcal{X} \to \mathcal{X}$ *and* $X_2(\lambda): \mathcal{Y} \to \mathcal{X}$ *such that*

$$X_1(\lambda)A_1(\lambda)+X_2(\lambda)A_2(\lambda) = I, \quad \lambda \in \mathbb{C};$$

(ii) *the analytic operator function*

$$\begin{bmatrix} A_1(\lambda) \\ A_2(\lambda) \end{bmatrix}: \mathcal{X} \to \mathcal{X} \oplus \mathcal{Y}$$

is left invertible for every $\lambda \in \mathbb{C}$;

(iii) *for some positive integer* r, *the operator*

$$\mathrm{col}[ZT^i]_{i=0}^{r-1}$$

is left invertible, where $Z = \sum_{j=0}^{m} A_{2j}XT^j$ *and* (X,T) *is a right spectral pair for* $A_1(\lambda)$ *with respect to* $\mathbb{C}$.

For the proof of Theorem 8.3.2, it will be convenient to establish first a simple lemma.

LEMMA 8.3.3. *Let* $L(\lambda)$ *be an operator polynomial (with coefficients in* $L(\mathcal{X})$*) with compact spectrum and with spectral triple* (X,T,Y) *with respect to* $\mathbb{C}$, *and let* $F(\lambda) = \sum_{j=0}^{\infty} \lambda^j F_j$ *be an entire operator-valued function (with coefficients in* $L(\mathcal{X},\mathcal{Y})$*). Then there exists an entire operator-valued function* $G(\lambda)$ *such that* $F(\lambda) = G(\lambda)L(\lambda)$ *if and only if*

$$\sum_{j=0}^{\infty} F_j X T^j = 0. \tag{8.3.4}$$

Observe that the left-hand side of (8.3.4) converges in norm because the coefficients F_j satisfy

$$\lim_{j\to\infty} \|F_j\|^{1/j} = 0,$$

as it follows from the Cauchy-Hadamard formula for the radius of convergence of the power series $\sum_{j=0}^{\infty} \lambda^j F_j$.

PROOF. By the defining property of the spectral triple the function

$$L(\lambda)^{-1} - X(\lambda - T)^{-1} Y$$

is analytic in $\mathbb{C}$. Hence the coefficient of λ^{-k} (where k is a positive integer) in the Laurent expansion of $F(\lambda)L(\lambda)^{-1}$ at infinity coincides with the corresponding coefficient of $F(\lambda)X(\lambda-T)^{-1}Y$. The latter is equal to

$$\left(\sum_{j=0}^{\infty} F_j X T^j\right) T^{k-1} Y.$$

So $F(\lambda)L(\lambda)^{-1}$ is entire if and only if

$$\left(\sum_{j=0}^{\infty} F_j X T^j\right) T^{k-1} Y = 0 \text{ for } k = 1,2,\ldots\ . \tag{8.3.5}$$

However, the operator $[Y, TY, \ldots, T^{r-1}Y]$ is right invertible for some r, and therefore (8.3.5) is equivalent to (8.3.4). ∎

PROOF OF THEOREM 8.3.2. (i) ⇒ (ii) is evident.

(ii) ⇒ (iii). By Theorem 7.1.3, there exist entire operator functions $C(\lambda)$ and $D(\lambda)$ such that

$$C(\lambda)A_1(\lambda) + D(\lambda)A_2(\lambda) = I, \quad \lambda \in \mathbb{C}.$$

Therefore, by Lemma 8.3.3 (applied with $L(\lambda) = A_1(\lambda)$ and $F(\lambda) = I - D(\lambda)A_2(\lambda)$), we obtain

$$X = \sum_{j=0}^{\infty} D_j Z T^j, \tag{8.3.6}$$

where $D(\lambda) = \sum_{j=0}^{\infty} \lambda^j D_j$ and $Z = \sum_{j=0}^{m} A_{2j} X T^j$ (recall that A_{2j} are the coefficients of $A_2(\lambda)$). Let p be such that $\mathrm{col}[XT^j]_{j=0}^{p-1}$ is left invertible; then (8.3.6) leads to

$$\begin{bmatrix} X \\ XT \\ \vdots \\ XT^{p-1} \end{bmatrix} = \sum_{j=0}^{\infty} D_j^{(p)} \begin{bmatrix} X \\ XT \\ \vdots \\ XT^{p-1} \end{bmatrix} T^j,$$

where $D_j^{(p)} = \mathrm{diag}[D_j, D_j, \dots, D_j]$ (p times). Hence

$$\sum_{j=0}^{q-1} D_j^{(p)} \, \mathrm{col}[ZT^k]_{k=0}^{p-1} T^j$$

is left invertible for some q, and the left invertibility of $\mathrm{col}[ZT^k]_{k=0}^{p+q-2}$ follows.

(iii) ⇒ (i). By assumption, there exist operators $D_0, \dots, D_{r-1}$ such that

$$\sum_{j=0}^{r-1} D_j Z T^j = X.$$

Letting $X_2(\lambda) = \sum_{j=0}^{r-1} \lambda^j D_j$, it follows by Lemma 8.3.3 that the operator function

$$X_1(\lambda) = A_1(\lambda)^{-1}(I - X_2(\lambda) A_2(\lambda))$$

is entire. Now the condition (8.3.3) ensures that $X_1(\lambda)$ is in fact a polynomial, and (i) follows. ∎

The assumption (8.3.3) is essential in Theorem 8.3.2, as Example 8.3.1 shows. On the other hand, the proof of Theorem 8.3.2 shows that the implications (i) ⇒ (ii) ⇒ (iii) do not depend on the assumption (8.3.3).

8.4 Analytic behavior of common multiples

In the section we consider the given finite family of operator polynomials $L_1(\lambda),\ldots,L_r(\lambda)$ as depending on a parameter, and study the behavior of common multiples and least common multiples for $L_1(\lambda),\ldots,L_r(\lambda)$ as functions of this parameter.

In this section, all common multiples are with respect to $\mathbb{C}$.

We choose the analytic dependence on a parameter (analogous results can be proved for other types of dependence, for instance, continuous); so we assume that $L_i(\lambda) = L_i(\lambda,\varepsilon)$ depend analytically on the parameter ε, i.e.,

$$L_i(\lambda) = L_i(\lambda,\varepsilon) = \sum_{j=0}^{\ell_i} \lambda^j A_{ij}(\varepsilon), \quad i = 1,\ldots,r,$$

where $A_{ij}(\varepsilon) \in L(\mathcal{X})$ are operator functions analytic on ε in some domain $\Delta \subset \mathbb{C}$. (Here $\mathcal{X}$ is a Banach space.)

First we prove the following theorem on existence of an analytic common multiple inside Δ.

THEOREM 8.4.1. *Assume $\mathcal{X}$ is a Hilbert space. Let $L_1(\lambda,\varepsilon),\ldots,L_r(\lambda,\varepsilon)$ depend analytically on $\varepsilon \in \Delta$, and suppose that $\Sigma(L_i(\lambda,\varepsilon)) \cap \Sigma(L_j(\lambda,\varepsilon)) = \phi$ for $i \neq j$ and every $\varepsilon \in \Delta$. Suppose also that the set*

$$\bigcup_{\varepsilon\in K} \Sigma(L_j(\lambda,\varepsilon)) \tag{8.4.1}$$

is bounded for every compact $K \subset \Delta$ and $j = 1,\ldots,r$. Then for every bounded domain E such that $\overline{E} \subset \Delta$ there exists a positive integer $\ell_0 = \ell_0(E)$ and a common multiple $A(\lambda,\varepsilon)$ of degree ℓ_0 of $L_1(\lambda,\varepsilon),\ldots,L_r(\lambda,\varepsilon)$, whose coefficients are analytic in $\varepsilon \in E$.

In particular, the compactness of (8.4.1) implies that $L_j(\lambda,\varepsilon)$ has compact spectrum for every $\varepsilon \in \Delta$ and $j = 1,\ldots,r$. As the following example shows, the compactness of $\Sigma(L(\lambda,\varepsilon))$ for every $\varepsilon \in \Delta$, where $L(\lambda,\varepsilon)$ is an operator polynomial with coefficients analytically depending on $\varepsilon \in \Delta$, does not generally imply boundedness of $\bigcup_{\varepsilon\in K} \Sigma(L(\lambda,\varepsilon))$ for every compact $K \subset \Delta$: $L(\lambda,\varepsilon) = I+\varepsilon\lambda I$.

PROOF. Let E_0 be a bounded domain such that $\overline{E} \subset E_0$ and $\overline{E}_0 \subset \Delta$.

We show first that right spectral pairs for $L_1(\lambda,\varepsilon),\ldots,L_r(\lambda,\varepsilon)$ can be chosen to depend analytically on $\varepsilon \in E_0$. More exactly, there exist Hilbert spaces $\mathcal{Y}_1,\ldots,\mathcal{Y}_r$ and right spectral pairs $(X_j(\varepsilon),T_j(\varepsilon))$ $(j = 1,\ldots,r)$ of $L_j(\cdot,\varepsilon)$ with respect to $\mathbb{C}$ for every $\varepsilon \in E_0$, where $X_j(\varepsilon) \in L(\mathcal{Y}_j,\mathcal{X})$, $T_j(\varepsilon) \in L(\mathcal{Y}_j)$, such that the operator-valued functions $X_j(\varepsilon)$ and $T_j(\varepsilon)$ are analytic in E_0.

By assumption, the set $W = \bigcup_{j=1}^{r} \bigcup_{\varepsilon\in\overline{E}_0} \Sigma(L_j(\lambda,\varepsilon))$ is bounded. Let Γ_0 be a simple rectifiable contour such that $\overline{W}$ is inside Γ_0; pick $\alpha \in \mathbb{C}$ outside Γ_0 and define monic operator polynomials $M_j(\lambda,\varepsilon) = \lambda^{\ell_j}(L_j(\alpha,\varepsilon))^{-1}L_j(\lambda^{-1}+\alpha,\varepsilon)$, $\varepsilon \in \overline{E}_0$, where ℓ_j is the degree of $L_j(\lambda,\varepsilon)$, $j = 1,\ldots,r$. Note that the coefficients of $M_j(\lambda,\varepsilon)$ are analytic functions of $\varepsilon \in E_0$. Let $(\tilde{X}_j(\varepsilon),\tilde{T}_j(\varepsilon))$ be a right spectral pair of $M_j(\lambda,\varepsilon)$ $(\tilde{X}_j(\varepsilon) \in L(\mathcal{X}^{\ell_j},\mathcal{X})$, $\tilde{T}_j(\varepsilon) \in L(\mathcal{X}^{\ell_j}))$ such that $\tilde{X}_j(\varepsilon)$ and $\tilde{T}_j(\varepsilon)$ depend analytically on $\varepsilon \in E_0$. For instance, we can choose $\tilde{X}_j(\varepsilon) = [I \ \ 0 \ \cdots \ 0]$ and $\tilde{T}_j(\varepsilon)$ be the companion operator for $M_j(\lambda,\varepsilon)$ (see proof of Theorem 2.1.1). Let $\tilde{\Gamma} = \{(\lambda-\alpha)^{-1} \mid \lambda \in \Gamma\}$; by Theorem 6.7.2, the pair $\hat{X}_j(\varepsilon),\hat{T}_j(\varepsilon)$ defined by the equalities

$$\hat{X}_j(\varepsilon) = \tilde{X}_j(\varepsilon)|_{\Lambda_j(\varepsilon)}, \quad \hat{T}_j(\varepsilon) = (\tilde{T}_j(\varepsilon)|_{\Lambda_j(\varepsilon)})^{-1}+\alpha I$$

is right spectral pair for $L_j(\lambda,\varepsilon)$, where

$$\Lambda_j(\varepsilon) = \operatorname{Im}(2\pi i)^{-1} \int_{\tilde{\Gamma}} (\lambda I-\tilde{T}_j(\varepsilon))^{-1}d\lambda. \tag{8.4.2}$$

At this point we need some information concerning analytic families of subspaces. Let $\mathcal{X}$ be a Banach space, and let be given a subspace $\mathcal{Z}(\varepsilon) \subset \mathcal{X}$ for every $\varepsilon \in \Delta$, where Δ is a domain in the complex plane. The family of subspaces $\{\mathcal{Z}(\varepsilon)\}_{\varepsilon\in\Delta}$ is called *analytic* in Δ if for every ε_0 there exists a subspace $\mathcal{M} \subset \mathcal{X}$ and a neighborhood $U \subset \Delta$ of ε_0 such that

$$\mathcal{Z}(\varepsilon) = A(\varepsilon)\mathcal{M}, \quad \varepsilon \in U$$

for some analytic and invertible operator-valued function $A: U \to L(\mathcal{X})$. A basic result on analytic families of subspaces (see, Bungart [1], Gohberg-Leiterer [1]) says that if $\mathcal{X}$ is a Hilbert space, then every analytic (in Δ) family of subspaces $\{\mathcal{Z}(\varepsilon)\}_{\varepsilon\in\Delta}$ is *analytically trivial*, i.e., there is a subspace $\mathcal{M}_0 \subset \mathcal{X}$ and an analytic and invertible operator-valued function $A: \Delta \to L(\mathcal{X})$ such that

$$\mathcal{Z}(\varepsilon) = A(\varepsilon)\mathcal{M}_0, \quad \varepsilon \in \Delta.$$

Returning to the proof of Theorem 8.4.1, observe that (8.4.2) is an analytic family of subspaces in E_0. By the just-quoted result there exists a subspace $\mathcal{M}_j \subset \mathcal{X}^{\ell_j}$ and an analytic and invertible operator function $Q_j(\varepsilon) \in L(\mathcal{X}^{\ell_j})$, $\varepsilon \in E_0$ such that $\Lambda_j(\varepsilon) = Q_j(\varepsilon)\mathcal{M}_j$ for every $\varepsilon \in E_0$. Evidently, the pair $(X_j(\varepsilon), T_j(\varepsilon))$ given by

$$X_j(\varepsilon) = \hat{X}_j(\varepsilon)Q_j(\varepsilon)|_{\mathcal{M}_j}, \quad T_j(\varepsilon) = (Q_j(\varepsilon))^{-1}\hat{T}_j(\varepsilon)Q_j(\varepsilon)|_{\mathcal{M}_j},$$

is a right spectral pair of $L_j(\lambda,\varepsilon)$, analytic on $\varepsilon \in E_0$.

Consider now the analytic operator functions in E_0:

$$X(\varepsilon) = [X_1(\varepsilon)\cdots X_r(\varepsilon)]; \; T(\varepsilon) = \operatorname{diag}[T_1(\varepsilon),\ldots,T_r(\varepsilon)].$$

We claim that for every $\varepsilon \in E_0$ there is $\ell = \ell(\varepsilon)$ such that the operator

$$\operatorname{col}[X(\varepsilon)T(\varepsilon)^i]_{i=0}^{\ell-1} \tag{8.4.3}$$

is left invertible. Indeed, by Theorem 8.2.3 for each $\varepsilon \in \Delta$, there exists a least common left multiple $M(\lambda;\varepsilon)$ of $L_1(\lambda;\varepsilon),\ldots,L_r(\lambda;\varepsilon)$; let $(\hat{X}(\varepsilon),\hat{T}(\varepsilon))$ be a right spectral pair of

$M(\lambda;\varepsilon)$. Then $(X_j(\varepsilon),T_j(\varepsilon))$ is similar to a restriction $(\hat{X}(\varepsilon)|_{\mathcal{N}_j(\varepsilon)},\hat{T}(\varepsilon)|_{\mathcal{N}_j(\varepsilon)})$ where $\mathcal{N}_j(\varepsilon)$ is certain $\hat{T}(\varepsilon)$-invariant subspace. By Lemma 3.6.2 (applied with $S = \hat{T}(\varepsilon)$) the sum $\mathcal{N}(\varepsilon) = \mathcal{N}_1(\varepsilon)\dot{+}\cdots\dot{+}\mathcal{N}_r(\varepsilon)$ is a direct sum. It is easy to see now that $(X(\varepsilon),T(\varepsilon))$ is similar to the restriction $(\hat{X}(\varepsilon)|_{\mathcal{N}(\varepsilon)},\hat{T}(\varepsilon)|_{\mathcal{N}(\varepsilon)})$. As $\mathrm{col}[\hat{X}(\varepsilon)\hat{T}(\varepsilon)^i]_{i=0}^{\ell-1}$ is left invertible for some ℓ, the left invertibility of (8.4.2) follows.

Using the compactness of $\bar{E} \subset E_0$, and stability of the left invertibility under small perturbations, we deduce that there exists $\ell_0 = \ell_0(E)$ such that $\mathrm{col}[X(\varepsilon)T(\varepsilon)^i]_{i=0}^{\ell_0-1}$ is left invertible for every $\varepsilon \in E$. Let $V(\varepsilon) = [V_1(\varepsilon),\ldots,V_{\ell_0}(\varepsilon)]$ be an analytic left inverse of $\mathrm{col}[X(\varepsilon)(T(\varepsilon))^i]_{i=0}^{\ell_0-1}$, $\varepsilon \in E$ (the existence of which is ensured by Theorem 7.1.3). Put

$$A(\lambda,\varepsilon) = \lambda^{\ell_0} - X(\varepsilon)(T(\varepsilon))^{\ell_0}\left(V_1(\varepsilon)+V_2(\varepsilon)\lambda+\cdots+V_{\ell_0}(\varepsilon)\lambda^{\ell_0-1}\right).$$

By Theorem 2.4.1, $(X(\varepsilon),T(\varepsilon))$ is a right restriction of the right spectral pair of $A(\lambda,\varepsilon)$, for every $\varepsilon \in E$. By Theorem 6.5.2, $A(\lambda,\varepsilon)$ is a common multiple of $L_1(\lambda,\varepsilon),\ldots,L_r(\lambda,\varepsilon)$. ∎

PROBLEM 8.4.1. *Extend (if possible) the result of Theorem* 8.4.1 *to the framework of Banach space*. In this regard, it is worth noting that not every analytic (in Δ) family of subspaces in a Banach space $\mathcal{X}$ is analytically trivial, even if the subspaces in this family are assumed to have direct complements in $\mathcal{X}$ (an example of such analytic family of subspaces is given in Gohberg-Leiterer [1]). However, if Δ is simply connected or the group of invertible operators in $L(\mathcal{X})$ is connected, then it is true that every analytic (in Δ) family of subspaces in $\mathcal{X}$ is analytically trivial (see Gohberg-Leiterer [1]; an example of a (complex) Banach space with the disconnected group of invertible operators is given in Douady [1], and many examples of Banach spaces for which this group is connected are found in Mityagin [3]).

PROBLEM 8.4.2. *Is it possible to replace "common multiple" by "least common multiple" in Theorem* 8.4.1?

In connection with problem 8.4.2, observe that a result of local type is valid for the least common multiples.

THEOREM 8.4.2. *Let* $L_1(\lambda,\varepsilon),\dots,L_r(\lambda,\varepsilon)$ *satisfy the assumptions of Theorem* 8.4.1. *Then for every* $\varepsilon_0 \in \Delta$ *there exists a neighborhood* U *of* ε_0 *and a least common multiple* $A(\lambda,\varepsilon)$ *of* $L_1(\lambda,\varepsilon),\dots,L_r(\lambda,\varepsilon)$ *such that* $A(\lambda,\varepsilon)$ *is analytic in* U.

PROOF. As in the proof of Theorem 8.4.1, construct an analytic right spectral pair $(X_j(\varepsilon),T_j(\varepsilon))$ of $L_j(\lambda,\varepsilon)$ in a neighborhood U_1 of ε_0. We also know from the proof of Theorem 8.4.1 that there exists a positive integer ℓ such that the operator

$$\operatorname{col}[X(\varepsilon)T(\varepsilon)^i]_{i=0}^{\ell-1}$$

is left invertible for every $\varepsilon \in U_1$, where $X(\varepsilon) = [X_1(\varepsilon),\dots,X_r(\varepsilon)]$ and $T(\varepsilon) = T_1(\varepsilon) \oplus\cdots\oplus T_r(\varepsilon)$. Let $\alpha \in \mathbb{C}\setminus \bigcup_{j=1}^{r} \bigcup_{\varepsilon\in U_1} \sigma(T_j(\varepsilon))$; the existence of such number α follows from the upper semicontinuity of the spectrum (cf. the proof of Theorem 8.4.1). By Theorem 7.4.2 there exists a pair of operators (X_0,T_0) with $\sigma(T_0) = \{0\}$ such that the operator $\operatorname{col}[X(\varepsilon_0)(T(\varepsilon_0)-\alpha I)^{-i},X_0T_0^i]_{i=0}^{\ell-1}$ is invertible. Then $\operatorname{col}[X(\varepsilon)(T(\varepsilon)-\alpha I)^{-i},X_0T_0^i]_{i=0}^{\ell-1}$ is also invertible for ε in some neighborhood $U \subset U_1$ of ε_0. Let $\tilde{A}(\lambda,\varepsilon)$ be the monic operator polynomial of degree ℓ with right spectral pair $([X(\varepsilon),X_0],(T(\varepsilon)-\alpha I)^{-1}\oplus T_0)$, the existence of $\tilde{A}(\lambda,\varepsilon)$ is ensured by Corollary 2.2.2. Theorem 2.2.1 ensures also that $\tilde{A}(\lambda,\varepsilon)$ is analytic on $\varepsilon \in U$. Now Theorem 8.4.2 follows from the fact that the operator polynomial $A(\lambda,\varepsilon) = (\lambda-\alpha)^{\ell}\tilde{A}((\lambda-\alpha)^{-1})$ is a least common left multiple for $L_1(\lambda,\varepsilon),\dots,L_r(\lambda,\varepsilon)$. ∎

PROBLEM 8.4.3. *Study the analytic behavior of common multiples and least common multiples for operator polynomials with coefficients depending analytically on a parameter, but not necessarily with disjoint spectra (the disjointness of spectrum was assumed in Theorems* 8.4.1 *and* 8.4.2). In the framework of monic operator polynomials such an investigation was carried out in Gohberg-Kaashoek-van Schagen [1].

8.5 Notes

The contents and exposition of Sections 8.1 and 8.2 (except for Theorem 8.2.4) are based on Kaashoek-van der Mee-Rodman [3], where the setup is more general, namely, analytic operator functions with compact spectrum. Lemma 8.3.1 is a well-known and widely used fact (cf. Section IV.4 in Gohberg-Kreĭn [3]). Theorem 8.3.2 is a generalized version of a result due to Takahashi [1]. Necessary and sufficient conditions of solvability of the Bezout equation in terms of Vandermonde operators are given in Rodman [3]. Results of Section 8.4 are new. Analytic behavior of common multiples of monic matrix and operator polynomials was studied in Gohberg-Kaashoek-van Schagen [1], where many examples of singular behavior of analytic least common multiples are found.

CHAPTER 9: RESULTANT AND BEZOUTIAN OPERATORS

We introduce and study operator generalizations of the classical resultant and Bezoutian matrices. It turns out that the theory of spectral pairs developed in Chapter 6 provides a very convenient tool in analysis of the operator resultant and Bezoutian. Many well-known properties of the classical Bezoutian and resultant can be extended to the operator theory framework. Subsequently, the Bezoutian operator is used to prove certain spectrum separation theorems for operator polynomials. In addition to the Bezoutian operator, we need also inertia theorems for Hilbert space operators. These are exposed in Section 9.5.

9.1 Resultant operators and their kernel

Let $L(\lambda) = \sum_{i=0}^{\ell} A_i\lambda^i$ be an operator polynomial with coefficients in $L(\mathcal{X})$, where $\mathcal{X}$ is a Banach space. The following operators

$$R_q(L) = \begin{bmatrix} A_0 A_1 \cdots A_\ell & & & 0 \\ & A_0 A_1 \cdots A_\ell & & \\ & & \ddots & \\ 0 & & & A_0 A_1 \cdots A_\ell \end{bmatrix} : \mathcal{X}^q \to \mathcal{X}^{q-\ell}, \quad q = \ell+1, \ell+2, \ldots$$

will be called the *resultant operators* of L. Given a finite family of operator polynomials $L_1, \ldots, L_r$, we define the resultant operators as follows:

$$R_q(L_1, L_2, \ldots, L_r) = \begin{bmatrix} R_q(L_1) \\ R_q(L_2) \\ \vdots \\ R_q(L_r) \end{bmatrix},$$

where q is any integer bigger than the maximal degree of the polynomials $L_1, \ldots, L_r$.

In this section, we describe the kernel of $R_q(L_1,\dots,L_r)$ in terms of the spectral properties of $L_1,\dots,L_r$ and their greatest common divisor. The interest in studying Ker $R_q(L_1,\dots,L_r)$ is motivated mainly by the subsequent applications, but also because of the following simple statement:

PROPOSITION 9.1.1. *If* $G(\lambda) = \sum_{j=0}^{p} \lambda^j G_j$ *is another operator polynomial, and if* $G(\lambda) = P(\lambda)L(\lambda)$ *for some operator polynomial* $P(\lambda)$, *then*

$$\text{Ker}[G_0 G_1 \cdots G_p 0 \cdots 0] \supset \text{Ker } R_q(L) \tag{9.1.1}$$

for sufficiently large integers q. *Conversely, if* (9.1.1) *holds for some* $q \geq \max(p,\ell)+1$, *where* ℓ *is the degree of* $L(\lambda)$, *and if* $R_q(L)$ *has a generalized inverse, then* $G(\lambda) = P(\lambda)L(\lambda)$ *for some operator polynomial* $P(\lambda)$.

PROOF. For the converse statement, define $P_0,\dots,P_{q-\ell-1}$ by the formula

$$[P_0 P_1 \cdots P_{q-\ell-1}] = [G_0 \cdots G_p 0 \cdots 0](R_q(L))^+,$$

where $(R_q(L))^+$ is a generalized inverse of $R_q(L)$. Then (using (9.1.1)), we have

$$[P_0 \cdots P_{q-\ell-1}]R_q(L) = [G_0 \cdots G_p \quad 0 \cdots 0].$$

Hence $L(\lambda)$ is a right divisor of $G(\lambda)$ and the quotient is given by $\sum_{j=0}^{q-\ell-1} \lambda^j P_j$. Conversely, if $G(\lambda) = (\sum_{j=0}^{s} \lambda^j P_j)L(\lambda)$, put $q = \ell+s+1$. Then $q \geq p+1$ and $[P_0 \cdots P_s]R_q(L) = [G_0 \cdots G_p 0 \cdots 0]$, which implies (9.1.1). ■

We shall express the kernel of the resultant operators in terms of common restrictions of admissible pairs.

Let $\mathcal{X}$ be a fixed Banach space. A pair of operators (C,A), where $A \in L(\mathcal{Y})$ (here $\mathcal{Y}$ is a Banach space which may be different for different admissible pairs) and $C \in L(\mathcal{Y},\mathcal{X})$ is called an *admissible pair* with the base space $\mathcal{Y}$ if the operator $\text{col}[CA^i]_{i=0}^{\ell-1}$ is left invertible for some integer ℓ. The least integer ℓ for which $\operatorname{col}[CA^i]_{i=0}^{\ell-1}$ is left invertible will be called the *index of stabilization* of the admissible pair (C,A). For us the most important examples of admissible pairs are right Ω-spectral pairs of operator polynomials.

If the equalities

$$C_1 = CS, \quad SA_1 = AS \tag{9.1.2}$$

hold for admissible pairs (C_1,A_1) and (C,A) with left invertible S, we say that (C_1,A_1) is a *restriction* of (C,A), or (C,A) is an *extension* of (C_1,A_1).

We pass now to the notions of common restrictions and common extensions of admissible pairs. An admissible pair (C,A) is a *greatest common restriction* of admissible pairs $(C_1,A_1),\ldots,(C_r,A_r)$ if (C,A) is a restriction of each (C_i,A_i), and if (C',A') is a restriction of each (C_i,A_i), then (C',A') is a restriction of (C,A).

THEOREM 9.1.2. *Assume $\mathcal{X}$ is a Hilbert space. Then for any finite set of admissible pairs there exists a greatest common restriction, which is unique up to similarity.*

PROOF. For any admissible pair (C,A) with a base space $\mathcal{Y}$, the space $\mathcal{Y}$ is isomorphic to a subspace in the Hilbert space $\mathcal{X}^\ell$, so, without loss of generality, we can assume that $\mathcal{Y}$ itself is a Hilbert space. Further, given admissible pairs $(C_1,A_1),\ldots,(C_r,A_r)$, let $L_i(\lambda)$ be an operator polynomial with the right spectral pair (C_i,A_i) with respect to $\mathbb{C}$ (Theorem 7.4.1), and let $L(\lambda)$ be the greatest right common divisor of $L_1,\ldots,L_r$ with respect to $\mathbb{C}$ (Theorem 8.1.1). By the divisibility theorem 6.5.2, the right spectral pair of $L(\lambda)$ is a greatest common restriction of (C_i,A_i), $i = 1,\ldots,r$. ■

In contrast with common restriction, a common extension of a finite set of admissible pairs does not always exist (see the example in Section 8.2). If, however, a common extension exists, we have the following property of a greatest common restriction.

THEOREM 9.1.3. *Let* (C_1,A_1) *and* (C_2,A_2) *be admissible pairs for which a common extension* $(\tilde{C},\tilde{A})$ *exists, and let* m *be the index of stabilization of* $(\tilde{C},\tilde{A})$. *Then*

$$\text{(9.1.3)}\qquad \operatorname{Im}\left(\operatorname{col}[C_1A_1{}^i]_{i=0}^{p-1}\right) \cap \operatorname{Im}\left(\operatorname{col}[C_2A_2{}^i]_{i=0}^{p-1}\right) = \operatorname{Im}\left(\operatorname{col}[CA^i]_{i=0}^{p-1}\right).$$

for every integer $p \geq m$, *where* (C,A) *is a greatest common restriction of* (C_1,A_1) *and* (C_2,A_2).

PROOF. We can assume $(C_i,A_i) = (\tilde{C}_{|\mathcal{M}_i},\tilde{A}_{|\mathcal{M}_i})$, $i = 1,2$ where $\mathcal{M}_1$ and $\mathcal{M}_2$ are $\tilde{A}$-invariant subspaces. One can show (cf. the proof of Theorem 8.1.1) that (C,A) is similar to $(\tilde{C}_{|\mathcal{M}},\tilde{A}_{|\mathcal{M}})$, where $\mathcal{M} = \mathcal{M}_1 \cap \mathcal{M}_2$.

The inclusion $\supset$ in (9.1.3) is clear. Let us prove the opposite inclusion. Let

$$z = \operatorname{col}[\tilde{C}\tilde{A}^i x]_{i=0}^{p-1} = \operatorname{col}[\tilde{C}\tilde{A}^i y]_{i=0}^{p-1}$$

for some $x \in \mathcal{M}_1$, $y \in \mathcal{M}_2$. Then $x-y \in \operatorname{Ker} \operatorname{col}[\tilde{C}\tilde{A}^i]_{i=0}^{p-1}$. But $\operatorname{col}[\tilde{C}\tilde{A}^i]_{i=0}^{p-1}$ is left invertible; so $x = y \in \mathcal{M}_1 \cap \mathcal{M}_2$, and $z \in \operatorname{Im} \operatorname{col}[\tilde{C}(\tilde{A}_{|\mathcal{M}})^i]_{i=0}^{p-1}$. ■

We state now the main result concerning the kernels of resultant operators.

Theorem 9.1.4. *Assume* $\mathcal{X}$ *is a Hilbert space. Let* $L_1,\dots,L_r$ *be operator polynomials (with coefficients in* $L(\mathcal{X})$*) with compact spectrum of degrees less than or equal to* $\ell_1,\dots,\ell_r$, *respectively. Assume* $L_1,\dots,L_r$ *have a left common multiple (also with compact spectrum). Let* (X_F,T_F) *be the greatest common restriction of the right spectral pairs of* $L_1,\dots,L_r$. *Further, let* (X_∞,T_∞) *be the greatest common restriction of the right spectral pairs of the operator polynomials* $\lambda^{\ell_i}L_i(\lambda^{-1})$, $i = 1,\dots,r$ *corresponding to a small disc* $|\lambda| < \delta$. *Then there* *exists an integer* $q(L_1,\ldots,L_r) > \max_j \ell_j$ *such that for each* $q \geq q(L_1,\ldots,L_r)$, *the following representation holds:*

$$\text{(9.1.4)} \qquad \operatorname{Ker} R_q(L_1,L_2,\ldots,L_r) = \operatorname{Im} \operatorname{col}[X_F T_F^{i-1}]_{i=1}^{q} \dotplus \operatorname{Im} \operatorname{col}[X T^{q-i}]_{i=1}^{q}.$$

We stress that the right-hand side here is a direct sum.

The proof of Theorem 9.1.4 is rather long and will be done in the next section.

Before we embark on the proof of Theorem 9.1.4, some remarks are in order. First, if at least for one index i, it happens that L_i is monic (or, more generally, with invertible leading coefficient) of degree precisely ℓ_i, then the pair (X_∞, T_∞) is empty, and (9.1.4) becomes

$$\operatorname{Ker} R_q(L_1,L_2,\ldots,L_r) = \operatorname{Im} \operatorname{col}[X_F T_F^{i-1}]_{i=1}^{q}.$$

Secondly, the formula (9.1.4) can be regarded as a far-reaching generalization of the classical property of resultant of two scalar polynomials (see, e.g., Uspensky [1]). Namely, for the scalar polynomials $p(\lambda) = \sum_{j=0}^{\alpha} a_j\lambda^j$, $q(\lambda) = \sum_{j=0}^{\beta} b_j\lambda^j$ with $a_\alpha b_\beta \neq 0$, the matrix

$$\begin{bmatrix} a_0 & a_1 & \cdots & a_\alpha & & & 0 \\ & a_0 & a_1 & \cdots & a_\alpha & & \\ & & \ddots & & & \ddots & \\ 0 & & & a_0 & a_1 & \cdots & a_\alpha \\ b_0 & b_1 & \cdots & b_\beta & & & 0 \\ & b_0 & b_1 & \cdots & b_\beta & & \\ & & \ddots & & & \ddots & \\ 0 & & & b_0 & b_1 & \cdots & b_\beta \end{bmatrix}$$

is invertible if and only if $p(\lambda)$ and $q(\lambda)$ are coprime.

9.2 Proof of Theorem 9.1.4.

It is convenient to prove first the following lemma which is actually a particular case of Theorem 9.1.4.

LEMMA 9.2.1. *Let* $L(\lambda) = \sum_{j=0}^{\ell} \lambda^j A_j$ *be an operator polynomial with* $A_0 = I$, *and let* (X,T) *be a right spectral pair (with respect to* $\mathbb{C}$*) of the monic operator polynomial* $\lambda^{\ell} L(\lambda^{-1})$. *Then for* $q > \ell$ *we have*

$$\operatorname{Ker} R_q(L) = \operatorname{Im} \operatorname{col}[XT^{q-\alpha}]_{\alpha=1}^{q}.$$

PROOF. Put $F_{\alpha\beta} = XT^{\alpha}Z_{\beta}$ $(\alpha \geq 0,\ 1 \leq \beta \leq \ell)$, where $[Z_1 Z_2 \cdots Z_{\ell}] = [\operatorname{col}[XT^{i-1}]_{i=1}^{\ell}]^{-1}$. Then (Theorem 2.2.1)

$$R_q(L) = \begin{bmatrix} I & -F_{\ell\ell} & \cdots & -F_{\ell 1} & 0 & \cdots & 0 \\ 0 & \ddots & & \ddots & & \ddots & \vdots \\ \vdots & & \ddots & & \ddots & & 0 \\ 0 & \cdots & 0 & I & -F_{\ell\ell} & \cdots & -F_{\ell 1} \end{bmatrix}.$$

Introduce

$$\begin{bmatrix} 0 & \cdots & 0 & -I \\ \vdots & & . \cdot & -F_{\ell\ell} \\ 0 & . \cdot & & \vdots \\ -I & -F_{\ell\ell} & \cdots & -F_{q-2,\ell} \end{bmatrix}.$$

Using the equalities (see 2.6.5)) $F_{\alpha+1,\beta} = F_{\alpha\ell}F_{\ell\beta}+F_{\alpha,\beta-1}$; $1 \leq \beta \leq 1$; $\alpha \geq 0$ (where it is assumed $F_{\alpha 0} = 0$), we obtain $SR_q(L) = [U\ \Omega]$, where

$$U = \begin{bmatrix} 0 & & -I \\ & . \cdot & \\ -I & & 0 \end{bmatrix}, \quad \Omega = \begin{bmatrix} F_{\ell\ell} & \cdots & F_{\ell 1} \\ \vdots & \cdots & \vdots \\ F_{q-1,\ell} & \cdots & F_{q-1,1} \end{bmatrix}.$$

As S is invertible, we see that

$$\operatorname{Ker} R_q(L) = \left\{ \begin{bmatrix} \phi_1 \\ \phi_2 \end{bmatrix} \;\middle|\; U\phi_1 + \Omega\phi_2 = 0 \right\}$$

$$= \left\{ \begin{bmatrix} \phi_1 \\ \phi_2 \end{bmatrix} \;\middle|\; \phi_1 = -U^{-1}\Omega\phi_2 \right\} = \operatorname{Im} \begin{bmatrix} -U^{-1}\Omega \\ I \end{bmatrix} .$$

Now

$$-U^{-1}\Omega = \begin{bmatrix} F_{q-1,\ell} & \cdots & F_{q-1,1} \\ \vdots & \cdots & \vdots \\ F_{\ell\ell} & \cdots & F_{\ell 1} \end{bmatrix}, \quad I = \begin{bmatrix} F_{\ell-1,\ell} & \cdots & F_{\ell-1,1} \\ \vdots & \cdots & \vdots \\ F_{0\ell} & \cdots & F_{0,1} \end{bmatrix} .$$

As $F_{\alpha\beta} = XT^{\alpha}Z_{\beta}$, we obtain

$$\operatorname{Ker} R_q(L) = \operatorname{Im}\{\operatorname{col}[XT^{q-1-i}]_{i=0}^{q-1} [Z_{\ell-1}Z_{\ell-2}\cdots Z_0]\} .$$

But $[Z_{\ell-1}Z_{\ell-2}\cdots Z_0]$ is invertible, and hence the lemma is proved. ∎

We now prove Theorem 9.1.4.

First assume that $L_1,\dots,L_r$ are with lower coefficient I, so that the operator polynomials $\tilde{L}(\lambda) = \lambda^{\ell_j} L_j(\lambda^{-1})$, $j = 1,\dots,r$ are monic. Assume further that at least for one index i, the spectrum $\Sigma(L_i)$ does not separate zero from infinity, i.e., there exists a continuous curve $\Gamma(t)$, $0 \le t < \infty$ in the complex plane such that $\Gamma(0) = 0$, $\lim_{t\to\infty} \Gamma(t) = \infty$ and $\Gamma(t) \cap \Sigma(L_i) = \phi$ for all $0 \le t < \infty$. Let $(X_1,T_1),\dots,(X_r,T_r)$ be right spectral pairs of $\tilde{L}_1,\dots,\tilde{L}_r$, respectively with respect to the whole complex plane. From Lemma 9.2.1, we know that for $q > d_0 \overset{\text{def}}{=} \max_{1\le j\le r} \ell_j$ we have

$$\operatorname{Ker} R_q(L_1,\dots,L_r) = \bigcap_{j=1}^{r} \operatorname{Im} \operatorname{col}[X_j T_j^{q-\alpha}]_{\alpha=1}^{q}.$$

Further, for q sufficiently large, we conclude from Theorem 9.1.3 that

$$\operatorname{Ker} R_q(L_1,\ldots,L_r) = \operatorname{Im} \operatorname{col}[X_0T_0^{q-\alpha}]_{\alpha=1}^q,$$

where (X_0,T_0) is any greatest common restriction of the pairs $(X_1,T_1),\ldots,(X_r,T_r)$ (at this point we need the assumption concerning existence of a left common multiple of $L_1,\ldots,L_r$). We can assume that

$$X_i = [X_{iF}\ X_{i\infty}],\ T_i = T_{iF}^{-1} \oplus T_{i\infty},\quad 1 \le i \le r,$$

where (X_{iF},T_{iF}) is a right $\mathbb{C}$-spectral pair of $L_i(\lambda)$ and $(X_{i\infty},T_{i\infty})$ is a right spectral pair of $\tilde{L}_i(\lambda)$ corresponding to a small disc $|\lambda| < \delta$ (see Theorem 6.7.2). Let (X_F,T_F) be a greatest common restriction of the pairs $(X_{1F},T_{1F}),\ldots,(X_{rF},T_{rF})$, and similarly let (X_∞,T_∞) be a greatest common restriction of the pairs $(X_{1\infty},T_{1\infty}),\ldots,(X_{r\infty},T_{r\infty})$.

We prove now that T_F is invertible. It is sufficient to prove that a greatest common right $\mathbb{C}$-spectral divisor $D(\lambda)$ of $L_1(\lambda),\ldots,L_r(\lambda)$ has the property that $D(0)$ is invertible (recall that $L_1,\ldots,L_r$ are assumed to have the lower coefficient I). Indeed, let L_i be such that $\Sigma(L_i)$ does not separate zero from infinity, and let Γ be the continuous curve which appears in the definition of this notion. As $L_i(\Gamma(t)) = W_i(\Gamma(t))D(\Gamma(t))$ for $0 \le t < \infty$, where W_i is an entire operator function, it follows that $D(\Gamma(t))$ is left invertible for all t. Since the spectrum of D is compact, $D(\Gamma(t))$ is invertible for large t. It follows from Lemma 8.3.1 then that $D(\Gamma(t))$ is invertible for all t. In particular, $D(0) = D(\Gamma(0))$ is invertible.

It is not difficult to see that the pair $([X_F,X_\infty],T_F^{-1} \oplus T_\infty)$ is a greatest common restriction of pairs $(X_1,T_1),\ldots,(X_r,T_r)$.

It follows that for q large enough

$$\operatorname{Ker} R_q(L_1,\ldots,L_r) = \operatorname{Im} \operatorname{col}[X_FT_F^{\alpha-q},X_\infty T_\infty^{q-\alpha}]_{\alpha=1}^q.$$

Multiplying on the right by $T_F^{q-1} \oplus I$, one sees that

(9.2.1) $\operatorname{Ker} R_q(L_1,\ldots,L_r) = \operatorname{Im} \operatorname{col}[X_F T_F^{\alpha-1}]_{\alpha=1}^{q} + \operatorname{Im} \operatorname{col}[X_\infty T_\infty^{q-\alpha}]_{\alpha=1}^{q}$

for large q.

Lastly, we verify that the right-hand side in (9.2.1) is actually a direct sum. Indeed, let q be such that both $\operatorname{col}[X_F T_F^{\alpha-1}]_{\alpha=1}^{q}$ and $\operatorname{col}[X_\infty T_\infty^{q-\alpha}]_{\alpha=1}^{q}$ are left invertible. Choose distinct non-zero points $\beta_1, \beta_2 \in \mathbb{C}\backslash\sigma(T_F^{-1})$, and using Theorem 7.4.2, construct monic operator polynomials $N_1(\lambda)$ and $N_2(\lambda)$ of degree q with the following properties: $\Sigma(N_1) = \sigma(T_F^{-1}) \cup \{\beta_1\}$; $\Sigma(N_2) = \sigma(T_\infty) \cup \{\beta_2\}$ (observe that actually $\sigma(T_\infty) = \{0\}$) and the right spectral pairs of N_1 and N_2 corresponding to a neighborhood of $\sigma(T_F^{-1})$ and $\{0\}$ are (X_F, T_F^{-1}) and (X_∞, T_∞), respectively. As $\Sigma(N_1) \cap \Sigma(N_2) = \emptyset$, by Corollary 3.6.3, the Vandermonde operator $V_m(N_1, N_2)$ is regular for m large enough (actually, left invertible, because $\mathcal{X}$ is assumed to be a Hilbert space). Using Theorem 6.2.8, we can write right spectral pairs for $N_1(\lambda)$ and $N_2(\lambda)$ in the form

$$([X_F\ X'], T_F^{-1}\oplus T') \text{ and } ([X_\infty\ X''], T_\infty \oplus T''),$$

respectively, for suitable pairs of operators (X', T') and (X'', T''). Now from the regularity of $V_m(N_1, N_2)$ we deduce easily that

$$\operatorname{Im} \operatorname{col}[X_F T_F^{\alpha-1}]_{\alpha=1}^{m} + \operatorname{Im} \operatorname{col}[X_\infty T_\infty^{m-\alpha}]_{\alpha=1}^{m}$$

is a direct sum.

Consider now the general case. Choose $\alpha \in \mathbb{C}\backslash(\bigcup_{j=1}^{r} \Sigma(L_j))$ in such a way that at least for one polynomial L_j, the spectrum $\Sigma(L_j)$ does not separate α and infinity. Then the polynomials

$$\hat{L}_j(\lambda) = L_j(\lambda+\alpha) = \sum_{j=0}^{\ell_j} \lambda^k \hat{A}_{kj} \quad (j = 1, 2, \ldots, r)$$

have invertible lower coefficients and $\Sigma(\hat{L}_j)$ does not separate zero and infinity for at least one index j. So we can apply the already proved part of Theorem 9.1.4 to the polynomials $\hat{L}_1,\dots,\hat{L}_r$. It follows that for $q \geq q(L_1,L_2,\dots,L_r)$, we have

(9.2.2)
$$\operatorname{Ker} R_q(\hat{L}_1,\hat{L}_2,\dots,\hat{L}_r) = \operatorname{Im} \operatorname{col}(\hat{X}_F\hat{T}_F^{i-1})_{i=1}^q \dotplus \operatorname{Im} \operatorname{col}(\hat{X}_\infty\hat{T}_\infty^{q-i})_{i=1}^q,$$

where $(\hat{X}_F,\hat{T}_F)$ and $(\hat{X}_\infty,\hat{T}_\infty)$ are the greatest common restrictions of the right spectral pairs of $\hat{L}_1,\dots,\hat{L}_r$ corresponding to $\mathbb{C}$, and of the right spectral pairs of $\lambda^{\ell_i}\hat{L}_i(\lambda^{-1})$, $i = 1,\dots,r$ corresponding to a small disc $|\lambda| < \delta$, respectively.

A simple calculation shows that the operators $R_q(L_1,\dots,L_r)$ and $R_q(\hat{L}_1,\dots,\hat{L}_r)$ are related as follows:

(9.2.3)
$$R_q(\hat{L}_1,\hat{L}_2,\dots,\hat{L}_r) = \operatorname{diag}\left(U_{-\alpha}^{q-\ell_1}, U_{-\alpha}^{q-\ell_2},\dots,U_{-\alpha}^{q-\ell_r}\right) R_q(L_1,\dots,L_r)U_\alpha^q$$

where

(9.2.4)
$$U_\alpha^m = \left[\binom{j}{k}\alpha^{j-k}I\right]_{j,k=0}^{m-1},$$

and we assume $\binom{j}{k} = 0$ for $j < k$.

It follows from (9.2.3) that

$$\operatorname{Ker} R_q(L_1,L_2,\dots,L_r) = U_\alpha^q \operatorname{Ker} R_q(\hat{L}_1,\hat{L}_2,\dots,\hat{L}_r).$$

So, if $q > q(L_1,L_2,\dots,L_r)$, we may apply (9.2.2) to have

(9.2.5)
$$\operatorname{Ker} R_q(L_1,L_2,\dots,L_r) = U_\alpha^q \operatorname{Im} \operatorname{col}[\hat{X}_F\hat{T}_F^{i-1}]_{i=1}^q \dotplus U_\alpha^q \operatorname{Im} \operatorname{col}[\hat{X}_\infty\hat{T}_\infty^{q-i}]_{i=1}^q.$$

Denote $X_F = \hat{X}_F$, $T_F = \hat{T}_F+\alpha I$, $X_\infty = \hat{X}_\infty(I+\alpha\hat{T}_\infty)^{q-1}$, $T_\infty = (I+\alpha\hat{T}_\infty)^{-1}\hat{T}_\infty$. It is a matter of a simple calculation to show that formula (9.2.5) can be rewritten in the form

$$\text{(9.2.6)} \qquad \operatorname{Ker} R_q(L_1, L_2, \ldots, L_r)$$
$$= \operatorname{Im} \operatorname{col}[X_F T_F^{i-1}]_{i=1}^q \dotplus \operatorname{Im} \operatorname{col}[X_\infty T_\infty^{q-i}]_{i=1}^q.$$

Now let $(\hat{X}_{jF}, \hat{T}_{jF})$ designate the right spectral pair of $\hat{L}_j(\lambda)$ corresponding to $\mathbb{C}$, and let $(\hat{X}_{j\infty}, \hat{T}_{j\infty})$ be a right spectral pair of $\lambda^{\ell_j} \hat{L}_j(\lambda^{-1})$ corresponding to $|\lambda| < \delta$ (where $\delta > 0$ is small), for $j = 1, 2, \ldots, r$.

It follows from Lemma 9.2.2 below (see also Theorem 6.7.2) that the pair $(\hat{X}_{jF}, \hat{T}_{jF} + \alpha I)$ is a right spectral pair of $L_j(\lambda)$ while the pair

$$(\hat{X}_{j\infty}(I+\alpha\hat{T}_{j\infty})^{q-1}, \ (I+\alpha\hat{T}_{j\infty})^{-1}\hat{T}_{j\infty})$$

is the right spectral pair of $\lambda^{\ell_j} L_j(\lambda^{-1})$ corresponding to a small disc $|\lambda| < \delta$ $(j = 1, 2, \ldots, r)$. This implies easily that (X_F, T_F) and (X_∞, T_∞) are the greatest common restrictions of the right spectral pairs of $L_1, \ldots, L_r$ on $\mathbb{C}$ and of the right spectral pairs of $\lambda^{\ell_1} L_1(\lambda^{-1}), \ldots, \lambda^{\ell_r} L_r(\lambda^{-1})$ corresponding to $|\lambda| < \delta$. Moreover, (9.2.6) coincides with (9.1.4). This completes the proof of Theorem 9.1.4. ∎

It remains to prove the following lemma.

LEMMA 9.2.2. *Let* $L(\lambda)$ *be an operator polynomial with compact spectrum of degree* ℓ, *and let* (Y, F) *be the right spectral pair of the polynomial* $\hat{L}(\lambda) = \lambda^\ell L(\lambda^{-1})$ *corresponding to a sufficiently small disc* Δ *with center at* 0. *Denote* $L_\alpha(\lambda) = L(\lambda+\alpha)$, *where* $\alpha \in \mathbb{C}$ *is fixed. Then* $(Y, F(I-\alpha F)^{-1})$ *is a right spectral pair of the polynomial* $\hat{L}_\alpha(\lambda) = \lambda^\ell L_\alpha(\lambda^{-1})$ *corresponding to* $\{\lambda \mid |\lambda| < \delta\}$ *with* $\delta > 0$ *sufficiently small.*

PROOF. Write

$$L(\lambda) = \sum_{j=0}^{\ell} \lambda^j L_j; \quad L_\alpha(\lambda) = \sum_{j=0}^{\ell} \lambda^j L_{j\alpha}.$$

In view of Theorem 6.6.1, we have to prove the following statements:

(i) $\sum_{j=0}^{\ell} L_{j\alpha}Y(F(I-\alpha F)^{-1})^{\ell-j} = 0$;

(ii) $\mathrm{col}[YF^{j}(I-\alpha F)^{-j}]_{j=0}^{\ell-1}$ is left invertible;

(iii) every other pair of operators (Y',F') which satisfies (i) and (ii) and such that $\sigma(F') = \{0\}$, is a right restriction of $(Y,F(I-\alpha F)^{-1})$.

The property (i) is a consequence (in view of Theorem 6.6.1 applied to the right spectral pair (Y,F) of $\hat{L}(\lambda)$) of the following general fact: If $G \in L(\mathcal{Y})$, $Z \in L(\mathcal{Y},\mathcal{X})$ are such that $\sigma(G) = \{0\}$ and

$$\sum_{j=0}^{\ell} L_j Z G^{\ell-j} = 0,$$

then

$$\sum_{j=0}^{\ell} L_{j\alpha} Z(G(I-\alpha G)^{-1})^{\ell-j} = 0.$$

(As before, L_j and $L_{j\alpha}$ are the coefficients of $L(\lambda)$ and of $L_{\alpha}(\lambda)$, respectively.) The proof of this fact is elementary and is based on the equalities

$$L_{k\alpha} = \sum_{j=k}^{\ell} \binom{j}{j-k}\alpha^{j-k}L_j, \quad k = 0,\ldots,\ell.$$

The property (ii) follows from the easily verified equality

$$(9.2.7) \qquad \left[\binom{\ell-1-j}{k-j}\alpha^{k-j}I\right]_{j,k=0}^{\ell-1}\mathrm{col}[YF^{i}(I-\alpha F)^{-i}]_{i=0}^{\ell-1}\,(I-\alpha F)^{\ell-1}$$
$$= \mathrm{col}[YF^{i-1}]_{i=1}^{\ell}$$

(as usual, we assume $\binom{p}{q} = 0$ for $q < 0$), taking into account the left invertibility of $\mathrm{col}[YF^{i-1}]_{i=1}^{\ell}$. Assume now $\sum_{j=0}^{\ell} L_{j\alpha}Y'F'^{\ell-j} = 0$ for some pair of operators (Y',F') with $\sigma(F') = \{0\}$ and left invertible $\mathrm{col}[Y'F'^{j}]_{j=0}^{\ell-1}$. Then

$$\sum_{j=0}^{\ell} L_j Y'(F'(I+\alpha F')^{-1})^{\ell-j} = 0.$$

In view of equality (9.2.7) (with Y,F,α replaced by $Y',F',-\alpha$, respectively), the operator $\mathrm{col}[Y'(F'(I+\alpha F')^{-1})^j]_{j=0}^{\ell-1}$ is also left invertible. As (Y,F) is a right spectral pair of $\hat{L}(\lambda)$ corresponding to Δ, by Theorem 6.6.1 for some left invertible S we have $Y' = YS$, $SF'(I+\alpha F')^{-1} = FS$. But then $F(I-\alpha F)^{-1}S = SF'$; so (Y',F') is a right restriction of $(Y,F(I-\alpha F)^{-1})$. ■

9.3 Bezoutian operator

Consider the operator polynomials (not necessarily with compact spectrum)

$$A(\lambda) = \sum_{j=0}^{\ell} A_j\lambda^j, \qquad B(\lambda) = \sum_{j=0}^{q} B_j\lambda^j,$$

where $A_j, B_j \in L(\mathcal{X})$ and $A_\ell \neq 0$, $B_q \neq 0$ ($\mathcal{X}$ is a Banach space). We say that $A(\lambda)$ and $B(\lambda)$ are of degrees ℓ and q, respectively. It is assumed that $A(\lambda)$ and $B(\lambda)$ have a (left) common multiple in the class of operator polynomials. That is to say, there are operator polynomials (not necessarily with compact spectrum)

$$M(\lambda) = \sum_{j=0}^{r} M_j\lambda^j, \quad N(\lambda) = \sum_{j=0}^{s} N_j\lambda^j \tag{9.3.1}$$

with $M_r \neq 0$, $N_s \neq 0$, such that

$$M(\lambda)A(\lambda) = N(\lambda)B(\lambda). \tag{9.3.2}$$

Write $\Gamma(\lambda) = M(\lambda)A(\lambda) = N(\lambda)B(\lambda)$.

Given the existence of a common multiple, consider the polynomial in two variables $D(\lambda,\mu)$ defined by

$$\begin{aligned} D(\lambda,\mu) &= (\lambda-\mu)^{-1}(N(\lambda)B(\mu)-M(\lambda)A(\mu)) \\ &= \sum_{i,j=1}^{\sigma,p} D_{ij}\lambda^{i-1}\mu^{j-1} \end{aligned} \tag{9.3.3}$$

where $\sigma = \max(r,s)$ and $p = \max(\ell,q)$. The coefficients D_{ij} determine a $\sigma \times p$ matrix of operators

$$D = \begin{bmatrix} D_{11} & D_{12} & \cdots & D_{1p} \\ D_{21} & D_{22} & \cdots & D_{2p} \\ \vdots & \vdots & & \vdots \\ D_{\sigma 1} & D_{\sigma 2} & \cdots & D_{\sigma p} \end{bmatrix} : \mathcal{X}^p \to \mathcal{X}^\sigma \tag{9.3.4}$$

which is called the *Bezoutian operator* determined by A, B, M, and N. Note that D depends on A, B, M, and N, but frequently this dependence will not be reflected in the notation for the Bezoutian operator. If we wish to emphasize the dependence of D on A, B, M, and N, the notation $D_{M,N}(B,A)$ will be used for the Bezoutian operator. The reasons for its importance include the fact that, when A, B, M, and N have compact spectra, the kernel of T turns out to be independent of M and N and yields information on common divisors of A and B. We shall prove this fact in the next section.

In this section we develop some linear and non-linear equations satisfied by the Bezoutian D. It is convenient to adopt the convention that, if an operator has degree r, say $M(\lambda) = \sum_{j=0}^{r} M_j \lambda^j$ with $M_r \neq 0$, then the symbol M_k with $k > r$ is read as the zero operator. With this convention, the definition of the Bezoutian leads to the relation

$$D_{ij} = \sum_{k=0}^{j-1} (N_{i+k} B_{j-k-1} - M_{i+k} A_{j-k-1}), \tag{9.3.5}$$

for $1 \leq i \leq \sigma$ and $1 \leq j \leq p$. Note that the coefficients A_p, B_p, N_0, M_0 do not appear explicitly here. Because of (9.3.2), there is a dual form

$$D_{ij} = \sum_{k=0}^{i-1} (M_{i-k-1} A_{j+k} - N_{i-k-1} B_{j+k}) \tag{9.3.6}$$

for the same values of i and j, in which M_σ, N_σ, A_0, and B_0 do not appear explicitly.

We need certain operator equations satisfied by the Bezoutian. To set up the framework, we can (and will) assume without loss of generality that $\ell \geq q$ and $s \geq r$. Further, for an operator polynomial $E(\lambda) = \sum_{j=0}^{\alpha} \lambda^j E_j$ with coefficients in $L(\mathcal{X})$ and an integer $\beta \geq \alpha$ define

$$G_E^{(\beta)} = \begin{bmatrix} -E_1 & -E_2 & \cdots & -E_\beta \\ I & 0 & \cdots & 0 \\ 0 & I & \cdots & 0 \\ \vdots & \vdots & & \vdots \\ 0 & 0 & \cdots\ I & 0 \end{bmatrix} \in L(\mathcal{X}^\beta)$$

(it is assumed here $E_{\alpha+1} = \cdots = E_\beta = 0$).

In the following theorem we use the notation A^T to denote the formal transpose of an operator matrix

$$A = \begin{bmatrix} A_{11} & A_{12} & \cdots & A_{1n} \\ A_{21} & A_{22} & \cdots & A_{2n} \\ \vdots & & & \vdots \\ A_{n1} & A_{n2} & \cdots & A_{nn} \end{bmatrix}.$$

Thus

$$A^T = \begin{bmatrix} A_{11} & A_{12} & \cdots & A_{1n} \\ A_{21} & A_{22} & \cdots & A_{2n} \\ \vdots & & & \vdots \\ A_{n1} & A_{n2} & \cdots & A_{nn} \end{bmatrix}.$$

THEOREM 9.3.1. *Let* $A(\lambda)$, $B(\lambda)$, $M(\lambda)$, *and* $N(\lambda)$ *be comonic (i.e., with lower coefficients* $A(0)$, $B(0)$, $M(0)$, *and* $N(0)$ *equal to* I) *operator polynomials satisfying* (9.3.2). *Then*

$$DG_A^{(\ell)} - G_M^{(s)T} D = -D \begin{bmatrix} I_{\mathcal{X}} & 0 \\ 0 & 0 \end{bmatrix} D, \tag{9.3.7}$$

$$DG_B^{(\ell)} - G_N^{(s)T} D = D \begin{bmatrix} I_{\mathcal{X}} & 0 \\ 0 & 0 \end{bmatrix} D, \tag{9.3.8}$$

(9.3.9) $$DG_A^{(\ell)} - G_N^{(s)T} D = 0,$$

(9.3.10) $$DG_B^{(\ell)} - G_M^{(s)} D = 0.$$

These equalities are not difficult to verify using (9.3.5) and (9.3.6). We leave the verification to the reader.

We conclude this section with several formulas involving the Bezoutian that we will need later. The first formula is just a rewriting of (9.3.5):

(9.3.11)
$$D = \begin{bmatrix} N_1 & N_2 & \cdots & N_\sigma \\ N_2 & \vdots & ⋰ & \\ \vdots & N_\sigma & & \\ N_\sigma & & & 0 \end{bmatrix} \begin{bmatrix} B_0 & B_1 & \cdots & B_{p-1} \\ 0 & B_0 & \cdots & B_{p-2} \\ & & \ddots & \vdots \\ \vdots & \vdots & & B_0 \\ 0 & 0 & \cdots & 0 \end{bmatrix} - \begin{bmatrix} M_1 & M_2 & \cdots & M_\sigma \\ M_2 & \vdots & ⋰ & \\ \vdots & M_\sigma & & \\ M_\sigma & & & 0 \end{bmatrix} \begin{bmatrix} A_0 & A_1 & \cdots & A_{p-1} \\ 0 & A_0 & \ddots & \vdots \\ \vdots & & \ddots & A_0 \\ 0 & 0 & \cdots & 0 \end{bmatrix}$$

Analogously, one can rewrite (9.3.6) in the form

(9.3.12)
$$D = \begin{bmatrix} 0 & 0 & \cdots & 0 & M_0 \\ 0 & & & M_0 & M_1 \\ \vdots & & ⋰ & & \vdots \\ 0 & M_0 & & & \\ M_0 & M_1 & \cdots & & M_\sigma \end{bmatrix} \begin{bmatrix} 0 & 0 & \cdots & 0 \\ A_p & 0 & & 0 \\ \vdots & & \ddots & \vdots \\ A_2 & & & 0 \\ A_1 & A_2 & \cdots & A_p \end{bmatrix} - \begin{bmatrix} 0 & 0 & \cdots & 0 & N_0 \\ 0 & & & N_0 & N_1 \\ \vdots & & ⋰ & & \vdots \\ 0 & N_0 & & & \\ N_0 & N_1 & \cdots & & N_\sigma \end{bmatrix} \begin{bmatrix} 0 & 0 & \cdots & 0 \\ B_p & 0 & & 0 \\ \vdots & & \ddots & \vdots \\ B_2 & & & 0 \\ B_1 & B_2 & \cdots & B_p \end{bmatrix}$$

9.4 The kernel of a Bezoutian operator

We continue here to study the Bezoutian operator D introduced in Section 9.3.

Let $A(\lambda)$, $B(\lambda)$ be operator polynomials of degrees ℓ and q, respectively, and let

$$M(\lambda)A(\lambda) = N(\lambda)B(\lambda) \tag{9.4.1}$$

for some operator polynomials $M(\lambda)$ and $N(\lambda)$. Define the Bezoutian operator D by (9.3.4).

It turns out that the kernel of D is independent of the choice of $M(\lambda)$ and $N(\lambda)$ satisfying (9.4.1), and depends on $A(\lambda)$ and $B(\lambda)$ only. This remarkable fact, as well as description of Ker D in terms of $A(\lambda)$ and $B(\lambda)$, is the main result of this section.

We start with the case when A and B are *comonic*, i.e., with the lower coefficient I.

THEOREM 9.4.1. *Let* $A(\lambda) = \sum_{i=0}^{\ell} A_i\lambda^i$ *and* $B(\lambda) = \sum_{i=0}^{q} B_i\lambda^i$ *be comonic, and assume* $\ell \geq q$. *Let* $\hat{B}(\lambda) = \sum_{i=0}^{\ell} B_i\lambda^i$ *with* $B_i = 0$ *if* $i > q$, *and let*

(9.4.2)

$$G_A = \begin{bmatrix} -A_1 & -A_2 & \cdots & -A_\ell \\ I & 0 & \cdots & 0 \\ 0 & I & \cdots & 0 \\ \vdots & \vdots & & \vdots \\ 0 & 0 & \cdots I & 0 \end{bmatrix}, \qquad G_{\hat{B}} = \begin{bmatrix} -B_1 & -B_2 & \cdots & -B_\ell \\ I & 0 & \cdots & 0 \\ 0 & I & \cdots & 0 \\ \vdots & \vdots & & \vdots \\ 0 & 0 & \cdots I & 0 \end{bmatrix}.$$

Assume that $M(\lambda)$ *and* $N(\lambda)$ *are comonic as well. Then* Ker D *is the subspace* $\mathcal{S} \subset \mathcal{X}^p$ *which is maximal with respect to the properties that* $\mathcal{S}$ *is both* G_A*-invariant and* $G_{\hat{B}}$*-invariant, and*

$$G_A|_{\mathcal{S}} = G_{\hat{B}}|_{\mathcal{S}}.$$

PROOF. We show first that Ker D is $G_A, G_{\hat{B}}$-invariant and that the restrictions of $G_A, G_{\hat{B}}$ to Ker D agree. If $y \in$ Ker D, it follows immediately from (9.3.7) and (9.3.8) that

$$D(G_A y) = 0, \quad D(G_{\hat{B}} y) = 0$$

so that Ker D is $G_A, G_{\hat{B}}$-invariant. Then the definitions of G_A and $G_{\hat{B}}$ imply that

$$(G_{\hat{B}} - G_A)y = \begin{bmatrix} [A_1 - B_1, A_2 - B_2, \ldots, A_\ell - B_\ell]y \\ 0 \\ \vdots \\ 0 \end{bmatrix}.$$

But also (by (9.3.6))

$$[A_1 - B_1, A_2 - B_2, \ldots, A_\ell - B_\ell]y = [I \quad 0 \cdots 0]Dy = 0,$$

and it follows that $G_{\hat{B}} y = G_A y$. Thus, the restrictions of G_A and $G_{\hat{B}}$ to Ker D are the same and Ker D $\subset \mathcal{S}$.

To obtain the reverse inclusion, let $z \in \mathcal{S}$. Then $G_{\hat{B}} z = G_A z$ and, as above,

$$(9.4.3) \qquad 0 = (G_{\hat{B}} - G_A)z = \begin{bmatrix} (Dz)_1 \\ 0 \\ \vdots \\ 0 \end{bmatrix},$$

where we adopt the notation that x_k denotes the kth entry of a column vector x. Combining this relation with (9.3.7), it is found that

$$DG_A z = G_{\hat{M}}^T Dz = V_\sigma Dz,$$

where $V_\sigma = [\delta_{i+1,j} I]_{i,j=1}^{\sigma}$. This means that, for $k = 1, 2, \ldots, \sigma - 1$, the kth element of $DG_A z$ is the $(k+1)^{st}$ element of Dz. In particular, $(Dz)_2 = (DG_A z)_1$. But $G_A z \in \mathcal{S}$ and, as in (9.4.3), it is found that $(DG_A z)_1 = 0$. Thus, $(Dz)_2 = 0$ also. Repeating the argument, it follows that $Dz = 0$, and hence that $\mathcal{S} =$ Ker D. ∎

Next, we give a description of the kernel of the Bezoutian in terms of greatest common restrictions of right spectral pairs (see Section 9.1 for the definition of this notion).

THEOREM 9.4.2. *Assume $\mathcal{X}$ is a Hilbert space, and let*

$$A(\lambda) = \sum_{i=0}^{\ell} A_i \lambda^i, \qquad B(\lambda) = \sum_{i=0}^{q} B_i \lambda^i$$

be operator polynomials with coefficients in $L(\mathcal{X})$ and compact spectrum. Further, assume that $\ell \geq q$ and let $\hat{B}(\lambda) = \sum_{i=0}^{\ell} B_i \lambda^i$ with $B_i = 0$ for $i > q$.

Let (X_F, T_F) be the greatest common restriction of the right spectral pairs of $A(\lambda)$ and $B(\lambda)$ corresponding to $\mathbb{C}$. Denote by (X_∞, T_∞) the greatest common restriction of the right spectral pairs of the polynomials $\lambda^\ell A(\lambda^{-1})$ and $\lambda^\ell \hat{B}(\lambda^{-1})$ corresponding to a small disc $|\lambda| < \delta$. If $D = D_{M,N}(B,A)$ is the Bezoutian for $A(\lambda)$ and $B(\lambda)$, where $M(\lambda)$ and $N(\lambda)$ have compact spectrum, then

$$\operatorname{Ker} D = \operatorname{Im}[\operatorname{col}[X_F T_F^i]_{i=0}^{\ell-1}] \dotplus \operatorname{Im}[\operatorname{col}[X_\infty T_\infty^{\ell-i}]_{i=1}^{\ell}].$$

PROOF. Assume first that $A(\lambda)$, $B(\lambda)$, $M(\lambda)$, and $N(\lambda)$ are comonic (i.e., with $A(0) = B(0) = I$), and that at least one of $\Sigma(A)$ and $\Sigma(B)$ does not separate zero from infinity.

We begin by describing Ker D as in Theorem 9.4.1. With (X_F, T_F) defined as in the theorem statement, it follows from the hypotheses of the theorem that T_F is invertible (see the proof of Theorem 9.1.4). Then it is easily deduced from the property (Q2) of Theorem 6.6.1 that

$$G_A \begin{bmatrix} X_F \\ X_F T_F \\ \vdots \\ X_F T_F^{\ell-1} \end{bmatrix} = \begin{bmatrix} X_F \\ X_F T_F \\ \vdots \\ X_F T_F^{\ell-1} \end{bmatrix} T_F^{-1},$$

and a similar relation also holds with G_A replaced by $G_{\hat{B}}$ (the operators G_A and $G_{\hat{B}}$ are defined by (9.4.2)). Hence

$$\mathcal{S}_F \overset{\text{def}}{=} \operatorname{Im}[\operatorname{col}[X_F T_F^i]_{i=0}^{\ell-1}]$$

is invariant for both G_A and $G_{\hat{B}}$ and

$$G_{A}\big|_{\mathcal{S}_F} = G_{\hat{B}}\big|_{\mathcal{S}_F}.$$

Similarly, it is found that

$$G_A \begin{bmatrix} X_\infty T_\infty^{\ell-1} \\ \vdots \\ X_\infty T_\infty \\ X_\infty \end{bmatrix} = \begin{bmatrix} X_\infty T_\infty^{\ell-1} \\ \vdots \\ X_\infty T_\infty \\ X_\infty \end{bmatrix} T_\infty$$

with a similar relation for $G_{\hat{B}}$. Thus, $\mathcal{S}_\infty$ is $G_A, G_{\hat{B}}$-invariant and $G_{A}\big|_{\mathcal{S}_\infty} = G_{\hat{B}}\big|_{\mathcal{S}_\infty}$. It is easily seen that the sum of $\mathcal{S}_F$ and $\mathcal{S}_\infty$ is direct and so $\mathcal{S}$ is indeed $G_A, G_{\hat{B}}$-invariant subspace has the form $\mathcal{S}_F \dot{+} \mathcal{S}_\infty$ for some choice of admissible pairs (X_F, T_F) and (X_∞, T_∞). The maximality of $\mathcal{S}$ therefore follows from the property (Q4) of Theorem 6.6.1.

Consider now the general case. There exists $\alpha \in \mathbb{C}$ such that $B(\alpha)$, $A(\alpha)$, $M(\alpha)$, and $N(\alpha)$ are invertible, and α belongs to the unbounded connected component of at least one of $\mathbb{C}\backslash\Sigma(B)$ and $\mathbb{C}\backslash\Sigma(A)$. Put $B_\alpha(\lambda) = B(\lambda+\alpha)$, $N_\alpha(\lambda) = N(\lambda+\alpha)$, $M_\alpha(\lambda) = M(\lambda+\alpha)$, $A_\alpha(\lambda) = A(\lambda+\alpha)$. Then B_α and M_α have invertible constant terms and $M_\alpha(\lambda)A_\alpha(\lambda) = N_\alpha(\lambda)B_\alpha(\lambda)$. By the already proved part of the theorem,

$$\text{(9.4.4)} \qquad \operatorname{Ker} D_{M_\alpha,N_\alpha}(B_\alpha, A_\alpha) = \operatorname{Im}\begin{bmatrix} X_{0\alpha} \\ X_{0\alpha}T_{0\alpha} \\ \vdots \\ X_{0\alpha}T_{0\alpha}^{\ell-1} \end{bmatrix} \dot{+} \operatorname{Im}\begin{bmatrix} X_{\infty\alpha}T_{\infty\alpha}^{\ell-1} \\ \vdots \\ X_{\infty\alpha} \end{bmatrix},$$

where $(X_{0\alpha}, T_{0\alpha})$ is the greatest common restriction of the right spectral pairs of A_α and B_α corresponding to the whole complex plane, and $(X_{\infty\alpha}, T_{\infty\alpha})$ is the greatest common restriction of the right spectral pair of $\lambda^\ell A_\alpha(\lambda^{-1})$ and $\lambda^\ell B_\alpha(\lambda^{-1})$ corresponding to a small disc around zero.

At this point we need the relation between $D_{M,N}(B,A)$ and $D_{M_\alpha,N_\alpha}(B_\alpha,A_\alpha)$ which is given by the formula

$$D_{M_\alpha,N_\alpha}(B_\alpha,A_\alpha) = U_\alpha^{(m)} D_{M,N}(B,A) U_\alpha^{(\ell)^T}, \tag{9.4.5}$$

where $U_\alpha^{(k)} = \left[\binom{j}{i}\alpha^{j-i}I\right]_{i,j=0}^{k-1}$ (it is assumed that $\binom{j}{i} = 0$ for $j < i$). Recall that by superscript "T" we denote the formal transpose of a block operator matrix.

Combining (9.4.4) and (9.4.5), we obtain

$$\operatorname{Ker} D = U_\alpha^{(\ell)T} \operatorname{Ker} D_{M_\alpha,N_\alpha}(B_\alpha,A_\alpha)$$

$$= U_\alpha^{(\ell)T} \operatorname{Im} \begin{bmatrix} X_{0\alpha} \\ X_{0\alpha}T_{0\alpha} \\ \vdots \\ X_{0\alpha}T_{0\alpha}^{\ell-1} \end{bmatrix} + U_\alpha^{(\ell)T} \operatorname{Im} \begin{bmatrix} X_{\infty\alpha}T_{\infty\alpha}^{\ell-1} \\ X_{\infty\alpha}T_{\infty\alpha}^{\ell-2} \\ \vdots \\ X_{\infty\alpha} \end{bmatrix}.$$

By Lemma 9.2.2, we know that (up to similarity) $X_{\infty\alpha} = X_\infty$, $T_{\infty\alpha} = T_\infty(I-\alpha T_\infty)^{-1}$. Now it is easy to verify that

$$U_\alpha^{(\ell)T} \operatorname{Im} \begin{bmatrix} X_{\infty\alpha}T_{\infty\alpha}^{\ell-1} \\ X_{\infty\alpha}T_{\infty\alpha}^{\ell-2} \\ \vdots \\ X_{\infty\alpha} \end{bmatrix} = \operatorname{Im} \begin{bmatrix} X_\infty T_\infty^{\ell-1} \\ X_\infty T_\infty^{\ell-2} \\ \vdots \\ X_\infty \end{bmatrix}.$$

On the other hand, also

$$U_{\alpha}^{(\ell)T}\begin{bmatrix} X_{0\alpha} \\ X_{0\alpha}T_{0\alpha} \\ \vdots \\ X_{0\alpha}T_{0\alpha}^{\ell-1} \end{bmatrix} = \begin{bmatrix} X_{0\alpha} \\ X_{0\alpha}(T_{0\alpha}+\alpha I) \\ \vdots \\ X_{0\alpha}(T_{0\alpha}+\alpha I)^{\ell-1} \end{bmatrix},$$

and, taking into account that $(X_{0\alpha}, T_{0\alpha}+\alpha I)$ is the greatest common restriction of the spectral pairs of B and A (corresponding to the whole complex plane), we finish the proof of Theorem 9.4.2. ∎

9.5 Inertia theorems

This section plays an auxilliary role. Here we prove the basic inertia theorems which are needed in the next section for the study of the problems of spectrum location of operator polynomials (with respect to the unit circle).

We need some preliminaries. Denote $\pi_+ = \{z \in \mathbb{C} \mid \mathrm{Re}z > 0\}$, $\pi_- = \{z \in \mathbb{C} \mid \mathrm{Re}z < 0\}$, $\pi_0 = \{z \in \mathbb{C} \mid \mathrm{Re}z = 0\}$. Given an operator $A \in L(\mathcal{H})$, where $\mathcal{H}$ is a Hilbert space, denote by M_A the set of all triples of projectors (E_+, E_-, E_0) with the following properties:

$$E_\eta A = AE_\eta, \sigma(A|_{E_\eta H}) \subset \overline{\pi}_\eta, \ \eta = 0,+,-;$$

$$H = E_+H \dotplus E_-H \dotplus E_0H.$$

If the set M_A is not empty, denote

$$i_{\pm}^{\pi}(A) = \min\{\dim E_{\pm}H \mid (E_+, E_-, E_0) \in M_A\};$$

$$i_0^{\pi}(A) = \sup\{\dim E_0H \mid (E_+, E_-, E_0) \in M_A\}.$$

The triple $\mathrm{In}_\pi(A) = \{i_+^\pi(A), i_-^\pi(A), i_0^\pi(A)\}$ is called the *inertia* of A (relative to π_+, π_-, π_0).

THEOREM 9.5.1. *Let* $A \in L(\mathcal{H})$, *where* $\mathcal{H}$ *is a Hilbert space. Then there is a self-adjoint operator* $H \in L(\mathcal{H})$ *such that*

Re(HA) *is positive definite and invertible if and only if* $\sigma(A) \cap \pi_0 = \emptyset$. *(As usual, here* Re X *denotes the real part of an operator* X: $\mathrm{Re}\, X = \frac{1}{2}(X+X^*)$.)

PROOF. Assume $G \overset{\mathrm{def}}{=} 2\,\mathrm{Re}(HA)$ is positive definite and invertible, where $H = H^* \in L(\mathcal{H})$. Then

$$\lambda_0 \overset{\mathrm{def}}{=} \inf_{\|x\|=1} \langle Gx, x\rangle > 0. \tag{9.5.1}$$

On the other hand, arguing by contradiction, assume that $\sigma(A) \cap \pi_0 \neq \emptyset$, and let $\lambda = i\mu$ (μ is real) be a boundary point of $\sigma(A)$ which lies on the imaginary axis. By a well-known theorem on boundary points of the spectrum of a (linear bounded) operator (see, e.g., Daleckii-Kreĭn [1]), there is a sequence of vectors $\{f_n\}_{n=1}^{\infty}$ such that $\|f_n\| = 1$ for all n and

$$\lim_{n\to\infty}(Af_n - \lambda f_n) = 0. \tag{9.5.2}$$

Now

$$\begin{aligned} 0 < \lambda_0 \le \langle Gf_n, f_n\rangle &= \langle HAf_n, f_n\rangle + \langle Hf_n, Af_n\rangle \\ &= \langle H(Af_n - \lambda f_n), f_n\rangle + i\mu\langle Hf_n, f_n\rangle \\ &\quad + \langle Hf_n, (Af_n - \lambda f_n)\rangle - i\mu\langle Hf_n, f_n\rangle \\ &\le 2\|H\|\cdot\|Af_n - \lambda f_n\|, \end{aligned}$$

which tend to zero in view of (9.5.2), a contradiction.

Conversely, assume $\sigma(A) \cap \pi_0 = \emptyset$. Then $\mathcal{H}$ decomposes into a direct sum $\mathcal{H} = \mathcal{H}_+ \dot{+} \mathcal{H}_-$, where $\mathcal{H}_+, \mathcal{H}_-$ are A-invariant subspaces and $\sigma(A|_{\mathcal{H}_+})$ lies in the open right halfplane, while $\sigma(A|_{\mathcal{H}_-})$ lies in the open left half-plane. Denote by P_+ the projection on $\mathcal{H}_+$ along $\mathcal{H}_-$, and let $P_- = I - P_+$.

Let G be any positive definite invertible operator. Consider the operator

$$X = \int_0^{\infty} e^{-A^*t} P_+^* G P_+ e^{-At}\,dt. \tag{9.5.3}$$

We check first that the integral in (9.5.3) is correctly defined, i.e., converges. To this end, let ρ be any negative number such that $\sigma(-A_{|\mathcal{H}_+})$ lies on the left-hand side of the line $\{z \in \mathbf{C} \mid \mathrm{Re} z = \rho\}$. Pick a simple rectifiable contour Γ in the half plane $\{z \in \mathbf{C} \mid \mathrm{Re} z < \rho\}$ such that $\sigma(-A_{|\mathcal{H}_+})$ is inside Γ. Then

$$\exp(-A_{|\mathcal{H}_+} t) = \frac{1}{2\pi i} \int_\Gamma e^{\lambda t}(\lambda I + A_{|\mathcal{H}_+})^{-1} d\lambda,$$

and hence

$$\|P_+ e^{-At}\| \le N_1 e^{\rho t}, \tag{9.5.4}$$

where

$$N_1 = \frac{1}{2\pi}\left(\max_{\lambda\in\Gamma}\|(\lambda I + A_{|\mathcal{H}_+})^{-1}\|\right) \cdot (\text{the length of } \Gamma).$$

Analogously

$$\|e^{-A^* t} P_+^*\| \le N_2 e^{\rho t}, \tag{9.5.5}$$

where the constant N_2 is independent of t. The inequalities (9.5.4) and (9.5.5) imply convergence of the integral in (9.5.3).

Next, one verifies that

$$A^* X + XA = P_+^* G P_+.$$

In fact,

$$A^* X + XA = -\int_0^\infty [-A^* e^{-A^* t} P_+ G P_+ e^{-At} - e^{-A^* t} P_+ G P_+ e^{-At} A] dt$$

$$= -\int_0^\infty d(e^{-A^* t} P_+ G P_+ e^{-At}) = -[e^{-A^* t} P_+ G P_+ e^{-At}]_{t=0}^{t=\infty} = P_+ G P_+.$$

Define also

$$Y = -\int_0^\infty e^{A^* t} P_-^* G P_- e^{At} dt. \tag{9.5.6}$$

Analogously one can verify that

$$A^*Y+YA = P_-^*GP_-,$$

Now put $H = X+Y$. The operator H is clearly self-adjoint (because X and Y are such).

Further,

$$HA+A^*H = P_+^*GP_+ + P_-^*GP_-,$$

and for any $x \in \mathcal{H}$ we have (where $\lambda_0 = \inf_{\|y\|=1} \langle Gy,y\rangle > 0$):

$$\langle (P_+^*GP_+ + P_-^*GP_-)x,x\rangle = \langle GP_+x,P_+x\rangle + \langle GP_-x,P_-x\rangle$$

$$\geq \lambda_0(\|P_+x\|^2+\|P_-x\|^2) \geq \frac{1}{2}\lambda_0(\|P_+x\|+\|P_-x\|)^2 \geq \frac{1}{2}\lambda_0\|x\|^2,$$

where in the last inequality we have used $x = P_+x+P_-x$. Thus, $\mathrm{Re}(HA)$ is positive definite and invertible. ∎

It turns out that (in the notation in Theorem 9.5.1), the inertia of A coincides with the invertia of H. This fact, together with Theorem 9.5.1, is the main result of this section.

THEOREM 9.5.2. *Let* $A \in L(\mathcal{H})$, $H = H^* \in L(\mathcal{H})$ *be such that* $\mathrm{Re}(HA)$ *is positive definite and invertible. Then*

$$\mathrm{In}_\pi(A) = \mathrm{Im}_\pi(H).$$

It is convenient first to prove a lemma.

LEMMA 9.5.3. *Under the hypotheses of Theorem* 9.5.2, *the operator* H *is invertible.*

PROOF. Arguing by contradiction, assume $0 \in \sigma(H)$. Applying again the theorem on boundary points of the spectrum of an operator, we find sequence of vectors $\{f_n\}_{n=0}^\infty$, $\|f_n\| = 1$ such that $Hf_n \to 0$ as $n \to \infty$. Now, with λ_0 defined by (9.5.1), we have

$$0 < \lambda_0 \leq \langle Gf_n,f_n\rangle = \langle Af_n,Hf_n\rangle + \langle Hf_n,Af_n\rangle$$

$$\leq 2\|A\|\ \|Hf_n\|,$$

which tends to zero, a contradiction. ∎

PROOF OF THEOREM 9.5.2. Theorem 9.5.1 implies that $\sigma(A) \cap \pi_0 = \phi$. Let $H_0 = X+Y$, where X and Y are given by (9.5.3) and (9.5.6), respectively. We have

$$\langle Xf, f \rangle > 0$$

for every $f \in \mathcal{H}_+$, $f \neq 0$. Also

$$\langle Yg, g \rangle < 0$$

for every $g \in \mathcal{H}_-$, $g \neq 0$. Hence (in view of invertibility of H_0 which is ensured by Lemma 9.5.3),

$$In_\pi(H_0) = \{\dim \mathcal{H}_+, \dim \mathcal{H}_-, 0\},$$

which coincides with $In_\pi(A)$.

Let

$$H(t) = (1-t)H+tH_0, \quad 0 \leq t \leq 1.$$

Then $Re(H(t)A)$ is positive definite and invertible for all $t \in [0,1]$. By Lemma 9.5.3, the self-adjoint operator $H(t)$ is invertible for all $t \in [0,1]$, and hence $In_\pi(H(t))$ does not depend on t. In particular, $In_\pi(H) = In_\pi(H_0)$, and we are done. ■

PROBLEM 9.5.1. In the finite-dimensional case ($\dim \mathcal{H} < \infty$), there are many results stronger than Theorems 9.5.1 and 9.5.2. For example, the following theorem due to Carlson-Schneider [1] is well known. Let A be $n\times n$ matrix without eigenvalues on the imaginary axis and such that $AH+HA^*$ is positive semidefinite for some invertible Hermitian matrix H. Then $In_\pi(A) = In_\pi(H)$. This leads naturally to the problem of *obtaining inertia results stronger than Theorems* 9.5.1 *and* 9.5.2 *(at least for special classes of operators) in the infinite-dimensional case* as well.

Theorems 9.5.1 and 9.5.2 have counterparts concerning the inertia with respect to the unit circle, which we will present now.

Let $\Delta_+ = \{z \in \mathbb{C} \mid |z| < 1\}$, $\Delta_- = \{z \in \mathbb{C} \mid |z| > 1\}$, $\Delta_0 = \{z \in \mathbb{C} \mid |z| = 1\}$. Replacing π_η by Δ_η ($\eta = +,-,0$) in the definition of $In_\pi(A)$, we obtain the *inertia* $In_\Delta(A) = \{i_+^\Delta(A), i_-^\Delta(A), i_0^\Delta(A)\}$ of A (relative to $\Delta_+, \Delta_-, \Delta_0$).

THEOREM 9.5.4. *Let $C \in L(\mathcal{H})$ and suppose that $\Delta_0 \cap (\mathbb{C}\backslash\sigma(C)) \neq \phi$. There exists a self-adjoint operator $G \in L(\mathcal{H})$ such that $G-C^*GC$ is positive definite and invertible if and only if $\sigma(C) \cap \Delta_0 = \phi$. In this case, $In_\Delta(C) = In_\pi(G)$.*

PROOF. Replacing C by $\lambda_0 C$ for some $\lambda_0 \in \Delta_0$, we can assume without loss of generality that $1 \notin \sigma(C)$.

Assume that $G-C^*GC$ is positive definite and invertible for some $G = G^*$. Because of the assumption $1 \notin \sigma(C)$, the operator $A = (I+C)(I-C)^{-1}$ is well defined. Let us check that $Re(GA)$ is positive definite and invertible. Indeed,

$$GA+A^*G = (I-C^*)^{-1}\Big[(I-C^*)G(I+C)+(I+C^*)G(I-C)\Big](I-C)^{-1}$$

$$= 2(I-C^*)^{-1}[G-C^*GC](I-C)^{-1}.$$

Now apply Theorems 9.5.1 and 9.5.2, and use the spectral mapping theorem to verify that

$$In_\Delta(C) = In_\pi(A).$$

To prove the converse statement, use again the reduction to Theorem 9.5.1 applied to the operator A. ■

We remark that the condition

$$\Delta_0 \cap (\mathbb{C}\backslash\sigma(C)) \neq \phi$$

is essential in Theorem 9.5.4, as the following example shows.

EXAMPLE 9.5.1. Let $C \in L(\ell_2)$ be defined by

$$C(x_1, x_2, \ldots) = (0, 2x_1, 2x_2, \ldots).$$

For $G = -I$, we have

$$G-C^*GC = -I+C^*C = 3I,$$

which is positive definite and invertible. However,

$$\sigma(C) = \{z \in \mathbb{C} \mid |z| \leq 2\} \supset \Delta_0. \qquad \blacksquare$$

9.6 Spectrum separation

Using the notion of a Bezoutian, and the inertia reuslts developed in the preceding section, we study in this section the problem of location of the spectrum of an operator polynomial relative to the unit circle. This problem (for scalar polynomials) is commonly known as the Schur-Cohn problem.

Everywhere in this section the operator polynomials are assumed to have coefficients in $L(\mathcal{X})$, where $\mathcal{X}$ is a Hilbert space.

Let $M(\lambda) = \sum_{j=0}^{m} \lambda^j M_j$ be an operator polynomial (not necessarily with compact spectrum). Denote by $M^*(\lambda)$ the operator polynomial with adjoint coefficients:

$$M^*(\lambda) = \sum_{j=0}^{m} \lambda^j M_j^*.$$

For an operator polynomial $L(\lambda)$ of degree ℓ, let

$$L_\infty^*(\lambda) = \lambda^\ell (L(\bar{\lambda}^{-1}))^*.$$

For an operator polynomial $M(\lambda)$ with compact spectrum define its inertia $\{i_+(M), i_-(M), i_0(M)\}$ with respect to $\Delta_+, \Delta_-, \Delta_0$ as follows:

$$i_\eta(M) = i_\eta^\Delta(T), \quad \eta = +, -, 0 ,$$

where T is taken from a spectral pair (X, T) for M, provided $\mathrm{In}_\Delta T$ exists.

The following theorem is the main result of this section.

THEOREM 9.6.1. *Let $L_1(\lambda)$ and $L(\lambda)$ be operator polynomials with compact spectrum of degree ℓ such that L has invertible leading coefficient, $L_1(0) = I$, and*

$$L_{1\infty}^{*}(\lambda)L_{1}(\lambda) = L_{\infty}^{*}(\lambda)L(\lambda). \tag{9.6.1}$$

Let L_0 *be a greatest right common divisor of* L *and* L_1*. Denote* $\Theta = \{\lambda \mid |\lambda| = 1;$ *at least one of* $L(\lambda)$ *and* $L_0(\lambda)$ *is not invertible*$\}$*, and assume that the set* $\{\lambda^{\ell} \mid \lambda \in \Theta\}$ *is not the whole of the unit circle. Assume that the subspace* $\operatorname{Im} D$ *is closed, where* $D = D_{L_{1\infty}^{*}, L_{\infty}^{*}}(L, L_1)$*. Then* L *and* L_0 *are invertible at every point on the unit circle, and*

$$i_{+}(L) = i_{+}^{\pi}(-P_{\ell}D)+i_{+}(L_0), \tag{9.6.2}$$

$$i_{-}(L) = i_{-}^{\pi}(-P_{\ell}D)+i_{-}(L_0), \tag{9.6.3}$$

where $P_{\ell} = [\delta_{i,\ell-j+1}I]_{i,j=1}^{\ell}$.

Note that equality (9.6.1) ensures that $P_{\ell}D$ is self-adjoint. Indeed, letting $L(\lambda) = \sum_{j=0}^{\ell} \lambda^{j}L_j$ and $L_1(\lambda) = \sum_{j=0}^{\ell} \lambda^{j}L_j^{(1)}$, formula (9.3.11) shows that

$$P_{\ell}D = \begin{bmatrix} L_0^{*} & 0 & \cdots & 0 \\ L_1^{*} & L_0^{*} & \cdots & 0 \\ \vdots & \vdots & \ddots & \vdots \\ L_{\ell-1}^{*} & L_{\ell-2}^{*} & \cdots & L_0^{*} \end{bmatrix} \begin{bmatrix} L_0 & L_1 & \cdots & L_{\ell-1} \\ 0 & L_0 & \cdots & L_{\ell-2} \\ \vdots & \vdots & \ddots & \vdots \\ 0 & 0 & \cdots & L_0 \end{bmatrix}$$

$$- \begin{bmatrix} L_0^{(1)*} & 0 & \cdots & 0 \\ L_1^{(1)*} & L_0^{(1)*} & \cdots & 0 \\ \vdots & \vdots & \ddots & \vdots \\ L_{\ell-1}^{(1)*} & L_{\ell-2}^{(1)*} & \cdots & L_0^{(1)*} \end{bmatrix} \begin{bmatrix} L_0^{(1)} & L_1^{(1)} & \cdots & L_{\ell-1}^{(1)} \\ 0 & L_0^{(1)} & \cdots & L_{\ell-2}^{(1)} \\ \vdots & \vdots & \ddots & \vdots \\ 0 & 0 & \cdots & L_0^{(1)} \end{bmatrix}$$

and the self-adjointness of $P_{\ell}D$ is evident. In particular, $\operatorname{In}_{\pi}(P_{\ell}D)$ exists (as follows from the spectral theorem for $P_{\ell}D$).

The rest of this section will be devoted to the proof of Theorem 9.6.1, which requires some preparation. It will be convenient to introduce the notation

$$M(X,T) = \sum_{j=0}^{\ell} M_j X T^j,$$

where $M(\lambda) = \sum_{j=0}^{\ell-1} \lambda^j M_j$ is an operator polynomial and (X,T) is a suitable pair of operators.

We prove first the following result.

PROPOSITION 9.6.2. *Let* $L_1(\lambda) = \sum_{j=0}^{\ell} \lambda^j L_{1j}$ *and* $L(\lambda) = \sum_{j=0}^{\ell} \lambda^j L_j$ *be operator polynomials with compact spectrum of degree* ℓ, *such that* $L_1(0) = I$ *and*

$$L_{1\infty}^*(\lambda) L_1(\lambda) = L_{\infty}^*(\lambda) L(\lambda). \tag{9.6.4}$$

Put $D = D_{L_{1\infty}^* L_{\infty}^*}(L, L_1)$ *and define*

$$D_0 = (\operatorname{col}[XT^{i-1}]_{i=1}^{\ell})^* \begin{bmatrix} 0 & & I \\ & I \cdot^{\cdot^{\cdot}} & \\ & & 0 \\ I & & \end{bmatrix} D(\operatorname{col}[XT^{i-1}]_{i=1}^{\ell}),$$

where (X,T) *is a right spectral pair for* $L(\lambda)$ *corresponding to the whole complex plane. Then*

$$D_0 - (T^*)^{\ell} D_0 T^{\ell} = -\tilde{S}^* \tilde{S}, \tag{9.6.5}$$

where

$$\tilde{S} = \operatorname{col}[L_1(X,T) T^{i-1}]_{i=1}^{\ell}.$$

PROOF. Denote by $(L_{1\infty}^*)_j$ and $(L_{\infty}^*)_j$ the jth coefficient of $L_{1\infty}^*(\lambda)$ and $L_{\infty}^*(\lambda)$, respectively, and compute using formula (9.3.11):

$$
\begin{aligned}
&D\,\mathrm{col}[XT^{i-1}]_{i=1}^{\ell} \\
&= \begin{bmatrix} (L_\infty^*)_1 & (L_\infty^*)_2 & \cdots & (L_\infty^*)_\ell \\ (L_\infty^*)_2 & & \cdot^{\cdot^{\cdot}} & 0 \\ \vdots & \cdot^{\cdot^{\cdot}} & & \vdots \\ (L_\infty^*)_\ell & 0 & \cdots & 0 \end{bmatrix} \begin{bmatrix} L_0 & L_1 & \cdots & L_{\ell-1} \\ 0 & L_0 & \cdots & L_{\ell-2} \\ \vdots & \vdots & \ddots & \vdots \\ 0 & 0 & \cdots & L_0 \end{bmatrix} \begin{bmatrix} X \\ XT \\ \vdots \\ XT^{\ell-1} \end{bmatrix} \\
&- \begin{bmatrix} (L_{1\infty}^*)_1 & (L_{1\infty}^*)_2 & \cdots & (L_{1\infty}^*)_\ell \\ (L_{1\infty}^*)_2 & & \cdot^{\cdot^{\cdot}} & 0 \\ \vdots & \cdot^{\cdot^{\cdot}} & & \vdots \\ (L_{1\infty}^*)_\ell & 0 & \cdots & 0 \end{bmatrix} \begin{bmatrix} L_{10} & L_{11} & \cdots & L_{1,\ell-1} \\ 0 & L_{10} & \cdots & L_{1,\ell-2} \\ \vdots & \vdots & \ddots & \vdots \\ 0 & 0 & \cdots & L_{10} \end{bmatrix} \begin{bmatrix} X \\ XT \\ \vdots \\ XT^{\ell-1} \end{bmatrix} \\
&= \begin{bmatrix} (L_\infty^*)_1 & (L_\infty^*)_2 & \cdots & (L_\infty^*)_\ell \\ (L_\infty^*)_2 & & \cdot^{\cdot^{\cdot}} & 0 \\ \vdots & \cdot^{\cdot^{\cdot}} & & \vdots \\ (L_\infty^*)_\ell & 0 & \cdots & 0 \end{bmatrix} \begin{bmatrix} L(X,T) \\ L(X,T)T \\ \vdots \\ L(X,T)T^{\ell-1} \end{bmatrix} \\
&- \begin{bmatrix} (L_{1\infty}^*)_1 & (L_{1\infty}^*)_2 & \cdots & (L_{1\infty}^*)_\ell \\ (L_{1\infty}^*)_2 & & \cdot^{\cdot^{\cdot}} & 0 \\ \vdots & \cdot^{\cdot^{\cdot}} & & \vdots \\ (L_{1\infty}^*)_\ell & 0 & \cdots & 0 \end{bmatrix} \begin{bmatrix} L_1(X,T) \\ L_1(X,T)T \\ \vdots \\ L_1(X,T)T^{\ell-1} \end{bmatrix} \\
&+ \begin{bmatrix} (L_\infty^*)_1 & (L_\infty^*)_2 & \cdots & (L_\infty^*)_\ell \\ (L_\infty^*)_2 & & \cdot^{\cdot^{\cdot}} & 0 \\ \vdots & \cdot^{\cdot^{\cdot}} & & \vdots \\ (L_\infty^*)_\ell & 0 & \cdots & 0 \end{bmatrix} \begin{bmatrix} -L_\ell XT^\ell \\ -L_{\ell-1}XT^\ell - L_\ell XT^{\ell+1} \\ \vdots \\ -L_1XT^\ell - L_2XT^{\ell+1} - \cdots - L_\ell XT^{2\ell-1} \end{bmatrix}
\end{aligned}
$$

$$+\begin{bmatrix} (L^*_{1\infty})_1 & (L^*_{1\infty})_2 & \cdots & (L^*_{1\infty})_\ell \\ (L^*_{1\infty})_2 & & \cdot & 0 \\ \vdots & \cdot & & \vdots \\ (L^*_{1\infty})_\ell & 0 & \cdots & 0 \end{bmatrix} \begin{bmatrix} L_{1\ell}XT^\ell \\ L_{1,\ell-1}XT^\ell + L_{1\ell}XT^{\ell+1} \\ \vdots \\ L_{11}XT^\ell + L_{12}XT^{\ell+1} + \cdots + L_{1\ell}XT^{2\ell-1} \end{bmatrix}.$$

Let us analyze the terms in the right-hand side of this equality. The first summand is clearly zero, because $L(X,T) = 0$. The sum of the third and fourth is zero because of (9.6.4). Thus

$$\text{(9.6.6)} \qquad D\ \mathrm{col}[XT^{i-1}]_{i=1}^{\ell}$$

$$= -\begin{bmatrix} (L^*_{1\infty})_1 & (L^*_{1\infty})_2 & \cdots & (L^*_{1\infty})_\ell \\ (L^*_{1\infty})_2 & & \cdot & 0 \\ \vdots & \cdot & & \vdots \\ (L^*_{1\infty})_\ell & 0 & \cdots & 0 \end{bmatrix} \begin{bmatrix} L_1(X,T) \\ L_1(X,T)T \\ \vdots \\ L_1(X,T)T^{\ell-1} \end{bmatrix}.$$

Our next observation is that

$$\text{(9.6.7)} \qquad \mathrm{col}[L_1(X,T)T^{i-1}]_{i=1}^{\ell}T = C_{L^*_{1\infty}}\ \mathrm{col}[L_1(X,T)T^{i-1}]_{i=1}^{\ell},$$

where $C_{L^*_{1\infty}}$ is the companion operator of the monic operator polynomial $C_{L^*_{1\infty}}$. To verify (9.6.7), simply observe that $L(\lambda)$ is a right divisor of the operator polynomial $L^*_{1\infty}(\lambda)L_1(\lambda)$, and hence (see Theorem 6.6.1(Q2) and Theorem 6.5.2)

$$\sum_{j=0}^{2\ell} Z_j XT^j = 0,$$

where Z_j are the coefficients of $L^*_{1\infty}(\lambda)L_1(\lambda)$ (the degree of this operator polynomial clearly does not exceed 2ℓ). Now

$$0 = \sum_{j=0}^{2\ell} Z_j X T^j = \sum_{\substack{0\le j-k\le \ell \\ 0\le k\le \ell}} (L_{1\infty}^*)_{j-k} L_{1k} X T^j$$

$$= \sum_{p=0}^{\ell} (L_{1\infty}^*)_p \left(\sum_{k=0}^{\ell} L_{1k} X T^k\right) T^p$$

$$= \sum_{p=0}^{\ell} (L_{1\infty}^*)_p L_1(X,T) T^p.$$

This equality clearly implies (9.6.7).

We continue to collect pieces of information leading to the proof of Proposition 9.6.2. The next piece is the equality

$$\begin{bmatrix} M_\ell & 0 & \cdots & 0 \\ M_{\ell-1} & M_\ell & \cdots & 0 \\ \vdots & \vdots & & \vdots \\ M_1 & M_2 & \cdots & M_\ell \end{bmatrix} K^\ell = - \begin{bmatrix} M_0 & M_1 & \cdots & M_{\ell-1} \\ 0 & M_0 & \cdots & M_{\ell-2} \\ \vdots & \vdots & & \vdots \\ 0 & 0 & \cdots & M_0 \end{bmatrix}, \tag{9.6.8}$$

where $M_0, \ldots, M_\ell \in L(\mathcal{X})$ with $M_\ell = I$ and

$$K = \begin{bmatrix} 0 & I & 0 & \cdots & 0 \\ 0 & 0 & I & \cdots & 0 \\ \vdots & \vdots & & & \vdots \\ -M_0 & -M_1 & -M_2 & \cdots & -M_{\ell-1} \end{bmatrix}$$

is the companion operator. The easiest way to prove (9.6.8) is by straightforward verification of the equalities

$$S_{m+1} K = S_m \quad (m = 0, 1, \ldots, \ell-1),$$

where

$$S_i = \left[\begin{array}{cccc|cccc} 0 & 0 & \cdots 0 & M_0 & & & & \\ \vdots & \vdots & . \; M_0 & M_1 & & & 0 & \\ 0 & M_0 \cdot^{\cdot} & & \vdots & & & & \\ M_0 & M_1 & \cdots & M_{\ell-i-1} & & & & \\ \hline & & & & -M_{\ell-i+1} & -M_{\ell-i+2} & \cdots & -M_\ell \\ & & & & -M_{\ell-i+2} & & & 0 \\ & & 0 & & \vdots & & \cdot^{\cdot^{\cdot}} & \vdots \\ & & & & -M_\ell & 0 & \cdots & 0 \end{array}\right]$$

and by definition

$$S_\ell = \begin{bmatrix} -M_1 & -M_2 & \cdots & -M_\ell \\ -M_2 & & \cdot & 0 \\ \vdots & \cdot & & \vdots \\ & -M_\ell & & 0 \\ -M_\ell & 0 & \cdots & 0 \end{bmatrix}, \quad S_0 = \begin{bmatrix} 0 & 0 & \cdots & 0 & M_0 \\ \vdots & \vdots & & M_0 & M_1 \\ 0 & M_0 & \cdot^{\cdot^{\cdot}} & & \vdots \\ M_0 & M_1 & \cdots & & M_{\ell-1} \end{bmatrix}.$$

In view of equalities (9.6.6), (9.6.7), and (9.6.8) (the latter is applied with $M_j = (L_{1\infty}^*)_j$, $j = 0,\ldots,\ell-1$), the proof of (9.6.5) is reduced to the verification of the following equality:

$$\text{(9.6.9)} \qquad (\mathrm{col}[XT^{i-1}]_{i=1}^{\ell})^* \begin{bmatrix} (L_{1\infty}^*)_\ell & 0 & \cdots & 0 \\ (L_{1\infty}^*)_{\ell-1} & (L_{1\infty}^*)_\ell & \cdots & 0 \\ \vdots & \vdots & & \vdots \\ (L_{1\infty}^*)_1 & (L_{1\infty}^*)_2 & \cdots & (L_{1\infty}^*)_\ell \end{bmatrix}$$

$$+ T^{*\ell}(\mathrm{col}[XT^{i-1}]_{i=1}^{\ell})^{*}\begin{bmatrix}(L_{1\infty}^{*})_0 & (L_{1\infty}^{*})_1 & \cdots & (L_{1\infty}^{*})_{\ell-1} \\ 0 & (L_{1\infty}^{*})_0 & \cdots & (L_{1\infty}^{*})_{\ell-2} \\ \vdots & \vdots & & \vdots \\ 0 & 0 & \cdots & (L_{1\infty}^{*})_0\end{bmatrix}$$

$$= (\mathrm{col}[L_1(X,T)T^{i-1}]_{i=1}^{\ell})^{*}.$$

It is convenient to pass to adjoint operators in both sides of (9.6.9) and verify the equality

$$\begin{bmatrix}L_{10} & L_{11} & \cdots & L_{1,\ell-1} \\ 0 & L_{10} & \cdots & L_{1,\ell-2} \\ \vdots & \vdots & & \vdots \\ 0 & 0 & \cdots & L_{10}\end{bmatrix}\mathrm{col}[XT^{i-1}]_{i=1}^{\ell}$$

$$+ \begin{bmatrix}L_{1\ell} & 0 & \cdots & 0 \\ L_{1,\ell-1} & L_{1\ell} & \cdots & 0 \\ \vdots & \vdots & \ddots & \vdots \\ L_{11} & L_{12} & \cdots & L_{1\ell}\end{bmatrix}\mathrm{col}[XT^{i-1}]_{i=1}^{\ell}T^{\ell}$$

$$= \mathrm{col}[L_1(X,T)T^{i-1}]_{i=1}^{\ell}.$$

This equality, however, follows easily from the definition of $L_1(X,T)$.

Proposition 9.6.2 is proved completely. ∎

We start now the proof of Theorem 9.6.1. The notation introduced in the statement of this theorem will be used without further explanation.

PROOF OF THEOREM 9.6.1. Let (X_L, T_L) be a right spectral pair for $L(\lambda)$ (with respect to the whole complex plane), and let (X_0, T_0) be a right spectral pair for L_0 (also with respect to $\mathbb{C}$).

Since L_0 is a right divisor of L, the pair (X_0, T_0) is a right restriction of (X_L, T_L) (see the divisibility theorem 6.5.2). Passing, if necessary, to a pair which is similar to (X_L, T_L), we can assume that

$$X_L = [X_1 \; X_0], \quad T_L = \begin{bmatrix} T_1 & 0 \\ T_2 & T_0 \end{bmatrix} \tag{9.6.11}$$

for some X_1, T_1, T_2, and the matrices in (9.6.11) are with respect to an orthogonal decomposition of the domain of T_L. Write

$$\tilde{X}_L \overset{\text{def}}{=} \mathrm{col}[X_L T_L^{i-1}]_{i=1}^{\ell} = [\tilde{X}_1 \tilde{X}_0],$$

where the partition of $\tilde{X}_L$ in the right-hand side is consistent with the partition in (9.6.11). Using the equality (9.6.6) and the fact that $L_1(X_0, T_0) = 0$ (because $L_0(\lambda)$ is a right divisor of $L_1(\lambda)$ as well), we conclude that $P_\ell D\tilde{X}_0 = 0$, where

$$P_\ell = \begin{bmatrix} 0 & & & I \\ & & \cdot^{\cdot^{\cdot}} & \\ & I & & \\ I & & & 0 \end{bmatrix}.$$

Consequently, the self-adjoint operator $D_0 \overset{\text{def}}{=} \tilde{X}_L^* P_\ell D\tilde{X}_L$ has the form

$$D_0 = \begin{bmatrix} \tilde{X}_1^* P_\ell D\tilde{X}_1 & 0 \\ 0 & 0 \end{bmatrix}. \tag{9.6.12}$$

By Proposition 9.6.2, we have, denoting $\tilde{S} = \mathrm{col}[L_1(X_L, T_L)T_L^{i-1}]_{i=1}^{\ell}$

$$\tilde{S}^*\tilde{S} = -\begin{bmatrix} \tilde{X}_1^* P_\ell D \tilde{X}_1 & 0 \\ 0 & 0 \end{bmatrix} + \begin{bmatrix} T_1^{*\ell} & * \\ 0 & T_0^{*\ell} \end{bmatrix} \begin{bmatrix} \tilde{X}_1^* P_\ell D \tilde{X}_1 & 0 \\ 0 & 0 \end{bmatrix}$$

$$\cdot \begin{bmatrix} T_1^\ell & 0 \\ * & T_0^\ell \end{bmatrix} = \begin{bmatrix} -\tilde{X}_1^* P_\ell D \tilde{X}_1 + T_1^{*\ell} \tilde{X}_1^* P_\ell D \tilde{X}_1 T_1^\ell & 0 \\ 0 & 0 \end{bmatrix}.$$

Decompose $\tilde{S} = [\tilde{S}_1, \tilde{S}_0]$, where

$$\tilde{S}_0 = \mathrm{col}[L_1(X_0, T_0) T_0^{i-1}]_{i=1}^{\ell}.$$

Let us prove that $\tilde{S}_1{}^*\tilde{S}_1$ is invertible. To this end, it is sufficient to check that Ker $\tilde{S}_1 = \{0\}$ and Im $\tilde{S}_1$ is closed. Let $\tilde{S}_1\phi = 0$ for some ϕ; so $\tilde{S}\begin{bmatrix}\phi \\ 0\end{bmatrix} = 0$. Formula (9.6.6) implies that $D\ \mathrm{col}[X_L T_L^{i-1}]_{i=1}^{\ell}\begin{bmatrix}\phi \\ 0\end{bmatrix} = 0$, or $D\tilde{X}_1\phi = 0$. So $\tilde{X}_1\phi \in$ Ker D. Theorem 9.4.2 says that Ker D = Im $\mathrm{col}[X_0 T_0^i]_{i=1}^{\ell-1}$ (at this point we use the assumption that spectral pairs of L and L_1 at infinity have empty common part, because $L(\lambda)$ has invertible leading coefficient). So $\tilde{X}_1\phi \in$ Im $\mathrm{col}[X_0 T_0^i]_{i=0}^{\ell-1}$; but $\tilde{X}_L$ is invertible (because the leading coefficient of L is invertible), consequently $\phi = 0$.

We prove now that Im $\tilde{S}_1$ is closed. It is sufficient to show that $\|\tilde{S}_1\phi\| \geq \alpha\|\phi\|$ for all ϕ, where $\alpha > 0$ is independent on ϕ. Suppose the contrary; then there is a sequence $\{\phi_n\}_{n=1}^{\infty}$, $\|\phi_n\| = 1$ such that $\tilde{S}_1\phi_n \to 0$. Formula (9.6.6) implies that $D\tilde{X}_1\phi_n \to 0$. Since $\tilde{X}_1$ is left invertible, $\|\tilde{X}_1\phi_n\| \geq \gamma_1 > 0$ for some γ_1 independent on n. As $\tilde{X}_L = [\tilde{X}_1, \tilde{X}_0]$ is invertible, we see that the distance between $\tilde{X}_1\phi_n$ and Im $\tilde{X}_0$ is bounded below by a positive number γ_2 independent of n. Let V be a direct complement to Ker D = Im $\tilde{X}_0$ in $\mathcal{X}^\ell$, and let P_V be the projection on V along Ker D = Im $\tilde{X}_0$. Then

(9.6.13) $$\|P_V\tilde{X}_1\phi_n\| \geq \gamma_3 > 0;$$

on the other hand, $D\tilde{X}_1\phi_n \to 0$ implies

(9.6.14) $$DP_V\tilde{X}_1\phi_n \to 0.$$

But $D_{|V}$ is left invertible; so (9.6.13) and (9.6.14) are contradictory.

We have proved that $\tilde{S}_1{*}\tilde{S}_1$ is invertible. Now

$$\tilde{S}_1{*}\tilde{S}_1 = -\tilde{X}_1^*P_\ell D\tilde{X}_1 + T_1^{*\ell} \cdot (\tilde{X}_1^*P_\ell D\tilde{X}_1) \cdot T_1^\ell .$$

The condition on Θ implies that $\lambda_0 \notin \sigma(T_L^\ell) \cup \sigma(T_0^\ell)$ for some λ_0 on the unit circle. So $\lambda_0 \notin \sigma(T_1^\ell)$, Theorem 9.5.4 is applicable, and we obtain

(9.6.15) $$\sigma(T_1^\ell) \cap \{\text{unit circle}\} = \phi,$$

(9.6.16) $$\mathrm{In}_\Delta(T_1^\ell) = \mathrm{In}_\pi(-\tilde{X}_1^*P_\ell D\tilde{X}_1).$$

Because of (9.6.15), the equality $\mathrm{In}_\Delta(T_1^\ell) = \mathrm{In}_\Delta(T_1)$ holds. Further, formula (9.6.12) implies $i_\pm^\pi(\tilde{X}_1^*P_\ell DX_1) = i_\pm^\pi(D_0)$; since $\tilde{X}_L$ is invertible, we have also $i_\pm^\pi(D_0) = i_\pm^\pi(P_\ell D)$. So

(9.6.17) $$i_\pm^\Delta(T_1) = i_\pm^\pi(-P_\ell D).$$

But

$$i_\pm(L) = i_\pm^\Delta(T_L) = i_\pm^\Delta(T_1)+i_\pm^\Delta(T_0) = i_\pm^\Delta(T_1)+i_\pm(L_0).$$

If both $i_+(L)$ and $i_+(L_0)$ are infinite, then (9.6.2) is evident. If at least one of $i_+(L), i_+(L_0)$ is finite, then

(9.6.18) $$i_+^\Delta(T) = i_+(L)-i_+(L_0) \geq 0,$$

so in fact $i_+(L_0)$ must be finite, and (9.6.2) follows from (9.6.17) and (9.6.18). The same argument establishes (9.6.3). ∎

9.7 Spectrum separation problem: deductions and special cases

In this section we derive some useful general information and special cases from Theorem 9.6.1 and its proof.

Firstly, we observe that the hypotheses of Theorem 9.6.1 can be relaxed somewhat (with the expense of being more technical). Namely, the condition that L has invertible leading coefficient may be replaced by a weaker condition that L_1 and L have no common spectral pairs at infinity, i.e., the greatest right common divisor of $\lambda^{\ell}L_1(\lambda^{-1})$ and $\lambda^{\ell}L(\lambda^{-1})$ over a small neighborhood of zero is trivial (can be taken identically I).

Next, we state a result analogous to Theorem 9.6.1 which does not involve the notion of a greatest common divisor of L and L_1.

THEOREM 9.7.1. *Let* $L_1(\lambda)$ *and* $L(\lambda)$ *be operator polynomials of degree* ℓ *such that* L *has invertible leading coefficient,* $L_1(0) = I$ *and* (9.6.1) *holds. Assume that there is a point* λ_0 *on the unit circle which belongs to the unbounded connected component of the set* $\mathbb{C}\backslash\{\lambda^{\ell} \mid L(\lambda)$ *is not invertible*$\}$. *Further assume that* Im D *is closed, where* $D = D_{L_{1\infty}^*, L_{\infty}^*}(L, L_1)$. *Then* L *is invertible at every point on the unit circle, and letting* $P_{\ell} = [\delta_{i,\ell-j+1}I]_{i,j=1}^{\ell}$ *the inequalities*

$$i_+(L) \geq i_+^{\pi}(-P_{\ell}D); \; i_-(L) \geq i_-^{\pi}(-P_{\ell}D)$$

hold.

The proof of Theorem 9.7.1 is the same as that of Theorem 9.6.1. Note that the hypothesis on λ_0 implies (but generally is not equivalent to) that the set

$$\{\lambda^{\ell} \mid |\lambda| = 1 \text{ and } L(\lambda) \text{ is not invertible}\}$$

does not cover the whole unit circle. We need this stronger hypothesis in order to be able to deduce $\lambda_0 \notin \sigma(T_1^{\ell})$ from $\lambda_0 \notin \sigma(T_L^{\ell})$; here T_L is a global linearization for L and

$$T_L = \begin{bmatrix} T_1 & 0 \\ T_2 & T_0 \end{bmatrix}$$

for some operators T_0 and T_2.

The following particular case of Theorem 9.6.1 deserves to be stated separately.

COROLLARY 9.7.2. *Let* $L(\lambda)$ *and* $L_1(\lambda)$ *be operator polynomials with compact spectrum of degree* ℓ *such that* (9.6.1) *holds. Assume that* L *has invertible leading coefficient,* $\Sigma(L) \cap \{\lambda \mid |\lambda| = 1\} = \emptyset$, $L_1(0) = I$, *and the polynomials* L *and* L_1 *are relatively coprime (i.e., the spectrum of their greatest right common divisor is empty). Assume further that* $\operatorname{Im} D_0$ *is closed, where*

$$D_0 = \begin{bmatrix} 0 & 0 & \cdots & I \\ 0 & 0 & \cdots & 0 \\ \vdots & \vdots & & \vdots \\ 0 & I & \cdots & 0 \\ I & 0 & \cdots & 0 \end{bmatrix} D_{L_1^*, L^*}(L, L_1).$$

Then the spectrum of L *is inside the unit circle if and only if* $\langle D_0 x, x\rangle \leq 0$ *for every* $x \in \mathcal{X}^\ell$. *The spectrum of* L *lies outside of the unit circle if and only if* $\langle D_0 x, x\rangle \geq 0$ *for every* $x \in \mathcal{X}^\ell$.

Consider now the spectrum separation problem for the class of operator polynomials of the form

$$M(\lambda) = \sum_{j=0}^{m} \lambda^j (a_j I + M_j), \tag{9.7.1}$$

where $a_0, \ldots, a_m$ are complex numbers, and M_j are compact operators acting in a Hilbert space $\mathcal{X}$. Denote by $\alpha_M(\lambda)$ the scalar polynomial $\sum_{j=0}^{m} \lambda^j a_j$. The following spectrum separation result is obtained using Theorem 9.6.1 (it will be assumed here that the Hilbert space $\mathcal{X}$ is infinite dimensional).

THEOREM 9.7.3. *Let* $L_1(\lambda)$ *and* $L(\lambda)$ *be operator polynomials with compact spectrum of the form* (9.7.1) *and of degree* ℓ, *such that* $L_1(0) = I$ *and the leading coefficient of* L *is invertible. Assume* (9.6.1) *holds. If* $a(\lambda) \overset{\text{def}}{=} \alpha_L(\lambda)$ *and*

$b(\lambda) \overset{\text{def}}{=} \alpha_{L_1}(\lambda)$ *have no common zeros, then* L *and the greatest right common divisor* L_0 *of* L *and* L_1 *are invertible at every point on the unit circle, and equalities* (9.6.2), (9.6.3) *hold*.

PROOF. We have to check the conditions of Theorem 9.6.1. First, Theorem 1.4.3 implies that the spectrum $\Sigma(L)$ is at most countable. Then, clearly, the set $\{\lambda^{\ell} \mid \lambda \in \Theta\}$ does not cover the whole unit circle. Further, the equality

$$b_{\infty}^{*}(\lambda)b(\lambda) = a_{\infty}^{*}(\lambda)a(\lambda)$$

follows easily from (9.6.1) (in view of the particular form (9.7.1) of L and L_1). Now, the Bezoutian $D = D_{L_{1\infty}^{*}, L_{\infty}^{*}}(L, L_1)$ can be written in the form

$$D = \tilde{D} \otimes I + \Phi,$$

where $\Phi: \mathcal{X}^{\ell} \to \mathcal{X}^{\ell}$ is compact and $\tilde{D} = D_{b_{\infty}^{*}, a_{\infty}^{*}}(a, b)$ is the $\ell \times \ell$ matrix. As invertibility of the operator $\tilde{D} \otimes I: \mathcal{X}^{\ell} \to \mathcal{X}^{\ell}$ is equivalent to that of $\tilde{D}$, we obtain that Im D is closed (even has finite codimension) provided the $\ell \times \ell$ matrix $\tilde{D}$ is invertible. Finally, observe that by Theorem 9.4.2 the invertibility of $\tilde{D}$ is equivalent to the relative primeness of the polynomials $a(\lambda)$ and $b(\lambda)$. ■

One can easily extend Theorem 9.7.3 to the framework of operator polynomials with coefficients in the block operator form

$$[\alpha_{ij} I + M_{ij}]_{i,j=1}^{m},$$

where $M_{ij} \in L(\mathcal{Y})$ are compact operators, and it is assumed that $\mathcal{X} = \mathcal{Y}^{m}$ for some Hilbert space $\mathcal{Y}$. We leave it to the reader to state and prove this result.

9.8 Application to difference equations

Consider the difference equation

$$A_\ell x_{i+\ell}+A_{\ell-1}x_{i+\ell-1}+\cdots+A_0x_i = 0, \quad i = 0,1,\dots, \tag{9.8.1}$$

where $\{x_i\}_{i=0}^{\infty}$ is a sequence of vectors in a Hilbert space $\mathcal{X}$, and $A_j \in L(\mathcal{X})$, $j = 0,1,\dots,\ell$. We shall assume that A_ℓ is invertible. Then a general solution $\{x_i\}_{i=0}^{\infty}$ of (9.8.1) is uniquely determined by the ordered ℓ-tuple of initial vectors $\{x_i\}_{i=0}^{\ell-1} \in \mathcal{X}^\ell$. We say that equation (9.8.1) admits *geometric dichotomy* if $\mathcal{X}^\ell$ can be decomposed into a direct sum of two subspaces $\mathcal{M}_+$ and $\mathcal{M}_-$ with the following property: there exists ρ, $0 < \rho < 1$, such that every solution $\{x_i\}_{i=0}^{\infty}$ of (9.8.1) with $\{x_i\}_{i=0}^{\ell-1} \in \mathcal{M}_+$ (resp. $\{x_i\}_{i=0}^{\ell-1} \in \mathcal{M}_-\backslash\{0\}$) satisfies the inequalities $\|x_i\| \le \rho^i$ (resp. $\|x_i\| \ge \rho^{-i}$) for i large enough. The ordered pair ($\dim \mathcal{M}_+$, $\dim \mathcal{M}_-$) will be called the *index* of the geometric dichotomy.

The results on the spectrum separation problem given in Sections 9.6 and 9.7 can be applied to prove the geometric dichotomy property for difference equation (9.8.1) under certain conditions. Let us state one theorem of this type.

THEOREM 9.8.1. *Let* $L_1(\lambda), L(\lambda) = \sum_{j=0}^{\ell} \lambda^j A_j$, $L_0(\lambda)$ *and* $D = D_{L_{1\infty}^*, L_\infty^*}(L,L_1)$ *be as in Theorem* 9.6.1. *Then the equation* (9.8.1) *admits geometric dichotomy with index*

$$(i_+^\pi(-P_\ell D)+i_+(L_0),\ i_-^\pi(-P_\ell D)+i_-(L_0)), \tag{9.8.2}$$

where $P_\ell = [\delta_{i,\ell-j+1}]_{i,j=1}^{\ell}$.

PROOF. It is not difficult to see that the general solution of (9.8.1) is given by the formula

$$x_i = XT^iz, \quad i = 0,1,\dots, \tag{9.8.3}$$

where (X,T) is a right spectral pair for the operator polynomial $L(\lambda) = \sum_{j=0}^{\ell} \lambda^j A_j$ (with respect to $\mathbb{C}$), and z is an arbitrary vector in the domain of T. Indeed, since A_ℓ is invertible, the operator $\mathrm{col}[XT^i]_{i=0}^{\ell-1}$ is invertible as well; so given the initial vectors $\{x_0, \ldots, x_{\ell-1}\}$ define z by

$$\mathrm{col}[x_i]_{i=0}^{\ell-1} = \mathrm{col}[XT^i]_{i=0}^{\ell-1} z.$$

This gives (9.8.3) for $i = 0,1,\ldots,\ell-1$. For the other values of i, (9.8.3) follows from the equality

$$\sum_{i=0}^{\ell} A_i X T^i = 0$$

(see formula (2.2.1)). By Theorem 9.6.1, the spectrum of T does not intersect the unit circle. The geometric dichotomy of (9.8.1) is given then by the subspaces $\mathcal{M}_\pm = [\mathrm{col}[XT^i]_{i=0}^{\ell-1}]|_{\mathcal{N}_\pm}$, where $\mathcal{N}_+$ (resp. $\mathcal{N}_-$) is the spectral subspace of T corresponding to the part of $\sigma(T)$ lying inside (resp. outside) the unit circle. The formula (9.8.2) follows again from Theorem 9.6.1. ∎

We leave to the reader the statements of other results on geometric dichotomy which can be obtained using one of the results of Section 9.7 in place of Theorem 9.6.1.

9.9 Notes

The notions of Bezoutian and resultant for scalar polynomials are classical. Recently, these notions were successfully extended to matrix and operator polynomials and even to more generalized functions. A strong impetus for this development, besides the inertia theory and the spectrum separation problems, came from the theory of multivariable linear control. List of relevant references (far from complete) includes Anderson-Jury [1], Barnett-Lancaster [1], Gohberg-Lerer [1], Gohberg-Heinig [2,3], Lerer-Rodman-Tismenetsky [1], Lerer-Tismenetsky [1,2], Clancey-Kon [1], and Lancaster-Maroulas [1].

Theorem 9.1.4 is proved using the same ideas as in the finite dimensional case (Gohberg-Kaashoek-Lerer-Rodman [1]). Lemma 9.2.2 is an infinite dimensional version of a result proved in Gohberg-Lerer-Rodman [3]. Various equalities (as in Section 9.3) satisfied by the Bezoutian are found in Clancey-Kon [1] and Lancaster-Maroulas [1]. Theorem 9.4.2 was proved in Lerer-Rodman-Timenetsky [1], and in Lerer-Tismenetsky [1] in the finite dimensional case. The proof of Theorem 9.5.1 follows Daleckii-Kreĭn [1]. Further results on inertia theorems in finite dimensional Hilbert spaces can be found in Cain [1,2]. Example 9.5.1 is taken from Cain [2].

For scalar polynomials Theorem 9.6.1 is a classical result (Fujiwara [1], Kreĭn-Naimark [1]); see also Datta [1,2] for modern treatment. This theorem for matrix polynomials was proved in Lerer-Tismenetsky [1] and in full generality in Lerer-Rodman-Tismenesky [1]. The exposition in Sections 9.7 and 9.8 follows Lerer-Rodman-Tismenetsky [1].

CHAPTER 10. WIENER-HOPF FACTORIZATION

10.1 Definition and the main result

Let Δ be the domain in the complex plane bounded by a simple closed rectifiable contour Γ. It will be assumed that $0 \in \Delta$. An operator function $W: \Gamma \to L(\mathcal{Y})$, where $\mathcal{Y}$ is a Banach space, is said to admit *left Wiener-Hopf factorization* with respect to Γ if the following representation holds:

$$W(\lambda) = E_-(\lambda)\Big(\sum_{i=1}^{r} \lambda^{\nu_i} P_i\Big)E_+(\lambda) \tag{10.1.1}$$

where the continuous operator function $E_-: (\mathbb{C}\cup\{\infty\})\backslash\Delta \to L(\mathcal{Y})$ is analytic on $(\mathbb{C}\cup\{\infty\})\backslash\Delta$ and all its values are invertible, the continuous operator function $E_+: \overline{\Delta} \to L(\mathcal{Y})$ is analytic in Δ and all its values are invertible, $P_1,\dots,P_r$ are non-zero projections with $P_iP_j = P_jP_i = 0$ for $i \neq j$ and $P_1+\cdots+P_r = I$, the numbers $\nu_1 < \nu_2 <\cdots< \nu_r$ are integers (positive, negative, or zero). Interchanging E_+ and E_- in (10.1.1), we obtain a *right Wiener-Hopf factorization*.

In the formula (10.1.1) the point $\lambda_0 = 0$ plays a special role (that is why we have assumed $0 \in \Delta$). One could consider the Wiener-Hopf factorization with any fixed point $\lambda_0 \in \Delta$ being special; in this case the right-hand side of (10.1.1) is replaced by

$$E_-(\lambda)\Big(\sum_{i=1}^{r} (\lambda-\lambda_0)^{\nu_i} P_i\Big)E_+(\lambda).$$

As the case when $\lambda_0 \in \Delta$ is special is reduced to the case when $\lambda_0 = 0$ is special by a simple change of variables, we will study throughout this chapter the Wiener-Hopf factorization (10.1.1) and its right counterpart (implicitly assuming that $0 \in \Delta$).

Clearly, a necessary condition for existence of left (or right) Wiener-Hopf factorization is that the function $W(\lambda)$ be continuous and invertible on Γ. It is well known that these conditions are not sufficient already in the scalar case ($\dim \mathcal{Y} = 1$); an example of a non-zero continuous scalar function on the unit circle does not admit Wiener-Hopf factorization (with respect to the unit circle) is

$$f(\lambda) = \sum_{n=2}^{\infty} \frac{1}{n \ell n\, n} (\lambda^{n} - \lambda^{-n}), \quad |\lambda| = 1.$$

See, e.g., Gohberg-Feldman [1], Section 1.5 for more details.

However, there are many large classes of continuous and invertible functions on Γ which admit left and right Wiener-Hopf factorizations with respect to Γ in the finite-dimensional case ($\dim \mathcal{Y} < \infty$), e.g., Hölder functions, and, of course, polynomials. In contrast with the finite-dimensional case, there are operator polynomials with invertible values on Γ that do not admit Wiener-Hopf factorizations, as the following example shows.

EXAMPLE 10.1.1. Let Γ be the unit circle, and $\mathcal{Y}$ be the Hilbert space of two-dimensional vector functions, each coordinate of which belongs to $L_2[0,1]$. The multiplication operators acting in $\mathcal{Y}$ will be naturally written as 2×2 matrix functions of $t \in [0,1]$. Let

$$W(\lambda) = \begin{bmatrix} 0 & 0 \\ t & 0 \end{bmatrix} + \lambda \begin{bmatrix} t-1 & 0 \\ t & t+1 \end{bmatrix} + \lambda^{2} \begin{bmatrix} 0 & t \\ 0 & 0 \end{bmatrix}. \tag{10.1.2}$$

This operator polynomial with coefficients in $L(\mathcal{Y})$ is invertible for every $|\lambda| = 1$ (because

$$\det \begin{bmatrix} \lambda(t-1) & \lambda^{2} t \\ t & \lambda(t+1) \end{bmatrix} = -\lambda^{2} \neq 0 \text{ for } |\lambda| = 1).$$

However, $W(\lambda)$ does not admit left Wiener-Hopf factorization with respect to Γ (but admits a right Wiener-Hopf factorization). This fact will be obtained as an application of the main result on existence of Wiener-Hopf factorizations for operator polynomials. ■

In this section we present criteria for the existence of Wiener-Hopf factorizations of an operator polynomial $W(\lambda)$ in terms of the moments of the inverse function $W(\lambda)$ with respect to Γ. For a continuous function $V: \Gamma \to L(\mathcal{Y})$ we define the operators of moments to be:

(10.1.3)

$$M_{pq}(V) = \frac{1}{2\pi i} \int_{\Gamma} \begin{bmatrix} V(\lambda) & \lambda V(\lambda) & \cdots & \lambda^{q-1}V(\lambda) \\ \lambda V(\lambda) & \lambda^2 V(\lambda) & \cdots & \lambda^q V(\lambda) \\ \vdots & \vdots & & \vdots \\ \lambda^{p-1}V(\lambda) & \lambda^p V(\lambda) & \cdots & \lambda^{p+q-2}V(\lambda) \end{bmatrix} d\lambda : \mathcal{Y}^q \to \mathcal{Y}^p.$$

Recall that an operator $Y \in L(\mathcal{X},\mathcal{Y})$ is called a generalized inverse of $X \in L(\mathcal{Y},\mathcal{X})$ if $YXY = Y$ and $XYX = X$; existence of a generalized inverse for X is equivalent to the conditions that $\operatorname{Ker} X$ is complemented and $\operatorname{Im} X$ is closed and complemented.

THEOREM 10.1.1. *Let $W(\lambda)$ be an operator polynomial with coefficient operators acting on a Banach space $\mathcal{Y}$, and such that $W(\lambda)$ is invertible for every $\lambda \in \Gamma$. Then $W(\lambda)$ admits a left Wiener-Hopf factorization with respect to Γ if and only if all the operators*

$$M_{m1}(W^{-1}), \ldots, M_{m,m-1}(W^{-1})$$

have generalized inverses, where m is the degree of $W(\lambda)$. The polynomial $W(\lambda)$ admits a right Wiener-Hopf factorization with respect to Γ if and only if all the operators

$$M_{1m}(W^{-1}), \ldots, M_{m-1,m}(W^{-1})$$

have generalized inverses.

Recall that we always assume $0 \in \Delta$.

It will be clear from the proof of Theorem 10.1.1 that one can replace m in Theorem 10.1.1 by any integer greater than or equal to the degree of $W(\lambda)$.

The proof of Theorem 10.1.1 will be given in the next section.

To illustrate Theorem 10.1.1, let us go back to Example 10.1.1. An easy calculation shows that for $W(\lambda)$ given by (10.1.2) the operators $M_{21}(W^{-1})$ and $M_{12}(W^{-1})$ are multiplication operators given by the matrices

$$\begin{bmatrix} -(t+1) & 0 \\ 0 & -(t-1) \\ 0 & 0 \\ t & 0 \end{bmatrix} \text{ and } \begin{bmatrix} -(t+1) & 0 & 0 & 0 \\ 0 & -(t-1) & t & 0 \end{bmatrix} .$$

As the multiplication operator by $-(t-1)$ on $L_2[0,1]$ does not have a generalized inverse, it follows from Theorem 10.1.1 that $W(\lambda)$ does not admit left Wiener-Hopf factorization with respect to the unit circle. On the other hand, the multiplication operator by $-(t+1)$ on $L_2[0,1]$ is invertible, and the multiplication operator by $[-(t-1) \quad t]$ (it maps $L_2[0,1] \oplus L_2[0,1]$ into $L_2[0,1]$) is right invertible, with one of its right inverses is the multiplication operator by

$$\begin{bmatrix} t+1 \\ t \end{bmatrix} .$$

So $M_{12}(W^{-1})$ is actually right invertible, and by Theorem 10.1.1 $W(\lambda)$ admits right Wiener-Hopf factorization with respect to the unit circle.

10.2 Pairs of finite type and proof of Theorem 10.1.1

For the proof of Theorem 10.1.1, we need auxiliary results that describe Wiener-Hopf factorization in terms of spectral pairs. To state these results, we introduce the notion of pairs of operators of finite type.

Let $\mathcal{X}$ and $\mathcal{Y}$ be Banach spaces, and let $A \in L(\mathcal{X})$, $B \in L(\mathcal{Y},\mathcal{X})$. We call the pair (A,B) of *finite type* if there exists a positive integer ℓ such that the operator

$$\Delta_j \overset{\text{def}}{=} [B, AB, \ldots, A^{j-1}B] \in L(\mathcal{Y}^j, \mathcal{X})$$

has a generalized inverse for $1 \leq j \leq \ell-1$ and is right invertible for $j = \ell$.

THEOREM 10.2.1. *Let* $L(\lambda)$ *be an operator polynomial invertible for* $\lambda \in \Gamma$, *and let* (A,B) *be a left* Δ*-spectral pair for* $L(\lambda)$. *Then* $L(\lambda)$ *admits with respect to* Γ *a left Wiener-Hopf factorization*

$$L(\lambda) = E_-(\lambda)\left(\Sigma_{i=1}^{r} \lambda^{\nu_i} P_i\right) E_+(\lambda) \tag{10.2.1}$$

if and only if the pair (A,B) *is of finite type.*

The proof of this result is given in Gohberg-Kaashoek-van Schagen [2]; it is too long to be reproduced here.

To formulate the analogous theorem for right Wiener-Hopf factorization, we have to consider right Δ-spectral pairs. A pair (C,A) of operators, where $C \in L(\mathcal{X},\mathcal{Y})$, $A \in L(\mathcal{X})$ is said to be of *finite type* if there exists a positive integer ℓ such that the operator

$$\Omega_j \overset{\text{def}}{=} \operatorname{col}[CA^{i-1}]_{i=1}^{j} : \mathcal{X} \to \mathcal{Y}^j$$

has a generalized inverse for $1 \leq j \leq \ell-1$ and is left invertible for $j = \ell$.

THEOREM 10.2.2. *Let* $L(\lambda)$ *be an operator polynomial invertible for* $\lambda \in \Gamma$, *and let* (C,A) *be a right* Δ*-spectral pair for* $L(\lambda)$. *Then* $L(\lambda)$ *admits with respect to* Γ *a right Wiener-Hopf factorization*

$$L(\lambda) = E_+(\lambda)\left(\Sigma_{i=1}^{r} \lambda^{\kappa_i} P_i\right) E_-(\lambda)$$

if and only if the pair (C,A) *is of finite type.*

For the proof of Theorem 10.2.2 we refer the reader again to Gohberg-Kaashoek-van Schagen [2].

With the help of these two theorems, we are ready to prove Theorem 10.1.1. Let $W(\lambda)$ be an operator polynomial of degree m as in Theorem 10.1.1, and let (C,A,B) be a spectral triple of $W(\lambda)$ with respect to Δ. Then the operators $M_{pq}(W^{-1})$ (defined by (10.1.3)) can be expressed in the form

$$M_{pq}(W^{-1}) = \begin{bmatrix} C \\ CA \\ \vdots \\ CA^{p-1} \end{bmatrix} [B, AB, \dots, A^{q-1}B]. \tag{10.2.3}$$

Assume now that $W(\lambda)$ admits left Wiener-Hopf factorization with respect to Γ. By Theorem 10.2.1, the pair (A,B) is of finite type. As the operator $\mathrm{col}[CA^j]_{j=0}^{m-1}$ is left invertible (Theorem 6.6.1), formula (10.2.3) shows that the operators $M_{mq}(W^{-1})$, q = 1,2,... have generalized inverses.

Conversely, assume the operators $M_{mq}(W^{-1})$, q = 1,2,...,m-1 have generalized inverses. Formula (10.2.3) shows that the same is true for the operators

$$[B, AB, \dots, A^{q-1}B], \quad q = 1,2,\dots,m-1.$$

As we know that $[B, AB, \dots, A^{m-1}B]$ is right invertible, it remains to appeal to Theorem 10.2.1.

The second part of Theorem 10.1.1 (concerning right Wiener-Hopf factorization) is proved analogously.

10.3 Finite-dimensional perturbations

Let Δ be a domain bounded by a contour Γ with the properties described in the beginning of Section 10.1. In this section we study the behavior of Wiener-Hopf factorizations of operator polynomials with respect to Γ under finite-dimensional perturbations. The basic result in this direction is the following.

THEOREM 10.3.1. *Let $W(\lambda)$ be an operator polynomial which is invertible on Γ and admits left (or right) Wiener-Hopf factorization with respect to Γ. If $\tilde{W}(\lambda)$ is an operator polynomial invertible on Γ for which*

$$\operatorname{Im}(\tilde{W}(\lambda)^{-1}-W(\lambda)^{-1}) \subset \mathcal{M}, \ \lambda \in \Gamma,$$

where $\mathcal{M}$ *is a fixed (i.e., independent on* $\lambda \in \Gamma$*) finite-dimensional subspace, then* $\tilde{W}(\lambda)$ *admits left (or right) Wiener-Hopf factorization with respect to* Γ *as well.*

The proof is immediate using Theorem 10.1.1 and Lemma 10.6.2 below (indeed, under the hypotheses of Theorem 10.1.1, the differences $M_{pq}(W^{-1})-M_{pq}(\tilde{W}^{-1})$ are finite rank operators for $p,q \geq 1$).

LEMMA 10.3.2. *Let* $A \in L(\mathcal{X},\mathcal{Y})$ *be an operator with generalized inverse (here* $\mathcal{X}$ *and* $\mathcal{Y}$ *are Banach spaces). Then for any finite rank operator* $K \in (\mathcal{X},\mathcal{Y})$ *the operator* $A+K$ *also has a generalized inverse.*

PROOF. With respect to the direct sum decomposition $\mathcal{X} = \mathcal{X}_1 \dot{+} \operatorname{Ker} A$, $\mathcal{Y} = \operatorname{Im} A \dot{+} \mathcal{Y}_1$, where $\mathcal{X}_1$ and $\mathcal{Y}_1$ are suitable complemented subspaces of $\mathcal{X}$ and $\mathcal{Y}$, respectively, write

$$A = \begin{bmatrix} A_{11} & 0 \\ 0 & 0 \end{bmatrix}, \quad K = \begin{bmatrix} K_{11} & K_{12} \\ K_{21} & K_{22} \end{bmatrix} .$$

Here $A_{11} \in L(\mathcal{X}_1, \operatorname{Im} A)$ is invertible, and K_{ij} $(1 \leq i,j \leq 2)$ are finite rank operators. Now clearly

$$(10.3.1) \qquad \operatorname{Im}(A_{11}+K_{11}) \subset \operatorname{Im}(A+K)$$
$$\subset \operatorname{Im}(A_{11}+K_{11}) + \operatorname{Im} K_{12} + \operatorname{Im} K_{21} + \operatorname{Im} K_{22}.$$

As $A_{11}+K_{11} \in L(\mathcal{X}_1, \operatorname{Im} A)$ is a Fredholm operator, the subspace $\operatorname{Im}(A_{11}+K_{11})$ is closed and has finite codimension in $\operatorname{Im} A$. Consequently, $\operatorname{Im}(A_{11}+K_{11})$ is complemented in $\operatorname{Im} A$, and hence it is complemented in $\mathcal{Y}$. Say

$$(10.3.2) \qquad \operatorname{Im}(A_{11}+K_{11}) \dot{+} \mathcal{Y}_2 = \mathcal{Y},$$

where $\mathcal{Y}_2$ is a (closed) subspace in $\mathcal{Y}$. The inclusions (10.3.1) show that

$$(10.3.3) \qquad \operatorname{Im}(A+K) = \operatorname{Im}(A_{11}+K_{11}) + \operatorname{Span}\{y_1, \ldots, y_m\},$$

for a finite set of vectors $\{y_1,\ldots,y_m\}$ which are linearly independent modulo $\mathrm{Im}(A_{11}+K_{11})$. Using (10.3.2), we choose $y_1,\ldots,y_m$ to belong to $\mathcal{Y}_2$. Further, (10.3.2) implies that

$$\|x+y\| \geq \alpha(\|x\|+\|y\|) \tag{10.3.4}$$

for any $x \in \mathrm{Im}(A_{11}+K_{11})$, $y \in \mathrm{Span}\{y_1,\ldots,y_m\}$, where the positive constant α is independent on x and y. Using (10.3.3) and (10.3.4), one easily shows that $\mathrm{Im}(A+K)$ is closed. The complementedness of $\mathrm{Im}(A+K)$ is clear from (10.3.2) and (10.3.3).

Now consider $\mathrm{Ker}(A+K)$. We have to prove that $\mathrm{Ker}(A+K)$ is complemented. As

$$\mathrm{Ker}(A+K) = \mathrm{Ker}[A_{11}+K_{11},K_{12}] \cap \mathrm{Ker}[K_{21},K_{22}]$$

and $\mathrm{Ker}[K_{21},K_{22}]$ has finite codimension, it remains to prove that $\mathrm{Ker}[A_{11}+K_{11},K_{12}]$ is complemented.

The operator $A_{11}+K_{11}$ is Fredholm, so we can write

$$A_{11}+K_{11} = \begin{bmatrix} B & 0 \\ 0 & 0 \end{bmatrix}$$

with respect to direct sum decompositions $\mathcal{X}_1 = \mathcal{X}'\dot{+}\mathrm{Ker}(A_{11}+K_{11})$, $\mathrm{Im}\, A = \mathrm{Im}(A_{11}+K_{11})\dot{+}\mathcal{Y}'$, where $\mathcal{Y}'$ and $\mathrm{Ker}(A_{11}+K_{11})$ are finite-dimensional and $B \in L(\mathcal{X}',\mathrm{Im}(A_{11}+K_{11}))$ is invertible. With respect to the same decomposition $\mathcal{X}_1 = \mathcal{X}'\dot{+}\mathrm{Ker}(A_{11}+K_{11})$ write

$$K_{12} = \begin{bmatrix} L_1 \\ L_2 \end{bmatrix} .$$

Then

$$\mathrm{Ker}[A_{11}+K_{11},K_{12}] = (\mathrm{Ker}(A_{11}+K_{11})\dot{+}\mathrm{Ker}[B,L_1]) \cap \mathrm{Ker}\, L_2 .$$

We have reduced the proof to the verification that $\mathrm{Ker}[B,L_1]$ is complemented. However, the operator $[B,L_1]$ is right invertible ($\begin{bmatrix} B^{-1} \\ 0 \end{bmatrix}$ is one of its right inverses), and the complementedness of $\mathrm{Ker}[B,L_1]$ follows. ∎

As an application of Theorem 10.3.1, we obtain the following result.

THEOREM 10.3.3. *Let $\mathcal{Y} = \mathcal{Z} \oplus \cdots \oplus \mathcal{Z}$ (n times) for a Banach space $\mathcal{Z}$, and let $W(\lambda)$ be an operator polynomial with coefficients in $L(\mathcal{Y})$ of the form $[\alpha_{pq} I_{\mathcal{Z}} + K_{pq}]_{p,q=1}^{n}$ with scalars α_{pq} and finite rank operators K_{pq}. If $W(\lambda)$ is invertible for $\lambda \in \Gamma$, then $W(\lambda)$ admits both left and right Wiener-Hopf factorizations with respect to Γ.*

PROOF. We leave aside the case when $\mathcal{Z}$ is finite dimensional (in this case the existence of the Wiener-Hopf factorizations of $W(\lambda)$ follows from Theorem 10.1.1).

Write

$$W(\lambda) = \sum_{j=0}^{m} \lambda^j [\alpha_{pq}^{(j)} I_{\mathcal{Z}} + K_{pq}^{(j)}]_{p,q=1}^{n},$$

where $K_{pq}^{(j)} \in L(\mathcal{Z})$ are finite rank operators. Introduce the operator polynomial

$$\tilde{W}(\lambda) = \sum_{j=0}^{m} \lambda^j [\alpha_{pq}^{(j)} I_{\mathcal{Z}}]_{p,q=1}^{n}.$$

Observe that

$$\det \left[\sum_{j=0}^{m} \lambda^j \alpha_{pq}^{(j)} \right]_{p,q=1}^{n} \neq 0$$

for $\lambda \in \Gamma$ (otherwise Im $W(\lambda)$ could not be the whole space $\mathcal{Y}$, a contradiction with the assumed invertibility of $W(\lambda)$, $\lambda \in \Gamma$). Hence, by Theorem 10.1.1, $\tilde{W}(\lambda)$ admits left and right Wiener-Hopf factorizations with respect to Γ. The inverse $\tilde{W}(\lambda)^{-1}$ has the form

$$\tilde{W}(\lambda)^{-1} = [b_{pq}^{(j)}(\lambda) I_{\mathcal{Z}}]_{p,q=1}^{n},$$

where $b_{pq}^{(j)}(\lambda)$ are rational functions on λ. So, for $\lambda \in \Gamma$, we have

$$\operatorname{Im}[\tilde{W}(\lambda)^{-1} - W(\lambda)^{-1}] = \operatorname{Im}[\tilde{W}(\lambda)^{-1}[W(\lambda) - \tilde{W}(\lambda)]W(\lambda)^{-1}]$$

$$= \operatorname{Im}\left[[b_{pq}^{(j)}(\lambda) I_{\mathcal{Z}}]_{p,q=1}^{n} [\sum_{j=0}^{m} \lambda^j K_{pq}^{(j)}]_{p,q=1}^{n} \right] \subset \mathcal{M},$$

where

$$\mathcal{M} = \begin{bmatrix} \sum\limits_{p,q,j} \operatorname{Im} K_{pq}^{(j)} \\ 0 \\ \vdots \\ 0 \end{bmatrix} + \begin{bmatrix} 0 \\ \sum\limits_{p,q,j} \operatorname{Im} K_{pq}^{(j)} \\ 0 \\ \vdots \\ 0 \end{bmatrix} + \cdots + \begin{bmatrix} 0 \\ \vdots \\ 0 \\ \sum\limits_{p,q,j} \operatorname{Im} K_{pq}^{(j)} \end{bmatrix}$$

is a finite-dimensional subspace in $\mathcal{Y}$. It remains to apply Theorem 10.3.1. ■

10.4 Notes

The notion of Wiener-Hopf factorization is fundamental in many parts of analysis. For example, it plays a fundamental role in the theories of singular integral equations (Gohberg-Krupnik [1], Clancey-Gohberg [1]) and of Toeplitz operators. Theorem 10.1.1 was proved in Rowley [1], and for the more general framework of analytic operator functions with compact spectrum in Kaashoek-van der Mee-Rodman [2]. Example 10.1.1 is taken from Gohberg-Leiterer [2]. Further results on Wiener-Hopf factorizations of operator polynomials and applications to Toeplitz operators are found in Gohberg-Lerer-Rodman [1,2,4].

For more information on Wiener-Hopf factorization of operator-valued functions, see Bart-Gohberg-Kaashoek [4,5], Gohberg-Kaashoek-van Schagen [2].

REFERENCES

G. Allan [1] Holomorphic vector-valued functions on a domain of holomorphy. J. London Math. Soc. 42(1967), 509-513.

B. D. O. Anderson and E. I. Jury [1] Generalized Bezoutian and Sylvester matrices in multivariable linear control, IEEE Trans. Autom. Control, AC-21(1976), 551-556.

T. Ando [1] Linear operators on Kreĭn spaces. Sapporo, Japan, 1979.

C. Apostol, K. Clancey [1] On generalized resolvents. Proc. Amer. Math. Soc. 58(1976), 163-168.

C. Apostol, L. A. Fialkow, D. A. Herrero, D. Voiculescu [1] Approximation of Hilbert space operators, Vol. II. Res. Notes in Math. 102, Pitman, 1984.

C. Apostol, C. Foias, C. Pearcy [1] That quasinilpotent operators are norm-limits of nilpotent operators revisited. Proc. Amer. Math. Soc. 73(1979), 61-64.

C. Apostol, C. Foias, N. Salinas [1] On stable invariant subspaces. Integral Equations and Operator Theory 8(1985), 721-750.

C. Apostol, D. Voiculescu [1] On a problem of Halmos. Rev. Roumaine Math. Pures Appl. 19(1974), 283-284.

N. G. Askerov, S. G. Kreĭn, G. I. Lapter [1] On a class of nonselfadjoint problems. Doklady AN USSR, 155(1964), 499-502 (Russian).

T. Ya. Azizov, I. S. Iohvidov [1] Basic theory of linear operators in spaces with indefinite metric. Nauka, Moscow, 1986 (Russian).

[2] Linear operators in spaces with indefinite metric and their applications. Mathematical Analysis 17, 113-205 (Itogi Nauki i Tekhniki), Moscow, 1979 (Russian).

S. Barnett and P. Lancaster [1] Some properties of the Bezoutian for polynomial matrices. Lin. and Multilin. Alg. 9(1980), 99-111.

H. Bart, I. Gohberg, M. A. Kaashoek [1] Stable factorization of monic matrix polynomials and stable invariant subspaces. Integral Equations and Operator Theory 1(1978), 496-517.

[2] Minimal factorization of matrix operator functions. Birkhäuser, Basel, 1979.

[3] Operator polynomials as inverses of characteristic functions. Integral Equations and Operator Theory 1(1978), 1-12.

[4] Invariants for Wiener-Hopf equivalence of analytic operator functions, in: Constructive Methods of Wiener-Hopf Factorization (eds. I. Gohberg, M. A. Kaashoek), Birkhäuser, Basel, 1986, pp. 317-355.

[5] Explicit Wiener-Hopf factorization and realization, in: Constructive Methods of Wiener-Hopf Factorization (eds. I. Gohberg, M. A. Kaashoek), Birkhäuser, Basel, 1986, pp. 235-316.

S. K. Berberian [1] Approximative proper vectors. Proc. Amer. Math. Soc. 13 (1962), 111-114.

M. S. Birman, M. Z. Solomjak [1] Spectral Theory of Self-Adjoint Operators in Hilbert Space. D. Reidel, Dordrecht, 1987.

H. den Boer [1] Linearization of operator functions on arbitrary open sets. Integral Equations and Operator Theory 1(1978), 19-27.

J. Bognar [1] Indefinite inner product spaces. Springer, Berlin, 1974.

M. S. Brodskii [1] Triangular and Jordan representations of linear operators. Transl. of Math. Monographs, Vol. 32, Amer. Math. Soc., Providence, RI, 1971.

M. S. Brodskii, M. S. Livsic [1] Spectral analysis of non-selfadjoint operators and intermediate systems. Uspehi Mat. Nauk 13(1958), 3-85; English transl.: Amer. Math. for Transl. 13(1960), 265-346.

L. Bungart [1] On analytic fiber bundles I. Holomorphic fiber bundles with infinite dimensional fibers. Topology 7(1968), 55-68.

B. E. Cain [1] An inertia theory for operators on a Hilbert space. J. Math. Anal. and Appl. 41(1973), 97-114.

[2] Inertia theory. Linear Algebra and Appl. 30(1980), 211-240.

S. Campbell, J. Daughtry [1] The stable solutions of quadratic matrix equations. Proc. Amer. Math. Soc. 74(1979), 19-23.

D. Carlson, H. Schneider [1] Inertia theorems for matrices: the semidefinite case. J. Math. Anal. and Appl. 6(1963), 430-446.

K. Clancey, I. Gohberg [1] Factorization of matrix functions and singular integral operators. Birkhäuser, Basel, 1981.

K. Clancey, B. A. Kon [1] The Bezoutian and the algebraic Riccati equation. Linear and Multilinear Algebra 15(1984), 265-278.

R. F. Curtain, A. J. Pritchard [1] Infinite dimensional linear systems theory. Springer-Verlag, Berlin-New York, 1978.

Ju. L. Daleckii, M. G. Krein [1] Stability of solutions of differential equations in Banach space. Transl. of Math. Monographs, Vol. 43, American Math. Soc., Providence, RI, 1974.

B. N. Datta [1] On the Routh-Hurwitz-Fujiwara and the Schur-Cohn-Fujiwara theorems for the root-separation problem. Linear Algebra and Appl. 22(1978), 235-246.

[2] Matrix equation, matrix polynomial and the number of zeros of a polynomial inside the unit circle. Linear and Multilinear Algebra 9(1980), 63-68.

A. Douady [1] Un espace de Banach dont le groupe lineaire n'est pas connexe. Indag. Math. 68(1965), 787-789.

R. G. Douglas, C. Pearcy [1] On a topology for invariant subspaces. J. Funct. Anal. 2(1968), 323-341.

R. J. Duffin [1] A minimax theory for overdamped networks. J. Rational Mech. and Analysis 4(1955), 221-233.

N. Dunford, J. T. Schwartz [1] Linear Operators. Part I: General Theory. John Wiley and Sons, New York, etc., 1957,1988.

[2] Linear Operators. Part II: Spectral Theory. John Wiley and Sons, New York, etc., 1963,1988.

G. Eckstein [1] Exact controllability and spectrum assignment. Topics in Modern Operator Theory, Operator Theory: Advances and Applications, Vol. 2, Birkhäuser, 1981, 81-94.

T. Figiel [1] An example of infinite dimensional reflexive Banach space non-isomorphic to its Cartesian square. Studia Mathematica 42(1972), 295-306.

C-K. Fong, D. A. Herrero, L. Rodman [1] Invariant subspaces lattices that complement every subspace. Illinois J. of Math. 32(1988), 151-158.

A. Friedman, M. Shinbrot [1] Nonlinear eigenvalue problems. Acta Mathematica 121(1968), 77-125.

P. Fuhrmann [1] On weak and strong reachability and controllability of infinite dimensional linear systems. J. Optim. Theory and Appl., 9(1972), 77-87.

[2] Linear systems on operators in Hilbert space. McGraw-Hill, New York, 1981.

M. Fujiwara [1] On algebraic equations whose roots lie in a circle or in a half-plane (in German). Math. Z. 24(1926), 161-169.

F. R. Gantmakher [1] The theory of matrices. 2 Vols. Chelsea, New York, 1959.

I. Gohberg [1] On linear operators that depend analytically on a parameter. Doklady AN USSR 78(1951), 629-632 (Russian).

I. Gohberg, I. A. Feldman [1] Convolution equations and projection methods for their solution. Amer. Math. Soc. Transl., Vol. 41, Providence, 1974.

I. Gohberg, S. Goldberg [1] Basic operator theory. Birkhäuser, Boston, etc., 1981.

I. Gohberg, G. Heinig [1] The resultant matrix and its generalizations I. The resultant operator for matrix polynomials. Acta Sc. Math., 37(1975), 41-61 (Russian).

I. Gohberg, M. A. Kaashoek, D. C. Lay [1] Equivalence, linearization and decomposition of holomorphic operator functions. J. Funct. Anal. 28(1978), 102-144.

I. Gohberg, M. A. Kaashoek, L. Lerer, L. Rodman [1] Common multiples and common divisors of matrix polynomials, II. Vandermonde and resultant, Linear and Multilinear Algebra, 12(1982), 159-203.

I. Gohberg, M. A. Kaashoek, L. Rodman [1] Spectral analysis of families of operator polynomials and a generalized Vandermonde matrix, I. The finite dimensional case, in: Topics in Functional Analysis (eds. I. Gohberg and M. Kac), Academic Press (1978), 91-128.

[2] Spectral analysis of families of operator polynomials and a generalized Vandermonde matrix, II. The infinite dimensional case, Journal of Functional Analysis, 30(1978), 359-389.

I. Gohberg, M. A. Kaashoek, F. van Schagen [1] Common multiples of operator polynomials with analytic coefficients. Manuscripta Math. 25(1978), 279-314.

[2] Similarity of operator blocks and canonical forms, II. Infinite dimensional case and Wiener-Hopf factorization, in: Topics in Modern Operator Theory. Operator Theory: Advances and Applications, Vol. 2. Birkhäuser, 1981, pp. 121-170.

I. Gohberg, M. G. Kreĭn [1] Introduction to the theory of linear nonselfadjoint operators in Hilbert space. Translations Math. Monographs, Vol. 18, Amer. Math. Soc., Providence, 1969.

[2] The basic propositions on defect numbers, root numbers and indices of linear operators. Uspehi Mat. Nauk 12(1957), 43-118; translation, Russian Math. Surveys 13(1960), 185-264.

[3] Theory of Volterra operators in Hilbert space and its applications. Amer. Math. Soc. Transl., Vol. 24, Providence, 1970.

I. Gohberg, N. Ya. Krupnik [1] Introduction to the theory of one-dimensional singular integral operators. Kishinev, Stiinca, 1973(Russian); German transl., Birkhäuser, Basel, 1979.

I. Gohberg, P. Lancaster, L. Rodman [1] Spectral analysis of matrix polynomials, I. Canonical forms and divisors. Linear Algebra and Applications, 20(1978), 1-44.

[2] Matrix Polynomials. Academic Press, New York, etc., 1982.

[3] Invariant Subspaces of Matrices with Applications. J. Wiley and Sons, New York, etc., 1986.

[4] Representation and divisibility of operator polynomials. Canadian Math Journal, 30, 5(1978), 1045-1069.

[5] Spectral analysis of selfadjoint matrix polynomials. Annals of Mathematics, 112(1980), 33-71.

[6] Matrices and indefinite scalar products. Operator Theory: Advances and Applications, Vol. 8, Birkhäuser Verlag, Basel, 1983.

[7] Perturbation theory for divisors of operator polynomials. SIAM Journal of Mathematical Analysis 10(1979), 1161-1183.

I. Gohberg, Ju. Leiterer [1] On cocycles, operator functions and families of subspaces. Matem. Issled. VIII: 2(28) (1973), 23-56 (Russian).

[2] General theorems on factorization of operator functions relative to a contour, I. Holomorphic functions. Acta Sci. Math. (Szeged), 34(1973), 103-120.

I. Gohberg, L. Lerer [1] Resultants of matrix polynomials. Bull. Amer. Math. Soc. 82(1976), 465-467.

I. Gohberg, L. Lerer, L. Rodman [1] On canonical factorization of operator polynomials, spectral divisors and Toeplitz matrices, Integral Equations and Operator Theory, 1(1978), 176-214.

[2] Stable factorization of operator polynomials, I. Spectral divisors simply behaved at infinity, Journal of Mathematical Analysis and Applications, 74(1980), 401-431.

[3] On factorization, indices and completely decomposable matrix polynomials. Technical Report 80-47(1980), 72 pp., Department of Mathematical Sciences, Tel Aviv University.

[4] Stable factorization of operator polynomials, II. Main results and applications to Toeplitz operators, Journal of Mathematical Analysis and Applications, 75(1980), 1-40.

I. Gohberg, A. S. Markus [1] Two theorems on the gap between subspaces of a Banach space. Uspehi Mat. Nauk 14(1959), 135-140 (Russian).

I. Gohberg, L. Rodman [1] On spectral analysis of non-monic matrix and operator polynomials, I. Reduction to monic polynomials, Israel Journal of Mathematics, 30(1978), 133-151.

I. Gohberg, E. I. Sigal [1] An operator generalization of the logarithmic residue theorem and the theorem of Rouché. Math. USSR, Sbornik 13(1971), 603-625.

[2] Global factorization of a meromorphic operator function and some of its applications. Matem. Issled. V1: 1(19)(1971), 63-82 (Russian).

D. Gurarie [1] On a geometric problem in Hilbert space, unpublished.

P. Halmos [1] Ten years in Hilbert space. Integral Equations and Operator Theory 2(1979), 529-564.

G. Heinig [1] Über ein kontinuierliches Analogon der Begleitmatrix eines Polynoms und die Linearisierung einiger Klassen holomorpher Operatorfunctionen. Beiträge zur Analysis 13(1979), 111-126.

[2] Generalized resultant operators and classification of linear operator pencils up to strong equivalence. Colloquia Math. Soc. János Bolyai 35(1980), 611-620.

[3] Bezoutiante, Resultante und Spektralverteilungsprobleme für Operatorpolynome. Math. Nachr. 91(1979), 23-43.

J. W. Helton, L. Rodman [1] Vandermonde and resultant matrices: an abstract approach. Mathematical System Theory 20(1987), 169-192.

D. A. Herrero [1] Approximation of Hilbert space operators, Vol. 1. Res. Notes in Math. 72, Pitman, Boston, etc., 1982.

[2] The Fredholm structure of a multicyclic operator. Indiana University Math. J. 36(1987), 549-566.

[3] A Rota universal model for operators with multiple connected spectrum. Rev. Roum. Math. Pures et Appl. 21(1976), 15-23.

R. A. Hirschfeld [1] On hulls of linear operators. Math. Zeitschrift 96(1967), 216-222.

I. S. Iohvidov, M. G. Kreĭn [1] Spectral theory of operators in spaces with indefinite metric, I. Trudy Mosc. Math. Society 5(1956), 367-432; II. Trudy Mosc. Math. Society 8(1959), 413-496.

I. S. Iohvidov, M. G. Kreĭn, H. Langer [1] Introduction to the spectral theory of operators in spaces with an indefinite metric. Akademie-Verlag, Berlin, 1982.

P. Jonas, H. Langer [1] Compact perturbations of definitizable operators. J. Operator Theory 2(1979), 311-325.

M. A. Kaashoek, C. V. M. van der Mee, L. Rodman [1] Analytic operator functions with compact spectrum, I. Spectral linearization and equivalence, Integral Equations and Operator Theory, 4(1981), 504-547.

[2] Analytic operator functions with compact spectrum, II. Spectral pairs and factorization, Integral Equations and Operator Theory, 5(1982), 791-827.

[3] Analytic operator functions with compact spectrum, III. Hilbert space case: inverse problem and applications, Journal of Operator Theory, 10(1983), 219-250.

M. A. Kaashoek, M. P. A. van de Ven [1] A linearization for operator polynomials with coefficients in certain operator ideals. Ann. Math. Pure Appl. IV, CXXV(1980), 329-336.

V. I. Kabak, A. S. Markus, V. I. Mereutsa [1] On a connection between spectral properties of a polynomial operator bundle and its divisors, in: Spectral Properties of Operators, Stiinca, Kishinev (1977), 29-57 (Russian).

T. Kailath [1] Linear Systems. Prentice-Hall, Englewood Cliffs, NJ, 1980.

T. Kato [1] Perturbation theory for linear operators, 2nd ed., Springer-Verlag, Berlin, etc., 1976.

M. V. Keldysh [1] On eigenvalues and eigenfunctions of some classes of nonselfadjoint equations. Doklady AN USSR 77(1951), 11-14 (Russian).

[2] On completeness of eigenfunctions of some classes of nonselfadjoint linear operators. Uspehi Mat. Nauk 27(1971), 15-47 (Russian).

H. König [1] A trace theorem and a linearization method for operator polynomials. Integral Equations and Operator Theory 5(1982), 828-849.

A. G. Kostyuchenko, A. A. Shkalikov [1] Selfadjoint quadratic operator bundles and elliptic problems. Functional Analysis and its Applications 17(1983), 38-61 (Russian).

M. G. Kreĭn [1] Introduction to the geometry of indefinite J-spaces and to the theory of operators in those spaces. AMS Transl. (2)93(1970), 103-176.

[2] On one new application of the fixed point principle in the theory of operators in a space with indefinite metric. Doklady Akad. Nauk USSR 154(1964), 1023-1026 (Russian).

M. G. Kreĭn, M. A. Krasnoselskii, D. P. Milman [1] On the defect numbers of linear operators in Banach space and on some geometric problems. Sbornik Trud. Inst. Mat. Akad. Nauk SSR 11(1948), 97-112 (Russian).

M. G. Kreĭn, H. Langer [1] On some mathematical principles in the linear theory of damped oscillations of continua I, II. Integral Equations and Operator Theory 1(1978), 364-399; 539-566 (transl. from Russian).

M. G. Kreĭn, M. A. Naimark [1] The method of symmetric and hermitian forms in the theory of separation of the roots of algebraic equations. Linear and Multilinear Alg. 10(1981), 265-308 (transl. from Russian).

E. Kreyszig [1] Introductory functional analysis with applications. J. Wiley and Sons, 1978.

R. Kühne [1] Über eine Klasse J-selbstadjungierter Operatoren. Math. Annalen 154(1964), 56-69.

P. Lancaster [1] Lambda-Matrices and Vibrating Systems. Pergamon Press, Oxford etc., 1966.

P. Lancaster, J. Maroulas [1] The kernel of the Bezoutian for operator polynomials. Linear and Multilinear Algebra 17(1985), 181-201.

P. Lancaster, M. Tismenetsky [1] The Theory of Matrices with Applications. Academic Press, 1985.

H. Langer [1] Factorization of operator pencils. Acta. Scient. Math. 38(1976), 83-96.

[2] Invariant Teilräume definisierbarer J-selbstadjungierter Operatoren. Suomalainen Tiede-Akatemia Ann., A-1, 471(1971).

[3] Spectral functions of definitizable operators in Kreĭn spaces. Functional Analysis, Proc. Dubrovnik Conference, Lecture Notes in Math., Springer-Verlag, Berlin 948(1982), 1-46.

[4] Eine Verallgemeinerung eines Satzes von L. S. Pontrjagin. Math. Annalen 152(1963), 434-436.

[5] Invariant subspaces of linear operators on a space with indefinite metric. Soviet Math. Doklady 7(1966), 849-852.

[6] Zur Spektraltheorie J-selbstadjungierter Operatoren. Math. Annalen 146(1962), 60-85.

[7] Über eine Klasse polynomialer Scharen selbstadjungierter Operatoren in Hilbertraum. J. of Funct. Anal. 12(1979), 13-29.

L. Lerer, L. Rodman, M. Tismetretsky [1] Bezoutian and the Schur-Cohn problem for operator polynomials. J. Math. Anal. Appl. 103(1984), 83-102.

L. Lerer, M. Tismenetsky [1] The eigenvalue separation problem for matrix polynomials. Integral Equations and Operator Theory 5(1982), 386-445.

[2] Generalized Bezoutian and matrix equations. Linear Algebra and its Applications 99(1988), 123-160.

V. B. Lidskii [1] Nonselfadjoint operators with trace. Doklady AN 125(1959), 485-488 (Russian).

J. C. Louis, D. Wexler [1] On exact controllability in Hilbert spaces. J. of Differential Equations 49(1983), 258-269.

A. I. Markushevich [1] Theory of functions of a complex variable, Vols. I-III. Prentice Hall, Englewood Cliffs, NJ, 1965.

A. S. Markus [1] Introduction to spectral theory of polynomial operator pencils. Stiinca, Kishinev, 1986 (Russian). English transl.: AMS Transl. of Math. Monographs, Vol. 71, 1988.

A. S. Markus, V. I. Matsaev [1] On spectral factorization of holomorphic operator functions, in: Operators in Banach spaces, Matem. Issledov. 47(1978), 71-100 (Russian).

A. S. Markus, V. I. Matsaev, G. I. Russu [1] On some generalizations of the theory of strongly damped bundles to the case of the bundles of arbitrary order. Acta Sci. Math. (Szeged) 34(1973), 245-271 (Russian).

A. S. Markus, I. V. Mereutsa [1] On complete sets of roots of the operator equations corresponding to an operator bundle. Izvestiya AN SSSR, Seriya Matem., 37(1973), 1108-1131 (Russian).

C. V. M. van der Mee [1] Realization and linearization, Rapport 109, Vrije Universiteit, Amsterdam, 1979.

I. V. Mereutsa [1] On factorization of an operator bundle into linear factors. Matem. Issledov. 8(1973), 102-114 (Russian).

J. Mikusinski [1] The Bochner Integral. Academic Press, 1978.

B. Mityagin [1] Linearization of holomorphic operator functions I. Integral Equations and Operator Theory 1(1978), 114-131.

[2] Linearization of holomorphic functions II. Integral Equations and Operator Theory 1(1978), 226-249.

[3] Homotopic structure of the linear group of a Banach space. Uspehi Matem. Nauk 25(1970), 63-106 (Russian).

J. R. Munkres [1] Topology: A First Course. Prentice Hall, Englewood Cliffs, NJ 1975.

A. Perelson [1] On trace and determinant for entire operator functions. Integral Equations and Operator Theory 7(1984), 218-230.

[2] Spectral representation of a generalized trace and determinant. Integral Equations and Operator Theory 9(1986).

[3] Generalized traces and determinants for compact operators. Ph.D. Thesis, Tel-Aviv University, 1987.

G. V. Radziyevskii [1] Problem of completeness of root vectors in the spectral theory of operator functions. Uspehi Matem. Nauk 37(1982), 81-145 (Russian).

R. Raghavendran [1] Toeplitz-Hausdorff theorem on numerical ranges, Proc. Amer. Math. Soc. 20(1969), 284-285.

A. C. M. Ran [1] unpublished notes.

[2] Minimal factorization of selfadjoint rational matric function. Integral Equations and Operator Theory 5(1982), 850-869.

J. R. Ringrose [1] Lectures on the trace in a finite von Neumann algebra. Lecture Notes in Mathematics 247, Springer-Verlag, Berlin etc., pp. 309-354 (1972).

L. Rodman [1] On existence of common multiples of monic operator polynomials. Integral Equations and Operator Theory 1(1978), 400-414.

[2] On analytic equivalence of operator polynomials. Integral Equations and Operator Theory 2(1979), 48-61.

[3] Bezout equation, Vandermonde operators, and common multiples of operator polynomials. Journal of Mathematical Analysis and Applications 133(1988), 68-78.

[4] On global geometric properties of the set of subspaces in Hilbert space. Journal of Functional Analysis 45(1986), 226-235.

[5] On factorization of operator polynomials and analytic operator functions. Rocky Mountain Journal of Mathematics 16(1986), 153-162.

[6] On exact controllability of operators, to appear in: Rocky Mountain Journal of Mathematics.

[7] On factorization of selfadjoint operator polynomials, preprint.

M. Rosenblum, J. Rovnyak [1] Hardy Classes and Operator Theory. Oxford University Press, New York, 1985.

G. C. Rota [1] On models for linear operators. Comm. Pure Appl. Math. 13(1960), 469-472.

B. Rowley [1] Wiener-Hopf factorization of operator polynomials. Integral Equations and Operator Theory 3(1980), 427-462.

W. Rudin [1] Real and Complex Analysis. McGraw-Hill, New York, 1966.

[2] Functional Analysis. McGraw-Hill, New York, 1973.

M. Schechter [1] Principles of Functional Analysis. Academic Press, New York, 1971.

M. Shinbrot [1] Note on a nonlinear eigenvalue problem. Proc. Amer. Math. Soc. 14(1963), 552-559.

E. I. Sigal [1] On the trace of an operator bundle. Mat. Issled. 4(1969), 148-151 (Russian).

B. Simon [1] Trace ideals and their applications. London Math. Soc. Lecture Notes 35, Cambridge Univ. Press, 1979.

A. Sourour [1] On strong controllability of infinite dimensional linear systems. J. Math. Anal. and Appl. 87(1982), 460-462.

B. Sz.-Nagy, C. Foias [1] Harmonic Analysis of Operators in Hilbert Space. North Holland, New York, 1970.

K. Takahashi [1] On relative primeness of operator polynomials. Linear Algebra and its Applications 50(1983), 521-526.

[2] Exact controllability and spectrum assignment. J. of Math. Anal. and Appl. 104(1984), 537-545.

J. Tamarkin [1] Some general problems of the theory of ordinary linear differential equations and expansion of an arbitrary function in series of fundamental functions. Math. Zeitschrift 27(1927), 1-54.

A. E. Taylor, D. C. Lay [1] Introduction to Functional Analysis, 2nd ed. Wiley, New York, 1980.

G. Ph. A. Thijsse [1] On solutions of elliptic differential equations with operator coefficients. Integral Equations and Operator Theory 1(1978), 567-579.

J. V. Uspensky [1] Theory of Equations. McGraw-Hill, New York, 1978.

J. Weidmann [1] Linear Operators in Hilbert Spaces. Springer Verlag, New York, 1980.

M. G. Zaidenberg, S. G. Kreĭn, P. A. Kuchment, A. A. Pankov [1] Banach fiber bundles and linear operators. Uspehi Mat. Nauk 30(1975), 101-157 (Russian).

J. Zemanek [1] The stability radius of a semi-Fredholm operator. Integral Equations and Operator Theory 8(1985), 137-144.

A. S. Zilbergleit, Yu. I. Kopilevich [1] On the properties of waves related to quadratic operator bundles. Doklady AN USSR 256(1981), 565-570.

NOTATION

$A \subset B$ inclusion (not necessarily proper); here A and B are sets.

$\mathbb{C}$ the field of complex numbers.

$\mathbb{R}$ the field of real numbers.

$\partial\Omega$ positively oriented boundary of a set $\Omega \subset \mathbb{C}$.

$\bar{\Omega}$ closure of a set $\Omega \subset \mathbb{C}$; norm closure of a set Ω in a Banach space.

Re z, Im z real and imaginary parts of a complex number z.

$$\delta_{jk} = \begin{cases} 1 & \text{if } j = k \\ 0 & \text{if } j \neq k \end{cases}$$ the Kronecker index.

$$\binom{j}{k} = \frac{j!}{(j-k)!k!} \quad (0 \leq k \leq j).$$

A subspace in a Banach space is assumed, by definition, to be norm-closed.

$\mathcal{X} \dot{+} \mathcal{Y}$ direct sum of subspaces in a Banach space.

For Banach spaces $\mathcal{X}_1, \ldots, \mathcal{X}_m$ define the Banach space

$$\mathcal{X}_1 \oplus \cdots \oplus \mathcal{X}_m = \{(x_1, x_2, \ldots, x_m) \mid x_i \in \mathcal{X}_i \text{ for } i = 1, \ldots, m\}$$

with the norm

$$\|(x_1, \ldots, x_m)\| = \left(\sum_{i=1}^{m} \|x_i\|^2\right)^{1/2}.$$

$\mathcal{X} \oplus \cdots \oplus \mathcal{X}$ (n times) is abbreviated to $\mathcal{X}^n$.

$\{0\}$ the zero subspace.

$\Sigma(L)$ spectrum of the operator polynomial $L(\lambda)$.

$\sigma(A)$ spectrum of the operator A.

$L(\mathcal{X}, \mathcal{Y})$ the set of all bounded linear operators acting from the Banach space $\mathcal{X}$ into the Banach space $\mathcal{Y}$.

$L(\mathcal{X}) = L(\mathcal{X}, \mathcal{X})$.

$0_{\mathcal{X}}$ (or 0) the zero operator acting on $\mathcal{X}$.

$I_{\mathcal{X}}$ (or I) the identity operator acting on $\mathcal{X}$.

λI is often abbreviated to λ (here $\lambda \in \mathbb{C}$).

$A|\mathcal{M}$, $A|_{\mathcal{M}}$ the restriction of the operator $A \in L(\mathcal{X},\mathcal{Y})$ to a subspace $\mathcal{M} \subset \mathcal{X}$.

Im A the image (range) of the operator A.

Ker A the kernel of the operator A.

Given $Z_1 \in L(\mathcal{X}_1),\ldots,Z_n \in L(\mathcal{X}_n)$, we use the notation

$$\operatorname{diag}[Z_i]_{i=1}^{n} = \operatorname{diag}\,[Z_1,Z_2,\ldots,Z_n] = Z_1\oplus\cdots\oplus Z_n$$

to designate the block diagonal operator

$$\begin{bmatrix} Z_1 & 0 & \cdots & 0 \\ 0 & Z_2 & \cdots & 0 \\ \vdots & \vdots & & \vdots \\ 0 & 0 & \cdots & Z_n \end{bmatrix} \in L(\mathcal{X}_1\oplus\cdots\oplus\mathcal{X}_n).$$

Given $Y_1 \in L(\mathcal{X},\mathcal{Y}_1)$, $Y_2 \in L(\mathcal{X},\mathcal{Y}_2),\cdots,Y_m \in L(\mathcal{X},\mathcal{Y}_m)$, we denote by

$$(1) \qquad \operatorname{col}[Y_i]_{i=1}^{m} = \operatorname{col}[Y_1,Y_2,\ldots,Y_m]$$

the operator

$$\begin{bmatrix} Y_1 \\ Y_2 \\ \vdots \\ Y_m \end{bmatrix} \in L(\mathcal{X},\mathcal{Y}_1\oplus\cdots\oplus\mathcal{Y}_m).$$

The notation (1) is also used to designate the vector $(Y_1,\ldots,Y_m) \in \mathcal{Y}_1\oplus\cdots\oplus\mathcal{Y}_m$ (here $Y_2 \in \mathcal{Y}_1$, ..., $Y_m \in \mathcal{Y}_m$).

INDEX